Excitation and Neural Control of the Heart

Excitation and Neural Control of the Heart

edited by MATTHEW N. LEVY

Department of Investigative Medicine, Mount Sinai Hospital,
and Departments of Medicine, Physiology,
and Biomedical Engineering,
Case Western Reserve University, Cleveland, Ohio

MARIO VASSALLE

Department of Physiology, State University of New York,
Downstate Medical Center, Brooklyn, New York

AMERICAN PHYSIOLOGICAL SOCIETY
Bethesda, Maryland

Library of Congress Catalog Card Number 81-22939

International Standard Book Number 0-683-04952-6

Printed in the United States of America by
Waverly Press, Inc., Baltimore, Maryland 21202

Distributed for the American Physiological Society by
The Williams & Wilkins Company, Baltimore, Maryland 21202

Preface

This book is a compilation of thirteen reviews on the excitation and neural control of the heart. The chapters on cardiac excitation include descriptions of the movements of several ions across the cardiac cell membranes. The latest concepts of the roles of such ion fluxes in the genesis of automaticity and in excitation-contraction coupling are presented. These basic concepts are then applied to the important clinical problem of the genesis of cardiac arrhythmias. Chapters on the development of reentrant circuits, on the effects of myocardial ischemia, and on the relation of repolarization to arrhythmia vulnerability discuss the practical applicability of such fundamental knowledge to the relevant clinical field.

In several of the chapters about cardiac excitation the effects of the autonomic neurotransmitters on ionic fluxes are described; other features of neural control are described in greater detail in the five chapters on the neural control of the heart. In two of these chapters some attention is directed toward the sympathetic and parasympathetic centers in the central nervous system that regulate the cardiovascular system. The other three chapters are devoted to the peripheral mechanisms of neurotransmitter release in the heart and blood vessels. The presynaptic mechanisms involved in peripheral sympathetic-parasympathetic interactions have been investigated intensively during the past several years, and these recent advances are described in detail in the chapters on the peripheral mechanisms.

These reviews are an outgrowth of our tenure as editors of the *American Journal of Physiology: Heart and Circulatory Physiology* from 1977 to 1981. During that period we initiated the policy of publishing a series of review articles on specific themes. Acknowledged authorities from the United States and abroad were invited to prepare the articles. The authors were requested to write the reviews with the aim of bringing the latest information and concepts not only to other workers in their specific fields but also to cardiovascular physiologists with different research interests and expertise. The authors were urged to use a style of writing that would be readily understood by most cardiovascular investigators and students regardless of their specific field of interest. These special articles were reviewed much more extensively than the regular scientific papers submitted to the Journal, to ensure scientific accuracy, balanced interpretation, and clarity of presentation.

The articles were so favorably received by the Journal readers that, with the encouragement and support of the Publications Committee of the American Physiological Society, we had them carefully updated and revised by the authors for this presentation. We trust that by collecting these reviews in a single publication we have provided a service to research scientists, teachers, and students in cardiovascular physiology.

Matthew N. Levy
Mario Vassalle

Contents

CHAPTER 1

Ionic Basis of Electrical Activity in Cardiac Tissues

EDOUARD CORABOEUF

Laboratoire de Physiologie comparée et de Physiologie cellulaire
associé au Centre National de la Recherche Scientifique,
Université Paris, France

Cardiac action potentials recorded from the various parts of the heart are very different in shape, amplitude, and duration; however, they are generally characterized by a rapid phase of depolarization followed by a long-lasting plateau. The early phase is highest and most rapid in Purkinje fibers, whereas the plateau is highest in the ventricular myocardium. In normal node cells, and in most abnormally depolarized cells, the early phase is smaller or absent, and the action potential may therefore be considered as being limited to a slowly developing, low-amplitude plateau. When cardiac tissues are, or tend to become, spontaneously active, their action potentials are preceded, in the pacemaker region, by a slow, local diastolic depolarization. This pacemaker potential can develop in different ranges of membrane potential, for example, in the range of -90 to -60 mV for the normal Purkinje fibers and in the range of -60 to -35 mV for the sinus node cells or depolarized Purkinje and myocardial fibers. Though the cardiac action potentials from the different tissues and/or different species show a large range of shapes, a few underlying ionic mechanisms are responsible for their development. Among these the most precisely described are passive ionic permeabilities due to the opening and closing of membrane "channels," although electrogenic ionic exchanges and active transports also participate in the development of the electrical activity.

1

Cardiac Resting Potential: Intracellular and Extracellular Ionic Activities

It was established 20 years ago (252) that the resting potential of Purkinje fibers is the result of a dominant permeability of the cell membrane to potassium ions, K (for recent review see ref. 222). If this permeability P_K is, as supposed, extremely high compared to permeabilities to other ions, the resting potential, E_r. lies close to the equilibrium potential for K ions, E_K, as given by the Nernst equation. If X is a cation of valency z

$$E_x = \frac{RT}{zF} \ln \frac{[X]_o}{[X]_i}$$

$$\text{at } 18°C, \; E_x(mV) = \frac{58}{z} \log \frac{[X]_o}{[X]_i}$$

$$\text{at } 37°C, \; E_x(mV) = \frac{61}{z} \log \frac{[X]_o}{[X]_i}$$

where R is the gas constant, T the absolute temperature, F the Faraday constant, and $[X]_o$ and $[X]_i$ the concentration of the extra- and intracellular ions.

Under such conditions a 10-fold increase in the external potassium concentration $[K]_o$ must lead to a decrease in E_r of 58 mV at 18°C or 61 mV at 37°C, a 10-fold decrease in $[K]_o$ having the opposite effect. A slope smaller than 58–61 mV per decade change in $[K]_o$ indicates that the membrane is appreciably permeable to ions other than K. In Purkinje fibers the resting membrane potential obeys almost perfectly the Nernst equation for K ions (that is, the membrane behaves as a nearly perfect potassium electrode) for values of $[K]_o$ greater than 2.7 mM K; whereas for lower values it does not (252). If it did, a 10-fold reduction of $[K]_o$, let us say from 5 to 0.5 mM, would lead to a 58-mV hyperpolarization; for example, from −85 to −143 mV at 18°C. In fact, in Purkinje fibers a strong depolarization instead of a hyperpolarization occurs in low-K or K-free solution, showing that the relative participation of K ions in membrane potential decreases compared to that of other ions, mainly sodium (186). In other tissues, such as the sinoatrial node and to a smaller extent the atrium of the rabbit (73), the diastolic membrane is noticeably permeable to ions other than K.

Determination of the equilibrium potential for a given ion may be achieved in some cases by measuring the potential at which an ionic membrane current reverses (reversal potential, see Fig. 3B) when this current is known to be carried by a single ionic species, but the usual method implies the measurement of external and internal activities for the given ion. Activity differs from concentration because of interactions between ions; the activity coefficient, γ (γ = activity/concentration), decreases when the ionic strength of the solution increases. Activity may be calculated from the concentration or directly measured using ion-selective electrodes. Using concentrations instead of activities, or even using the same activity coefficient for the intracellular and extracellular ions, leads to errors that may be very large if, for example, an appreciable number of internal ions are sequestered. Another source of error comes from the assumption that the extracellular concentration of a given ion is necessarily the same in the bulk solution and in the narrow clefts between the cardiac cells.

Measurement of intracellular activities and corresponding equilibrium potentials have been achieved in several tissues using cation-selective microelectrodes. In frog sinus venosus (247), E_K appears to be more negative than the maximum diastolic potential by as much as 32 mV; whereas in frog atrium and ventricle the difference between E_r and E_K is less than 3 mV. It has also been estimated (247) that the permeability ratio P_{Na}/P_K is of the same order of magnitude for the atrium and ventricle, but 1–2 orders of magnitude higher for the sinus venosus; whereas P_{Cl}/P_K would be of the same order of magnitude for the sinus venosus and atrium, but an order of magnitude lower for the ventricle. In rabbit papillary muscle the ratio P_{Cl}/P_K has been estimated at 0.11 (92). In the same tissue (158), the intracellular activity of K has been shown to be much less (82.6 mM) than the corresponding concentration (134.9 mM), giving an apparent intracellular activity coefficient for K, γ_K, of only 0.612. The difference is still greater for intracellular sodium, the concentration being 32.7 mM and the activity only 5.7 mM ($\gamma_{Na} = 0.175$). This means that much of the internal Na is compartmentalized or sequestered. It could be partly bound to extracellular structure and therefore incorrectly attributed to the cytoplasm, but it is probably also bound to sarcoplasmic macromolecules and compartmentalized at a high concentration within intracellular organelles such as the sarcoplasmic reticulum (158). In dog Purkinje fibers an intracellular K activity of 130.0 ± 2.3 mM was measured (177), wheras in sheep Purkinje fibers the internal Na activity was found to be 7.2 ± 2.0 mM at the normal $[Na]_o$ (140 mM, equivalent to an external Na activity of 105 mM), so that $E_{Na} = +70$ mV (87). The Na activity appears to be sensitive to changes in $[K]_o$, $[Na]_o$, and $[Ca]_o$. It can rise to more than 30 mM after inhibition of the Na pump by strophanthidin (88).

Accumulation of K ions in the narrow clefts between cardiac cells or in the T tubules has been thought to be responsible for the depolarization observed during rapid stimulation of cardiac tissues (101, 239). Such an increase in extracellular K activity was confirmed using K-sensitive microelectrodes inserted in the extracellular space of frog ventricular strips (146) and canine Purkinje fibers (145). It was shown that the loss of K by activated cells was sufficient to produce, during the course of each action potential, a reversible increase in K activity. In both frog and Purkinje fibers this increase can reach about 1 mM, that is, 25–30% of the normal K activity in the physiological solution. It is therefore no longer possible to consider the extracellular fluid composition as constant in the course of a cardiac cycle. Since in Purkinje bundles more than 80% of the cell membranes are facing narrow (average width 40 nm) extracellular clefts (114, 178), the extracellular space must be considered as being much smaller than the intracellular space. Depletion or accumulation of K ions can therefore occur in the clefts as a result of ion pumping and restricted diffusion from the bulk extracellular medium (46). The influence of the accumulation-depletion phenomena on ionic currents has been analyzed by different authors (11, 17, 30, 80, 191).

Membrane Ionic Currents and Ionic Channels

Action potentials are due to changes in transmembrane ionic currents. Inward currents, conventionally considered as negative, correspond to an entry of positive charges, that is, of cations into the cell; however, anions leaving the cell should also give rise to an inward current, a situation that was suggested in skeletal muscle fibers grown in tissue culture (94) but not in cardiac tissues. Outward (positive) currents correspond to cations leaving the cell or anions entering the cell (in principle, an outward current carried by anions entering the cell must be suppressed after replacement of these anions by impermeant ones). Taking a cell in which all areas of membrane undergo the same voltage changes, when the sum of inward currents (carried for example by Na or Ca ions) becomes greater than the sum of outward currents (carried, for example, by K ions), a net flux of positive charges enters the cell, and the membrane depolarizes. It repolarizes when inward currents become smaller than outward currents. When the membrane potential E_m is more negative (or less positive) than the equilibrium potential for a given cation or anion, this ion tends to cross the membrane in such direction that it gives rise to an inward depolarizing current; conversely, an outward current tends to develop when E_m is less negative (or more positive) than the equilibrium potential for the ion. In cardiac tissues several ionic currents are governed by conductances that are voltage and time dependent, such as those described in giant axon (119), and in this case the time course of membrane currents depends on the membrane potential. For example, sodium conductance, g_{Na}, increases when the membrane is depolarized (voltage dependence), but if the depolarization is established very abruptly, the increase in g_{Na} does not follow the potential instantaneously but develops more slowly, with a time course of a fraction of a millisecond (time dependence). Moreover, this time course depends on the potential to which the membrane has been brought. Several currents are also totally or partially controlled by quasi-instantaneous (time independent), inward-going rectifiers (188). By this mechanism the membrane passes inward current in response to negative potential changes more easily than it passes outward current in response to positive potential changes.

Voltage- and time-dependent conductances probably result from the opening of membrane channels that are controlled by gating mechanisms. For the sake of clarity such channels may be represented as in Figure 1A, in which the degree of opening or closure of the channel is indicated by gates that are more or less open, so that more or fewer ions can cross the membrane. In such a simplified representation, the channels shown in Figure 1A correspond in fact to the sum of all individual channels of a given type existing in the membrane, each individual channel having probably two positions, open or closed (see ref. 115). Therefore, the degree of opening of the gates shown in the figure actually corresponds to the proportion of individual channels with an open gate or the amount (fraction) of time spent in the open state by each gate. During the last few years, several investigators have succeeded in measuring very small membrane currents that were not ionic currents. These currents are known as gating currents. They are assumed to be due to the passive movement through the membrane, when the electric field changes, of charges triggering displacement of the gating structure responsible for the opening (activation) of the sodium channel (22, 143, 176). Inactivation is not associated with detectable gating current and may derive its voltage sensitivity from coupling to activation (see ref. 8). In such a case inactivation would not be independent from activation. Multistep models of activation have been proposed recently (7, 8). Two different

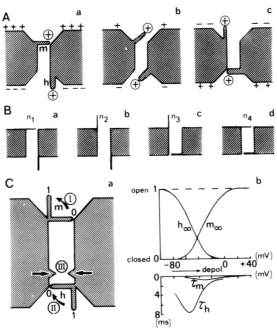

Fig. 1. Highly schematic diagram representing ionic channels and their control systems. A: channel shown corresponds to sum of individual channels of a given type (e.g., Na channels) existing in the membrane. A, a: at resting potential, E_r, all individual activation gates (m) are closed and inactivation gates (h) are open if E_r is high enough; as a result, m is shown closed and h completely open. A, b: for moderate steady-state membrane depolarization, a certain proportion of individual activation and inactivation gates are open so that m and h are shown partly open (a steady-state current flows through the membrane). A, c: for strong steady-state depolarization, all individual activation gates are open and inactivation gates are closed (no steady-state current). At the beginning of a strong depolarization (if it is abruptly established) m opens quickly, whereas h closes more slowly, so that current flows only transiently. B: individual channels are controlled by gates that probably have only 2 functional positions, open and closed. n_1, n_2, n_3, and n_4 represent the proportions of channels with a given position for each gate (see text). C, a: composite channel showing m (I) and h (II) gates with their 2 extreme positions, 0 (closed) and 1 (completely open) and a quasi-instantaneous inward-going rectifier (III). *Arrows* indicate evolution of control systems when membrane undergoes a depolarization. C, b: *upper curves,* steady-state values, from 0 to 1, of m (m_∞) and h (h_∞) as functions of membrane potential in millivolts (mV); *lower curves,* time constants of activation τ_m and inactivation τ_h, in milliseconds (ms), as functions of membrane potential. [Modified from Hodgkin and Huxley (119).]

gating currents associated with Na and Ca currents have been described in an *Aplysia* neuron (1, 3) and a snail neuron (152), suggesting that Na and Ca channels probably have similar gating mechanisms. In the case of nerve fibers there is evidence that the activation of each Na channel involves the displacement of several (three?) independent gating particles, but this refinement has been omitted in Figure 1. It has recently become possible to measure directly the current flowing through individual channels by using patch pipettes. Measurements were performed in giant axons in the case of the K channel (50) and in culture rat muscle cells in the case of the Na channel (221). When the conductance was open by applying depolarizing steps to the membrane, elementary events consisting of discrete current pulses of varying duration occurred. In the case of the Na channel, because of inactivation, elementary current pulses tend to occur only at the beginning of the depolarizing steps. From a macroscopic (statistical) point of view the Na channel may be considered as being controlled by two different and, at least partially, independent types of gating systems, m and h (1st-order kinetic variables). One, m, opens quickly (activation, labeled I in Fig. 1C, a) and the other, h, closes more slowly (inactivation, labeled II in Fig. 1C, a) when the membrane depolarizes. Consequently the Na current is brief, since it flows only during the time when m is already open and h not yet closed. The time constants of m and h processes, which depend on E_m, are schematically shown in Figure 1C, b. According to Hodgkin and Huxley's formulation (119)

$$i_{Na} = g_{Na} (E_m - E_{Na}) \text{ and } g_{Na} = \bar{g}_{Na} \cdot m^3 \cdot h$$

where i_{Na} is the sodium current, $E_m - E_{Na}$ is the driving force, i.e., the potential difference that generates the current, and $\bar{g}_{Na}$ is the maximum value reached by g_{Na} when m and h are fully opened ($m = h = 1$). It must be noticed that the model in Figure 1A does not account for the fact that g_{Na} is indeed the product of m and h, a point that can be understood by observing the channels in Figure 1B, in which the four possible position combinations for m and h gates, considered as independent systems, are represented with their corresponding probabilities n_1, n_2, n_3, and n_4. Assuming that the four probabilities are equal, a, b, c, and d represent the same number of channels. It can be seen in the figure that in this case the proportion of m and h gates in an open state is ½, whereas the proportion of open channels is only ½ × ½ = ¼. More generally, in the situation represented in the figure, the proportion of open channels would be $(n_2 + n_3) (n_1 + n_2)/(n_1 + n_2 + n_3 + n_4)^2$. K channels are also controlled by activation gates, but generally do not possess inactivation gates (119). When a gate opens or closes as a consequence of a long-lasting rectangular change in membrane potential, the gate reaches its new steady-state degree of opening (labeled m_∞, h_∞, Fig. 1C, b) according to an exponential kinetics, that is, theoretically, after an infinite time. The steady-state value of m (m_∞) and h (h_∞) as a function of the membrane potential (E_m) is schematically shown in Fig. 1C, b. When a membrane is submitted, as in Figure 4, to depolarizing steps of increasing amplitude, g_{Na} increases because m increases with depolarization (Fig. 1C, b); whereas ($E_m - E_{Na}$) decreases, reaches zero when $E_m = E_{Na}$, and then reverses. Therefore, i_{Na}, which is the product of g_{Na} and ($E_m - E_{Na}$) varies as a function of E_m, as shown in Figure 3B.

The presence of fixed charges in close proximity to an ionic channel (see refs. 19, 116, 166) is able to alter the local transmembrane potential across the channel without changing the transmembrane potential measured in the bulk solution, because the electric influence of such charges spreads over a very limited distance (2–3 nm). Since the gating mechanism is sensitive to the electric field in the membrane, it will be influenced by the presence of fixed charges in its neighborhood and also by ions that neutralize them, either by binding or by screening. Generally, divalent ions exert a much more efficient screening effect than monovalent ions at the same concentration. For example, a fivefold increase in Ca concentration in the external solution is able to shift the curves $m_\infty - E_m$ and $h_\infty - E_m$ toward less negative potentials by 10–15 mV (93). Changes in other divalent cations exert similar effects (116). Such a shift of the $m_\infty - E_m$ curve leads to an increase in threshold, while the same shift in the $h_\infty - E_m$ curve leads, for a given depolarization, to a smaller inactivation. The shifts come from the fact that after neutralization of external negative charges by Ca, a depolarization greater than the normal is necessary to influence the gating mechanisms. In the absence of divalent cations, monovalent ions become able to exert a marked screening of surface charges (116). A major source of negative charges that accounts for a considerable component of calcium binding on the cell surface are the ionized carboxyl groups of sialic acid (glycoproteins); those that are located in the very close vicinity of the membrane surface certainly influence the gating mechanisms. As was recently shown, the removal of sialic acid with neuraminidase markedly increases cellular calcium exchangeability of cultured heart cells (155). Lipids of the bilayer with charged polar head groups constitute another source of fixed negative charges (156).

Because the mechanisms responsible for the ionic conductances are inherently probabilistic, as demonstrated by single-channel recordings of K and Na currents in excitable membranes (50, 221), they can be tested by fluctuation analysis. Fluctuation analysis is a means of extracting information about the system under study from the spontaneous random fluctuations of variables being measured, that is, from the noise produced by the system (226). K and Na ion-current noise has been studied in the giant axon, and it has been shown that the data fit fairly well with the theory that allows only two possible channel conductance values (49). It has been shown that, in aggregates of embryonic heart cells, the increase in resistance in the range −70 to −50 mV (diastolic depolarization) is probably due to a change in the population of two conductance-state channels similar to those found in nerve axon (66), therefore that the current source for voltage noise can be predicted from membrane impedance (45).

Voltage Clamp in Cardiac Tissues

Most of our present knowledge about ionic conductances in cardiac tissues has been obtained from experiments using voltage clamp methods (for review, see refs. 38, 52, 90, 187, 207, 210, 234, 255). The different techniques, namely, the double-microelectrode technique, the single sucrose-gap, and the double sucrose-gap techniques (Fig. 2), have been extensively reviewed and criticized from a theoretical point of view (18, 134). Most of these criticisms are indeed justified, and it is clear that the limitations of the technique due to the structural complexity of cardiac tissues must be taken into consideration. Because membrane currents change with membrane voltage, one of the basic requirements of the voltage clamp is that the voltage be the same in each part of the membrane in which the current is measured. But this situation is difficult to obtain because of the multicellular and inherent cable properties of the fibers. One of the most serious limitations comes from the existence of an electrical resistance (R_s) that lies in series with the membrane resistance (R_m) (Fig. 3). R_s is primarily due to the resistance of the very narrow and tortuous clefts that connect any cell,

8   **Edouard Coraboeuf**

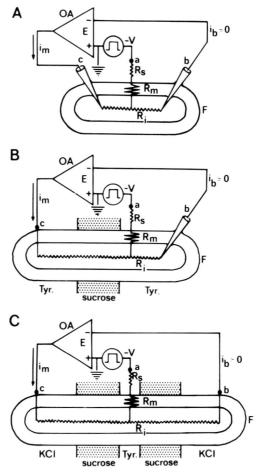

Fig. 2. Simplified diagram of arrangement for voltage clamp recording using 2 micro-electrodes (*A*), single sucrose gap (*B*), or double sucrose gap (*C*). The bundle of fibers, F, is assimilated to a single fiber. At the beginning of the experiment the resting potential recorded between a and b is balanced by an equal and opposite external potential source (not shown) so that the voltage E recorded by the operational amplifier OA (an amplifier with a very high gain and a very high input resistance) is zero; in such a case, the amplifier delivers no current. For the membrane to be depolarized by a given value, V, a potential step, −V (command pulse), must be applied as shown. This potential step tends to make E different from zero by making negative the negative input of the amplifier, through R_s, R_m, R_i, and the electrode b. As a consequence, the amplifier delivers, quasi-instantaneously, a current, i_m, so that this current produces a potential drop equal to V across R_m, whatever its value. Therefore the membrane potential is maintained at the chosen value. This is true only if R_s is very small compared with R_m (see text and Fig. 3). The current i_m delivered by the amplifier flows either through an intracellular electrode (*A*) or through an extracellular electrode separated from the central one by a nonconducting solution (sucrose, mannitol, etc.) (*B* and *C*); moreover, it cannot flow through electrode b since the amplifier has a very high input resistance (i_b = 0). Consequently, this current must cross the membrane to reach electrode a and is therefore a transmembrane current. R_m: membrane resistance; R_i: internal resistance; R_s: series resistance; Tyr: Tyrode solution; isotonic KCl is generally used in endchambers in the double sucrose-gap technique to decrease the membrane resistance and allow measurement of E_r between a and b.

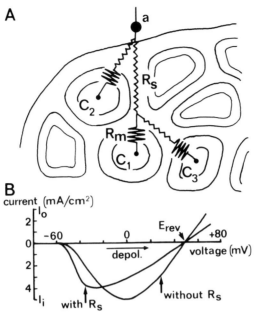

Fig. 3. *A:* schematic drawing of a bundle of fibers (C_1, C_2, C_3, etc.) showing membrane resistance (R_m) and series resistance (R_s) that connects each cell with the extracellular electrode a. *B:* transient current-voltage relationships resulting from a simulated voltage clamp of an equivalent membrane circuit (Hodgkin-Huxley model) with and without R_s. I_o: outward current; I_i: inward current; depol: depolarization; E_{rev}: reversal potential. [Adapted from Ramon et al. (204).]

situated in the depth of a bundle, with the physiological solution bathing the preparation and therefore with the electrode (a) and the voltage-clamp amplifier. When a rectangular voltage step (command pulse) is applied to the system (Fig. 2), the amplifier delivers a current, i_m, through the preparation, therefore across $R_m + R_s$, so that the resulting voltage drop $i_m (R_m + R_s)$ balances immediately and exactly the command pulse, whatever the value of R_m. By this operation the membrane potential should be clamped at any value chosen by the investigator. However, the current i_m produces a voltage drop $i_m \cdot R_m$ across the membrane that is necessarily different from the change in potential $i_m (R_m + R_s)$ applied to the preparation by the amplifier. This error in voltage clamp, which is equal to $i_m \cdot R_s$, is generally considered to be prohibitive when i_m becomes very large, as, for example, at the peak of rapid inward sodium current. The situation is better with thin preparations (100–200 µm in diameter), in which the maximum current cannot be very high and even the deepest cells are rather close to the periphery, provided that the fiber remains in good condition and the electrical coupling between the cells is satisfying. It must be noticed that the voltage clamp error due to R_s changes direction when the current flowing through the membrane is reversed: if we consider for example a rectangular depolarizing command pulse, the depolarization really applied to the membrane will be larger than the command pulse when the membrane current triggered by the pulse is inward and smaller when the membrane current is outward (204). Consequently, the current-voltage relationships corresponding to inward currents are shifted towards negative potentials, although the reversal potential of the current remains unchanged (Fig. 3B). This is true, however, only if the conductance of the membrane for other ions remains small (9). Computations corresponding to cardiac conductances have been done by Tarr et al. (229). Since R_s is distributed and has a different value for each cell, it is difficult to compensate for it, though its lumped value can be easily estimated (230). However, a partial compensation is possible (202). One means of decreasing the voltage clamp error $i_m \cdot R_s$ is to measure the membrane potential when $i_m = 0$, for example, after opening the current injection circuit for a very short time. Such a method has been proposed (26, 43, 262).

In spite of the criticism discussed above and others (see refs. 90 and 144) resulting from the lability of intercellular coupling (257, 263) diffusion of sucrose in the segment of the bundle placed in the test compartment (sucrose-gap technique) or point polarization (double-microelectrode technique), etc., voltage clamping remains one of the most useful techniques for the study of electrical phenomena in cardiac tissues. It has been shown by De Hemptinne (68) that the error factor resulting from the presence of the series resistance was relatively small in frog atrial fibers studied by the double sucrose-gap technique when the petrolatum-sealing method was used to separate the various fluid compartments. Transmembrane microelectrode recordings during voltage clamp experiments using this method indicated rather satisfactory voltage control even during the flow of the peak inward current (68, 217). Similar results have been obtained in model preparations in tissue culture (85, 182). Two different ways of measuring transmembrane currents in the same preparation have been recently used either by combining the single sucrose-gap technique and a second intracellular microelectrode (211) or by extending to cardiac tissues (137) the three-microelectrode technique of Adrian et al. (5). A last method using suction pipette to disrupt the membrane of neurons was recently applied to isolated adult cardiac cells and allowed voltage clamp measurements of the rapid Na current (159).

Ionic Currents Involved in Cardiac Electrical Activity

Depolarizing Currents. When depolarizing steps of increasing amplitude are applied to a cardiac preparation, for example, to a frog atrial fiber (215), current changes occur that are schematically represented in Figure 4. At the very beginning of the pulse, large and brief transients of current (t.c. in Fig. 4A) develop that are due to the charge of the membrane capacity. In the absence of R_s such transients would be infinitely short. For moderate depolarizing steps, large and brief peaks of inward current develop that disappear in Na-free Ringer solution or in the presence of tetrodotoxin (TTX), indicating that this early inward peak of current is carried by Na ions crossing the membrane through a TTX-sensitive channel. After suppression of this current, a slower and smaller peak of inward current still remains (Fig. 4B), which is suppressed either by adding manganese (2–4 mM) to the Ringer solution or by removal of both Ca and Na ions. Then only outward currents remain (Fig. 4C), which increase with time and are essentially carried by K ions.

From such experiments and from many other (see refs. 183, 184, 228) it may

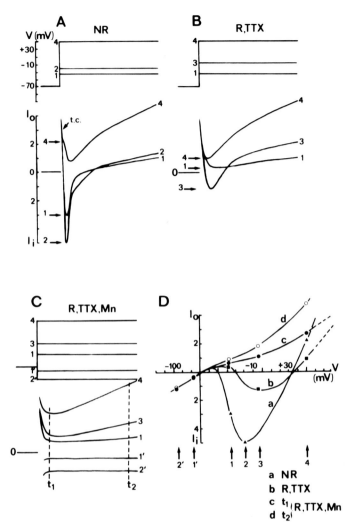

Fig. 4. *A, B,* and *C*: semischematic representation of transmembrane ionic currents (*lower curves*) elicited by depolarizing (1 to 4 in *A, B, C*) or hyperpolarizing (1′ and 2′ in *C*) steps (mV, *upper curves*) in a frog atrial fiber. *A*: in normal Ringer solution (NR), after a very short transient of capacity current (t.c.), a large and brief peak of inward (downward) current develops (1, 2, *arrows*), which becomes a peak of outward (upward) current (4, *arrow*) for strong depolarization. *B*: after tetrodotoxin (TTX) the brief peak of current (sodium current) is suppressed and a smaller and slower peak of inward current mainly carried by calcium ions remains. *C*: remaining inward current is blocked after further addition of manganese ions (Mn), unmasking outward potassium currents. *D*: current-voltage (*I-V*) relationships plotted at times corresponding to the peak of rapid inward current (*curve a*), peak of slow inward current (*curve b*), and at 2 different times indicated in C, t_1 (*curve c*) and t_2 (*curve d*). It can be seen that *curve a* is only a very approximate representation of the sodium current, since at the time of the rapid current peak the capacity current is not over, and slow inward current has already begun. Net slow inward current-voltage relationship is approximately given by b − c. I_o, outward current; I_i, inward current.

be concluded that two distinct inward currents contribute to the development of the action potential in most cardiac tissues. A rapid TTX-sensitive inward Na current, similar to the one described in nerve, is responsible for the initial rapid phase of depolarization and a secondary manganese-sensitive inward current (see ref. 265), mainly carried by Ca ions flowing through a slow channel, is responsible for the plateau of the action potential. The larger the inward current, the more rapid will be the depolarization of the membrane. For this reason the maximal rate of rise of the action potential (dV/dt_{max}) is proportional to the peak of net inward current measured on the I-V curves (Fig. 4D). Both rapid and slow channels are controlled by activation and inactivation kinetic variables, respectively labeled m and h for the rapid channel and d and f for the slow channel. Besides Ca ions, the slow channel can also accept Na ions (215, 218), mainly after removal of Ca ions. This accounts for long-lasting plateaus developing in Ca-free and Mg-free solution (98). Lithium cannot be substituted for Na through the slow channel in contrast to the rapid channel (42). There is evidence that in cat myocardium Na does not detectably contribute to the slow inward current (259), whereas it does in guinea pig myocardium (195) and rabbit sinus node (148, 163). It has been calculated that in cow papillary muscle the slow channel is about 100 times more permeable to Ca ions than to Na (or K) ions; however, because of the high $[Na]_o$, the relative contribution of i_{Na} in the slow inward current can reach about one-third of this inward current at $E_m = 0$ mV in this tissue and becomes still much greater at low $[Ca]_o$ (211). The slow channel also accepts some other divalent cations such as strontium and barium (149, 245), but magnesium to only a small extent (149) or not at all (42, 205). It is blocked by manganese, cobalt, nickel, lanthanum, and drugs like verapamil and D600 (see refs. 147, 149, 170, 200, 207, 209, 215, 255), though in certain tissues manganese can penetrate the cardiac cells (70, 196). On the other hand adrenaline increases g_{Ca} (205, 243), an effect that is probably mediated by intracellular cyclic AMP (208, 237, 248) and may be due to an increase in the number of functional slow channels (212). For that and other reasons it has been suggested that metabolic energy may be continuously required to maintain the slow channel (225). The decrease in slow inward current observed during metabolic inhibition [(181), see also ref. 55] can be attributed to a diminution in both g_{Ca} and driving force for Ca ions, the latter being due to an increase in intracellular free calcium produced by metabolic poisoning.

Recent analyses of the rapid Na channel have been performed in several tissues including frog atrial fibers (84, 216), rabbit Purkinje fibers (48), and isolated adult rat ventricular cells (159). The overlap of the $m_\infty - E_m$ and $h_\infty - E_m$ relationships (Fig. 5) seems to differ according to the tissue studied. In rabbit Purkinje fibers the overlap is obviously very small (47, 48), whereas in sheep Purkinje fibers it allows a sizeable steady-state inward current to enter the cell through the rapid Na channel (10). In the case of slow inward current the overlap of the $d_\infty - E_m$ and $f_\infty - E_m$ relationships is still larger (207, 236) and an appreciable steady-state slow inward current can therefore flow through the slow channel. Rather large values have generally been reported for the time constant of inactivation, τ_d, of the slow inward current (55 to 5–10 ms for E_m between −50 and −20 mV, see ref. 52). Recent measurements, however, made in isolated adult cardiac cells suggest values of τ_d as low as 0.6 ms at about 0 mV (G. Isenberg, unpublished data).

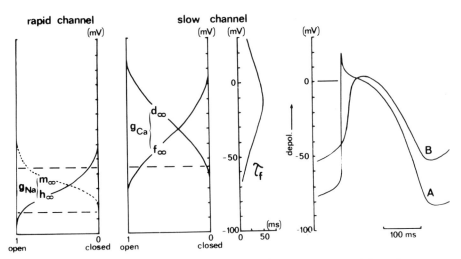

Fig. 5. Schematic composite diagram showing cardiac action potentials and parameters controlling depolarizing currents. Ordinate: membrane potential (mV). From *left* to *right*: steady-state conductance variables determining activation (m_∞) and inactivation (h_∞) of the rapid channel (g_{Na}); steady-state conductance variables determining activation (d_∞) and inactivation (f_∞) of the slow channel (g_{Ca}); time constant of inactivation (τ_f) of the slow inward current; and two types of spontaneous action potentials, one (A) resembling those of the normally polarized Purkinje fibers, the other (B) resembling those of the sinus node (comparative durations of action potentials have not been taken into consideration, and the corresponding time scale is only approximative). [*Slow channel, left* adapted from Reuter (207); *slow channel, right* from Horackova and Vassort (122), frog myocardium.]

The time constant of inactivation, τ_h, of the rapid Na channel (closure of h gate) has been estimated at 1.2 ms in frog atrial fibers at an E_m around 0 mV (215). For the slow current the corresponding time constant, τ_f, has been measured as a function of potential by several investigators with somewhat different results (see refs. 52, 122). It seems, however, that the relationship is similar to that obtained for other Hodgkin-Huxley conductances (Fig. 5). The fact that τ_f decreases when $[Ca]_o$ increases (150, 199) or during Ca injections (129) suggests that the inward movement of Ca or the resulting increase in $[Ca]_i$ can accelerate the inactivation of the slow inward current as suggested in other tissues (25). Voltage clamp experiments have clearly shown that the slow inward current has a more positive threshold than the rapid Na current. More precisely, as shown in Figure 5, the curves giving the steady-state values of activation and inactivation variables as a function of membrane potential are less negative by about 30–40 mV for g_{Ca} than for g_{Na} (207). This explains why, when a fiber undergoes partial depolarization, the rapid channel inactivates partially or even completely according to the value of the depolarization, whereas the slow Ca channel remains fully available. For this reason slowly developing and therefore slowly conducted action potentials occur (see ref. 61). The same situation occurs in true pacemaker cells of the sinoatrial node, in which the rapid channel is completely inactivated as a result of the low take-off potential (153). In subsidiary pacemakers excitation is due to the participation of both a small component of the fast sodium channel and a component of the slow channel (154, 162, 163). Since the rapid and slow channels are completely closed at high resting potentials, another source of inward current must be assumed to explain why E_r is less negative than E_K. This inward background current is probably carried mainly by sodium ions (see ref. 168) flowing through a potential-insensitive pathway. It is a point of interest to know whether or not rapid and slow channels undergo complete inactivation during large and long-lasting depolarizing pulses. If they do not, Na and Ca steady-state inward currents can participate in the development of long-lasting plateaus. According to Reuter (206) g_{Na} exhibits incomplete inactivation in Purkinje fibers, but similar observations have not been reported in myocardium. Moreover TTX does not shorten appreciably the plateau of the ventricular action potential (58), whereas it depresses that of the Purkinje fibers (37, 53). The fact that in dog Purkinje fibers the action potential plateau is about 100 times more sensitive to TTX than the spike strongly suggests that two populations of Na channels exist in this tissue with two different sensitivities for TTX. The most sensitive channels (about 0.1% of the total) seem to be deprived of inactivation gates (53). A noticeable fraction of the current flowing through the slow channel in Purkinje fibers also does not seem to be inactivated (136, 206, 210). A steady-state slow inward current inhibited by D600 can be observed at the end of 500 ms depolarizing pulses. The fact that this current is not depressed at $E_m \simeq 0$ mV (136), i.e., at a membrane potential corresponding to complete (or almost) inactivation of the slow channel suggests that this steady-state current cannot be due to the overlap of the activation and inactivation curves but rather to slow channels that do not possess inactivation gates.

An important feature of cardiac electrophysiology is the manner in which depolarizing currents recover from inactivation (recovery from inactivation, removal of inactivation, or reactivation). According to the Hodgkin-Huxley model, the time constant of inactivation (closing of h gate, see Fig. 1) and the time constant of reactivation (reopening of h gate) must be the same for a given potential. Recovery of Na conductance at the resting level has a time constant of about 10 ms in nerve. Weidmann (251) also showed that this recovery, measured as the recovery of the maximum rate of rise of the action potential (dV/dt_{max}) was rather rapid around the resting level in Purkinje fibers. In frog atrial muscle, however, much longer times (500–1,000 ms) were observed for full recovery of the dV/dt_{max} of an action potential following a previous response (time necessary for dV/dt_{max} to reach its maximum value again) as well as for the full recovery of the Na current (111). It was proposed that not one gate but two gates with different time constants control the inactivation process of the Na channel. If these gates are in series along the channel it can be easily understood that the faster gate controls inactivation, whereas the slower gate controls reactivation. Gettes and Reuter (99) have confirmed that the time constant of reactivation is very slow for dV/dt_{max} (i_{Na}) but depends on E_m being 10–20 ms for E_m from −90 to −80 mV and reaching 50 ms at $E_m = -70$ mV and 125–150 ms at $E_m = -60$ mV, whatever the tissue studied. In contrast, the time constants of inactivation and reactivation of the slow inward Ca current have been found to be identical by Gettes and Reuter (99) but not by Kohlardt et al. (150), who propose two different inactivation gates in cat myocardium. Interestingly, in *Aplysia* neuron recovery from Ca inactivation also consists of two phases (2). After a response, recovery from inactivation of the slow inward current determines recovery of the action potential plateau. Since the slow channel reopens more slowly than the rapid channel at large E_m (more negative than -70 mV) but more rapidly at low E_m, at least in guinea pig papillary muscle (99), it can be understood that in partially depolarized fibers the plateau recovers more rapidly than the spike, whereas the reverse situation is observed in highly polarized fibers, for example, in normal Purkinje fibers. Unexpectedly it has been observed in cat and dog myocardium that the time course of the plateau phase recovery can exhibit a transient period of increase in plateau amplitude above its control value together with enhanced contraction (15) and slow inward current (118). More recently oscillatory repriming of the slow inward current (associated with similar evolution of the twitch force) have been reported in calf Purkinje fibers treated with strophanthidin (258). Another kind of inward current develops in Purkinje fibers when they are exposed to cardiotonic steroids or to low-Na or Ca-rich media (135, 139, 157). These transient inward currents (TI), which occur when the membrane is repolarized after a pulse or an action potential, are responsible for the concomitent development of transient afterdepolarization (TD, see ref. 89) and aftercontractions. It has been proposed that TI is carried by a leak channel or reflects Ca extrusion by an electrogenic Ca-Na exchange (139).

Repolarizing Currents. Coming back to Figure 4C, where rapid and slow inward currents have been blocked, it can be seen that if we measure the remaining currents (mainly K currents) at the beginning of the steps (t_1 in Fig. 4C), they are larger for hyperpolarizing than for depolarizing steps of the same value (for example, pulses 1 and 2′ in Fig. 4C). As a consequence, the *filled-circles* curve in Figure 4D is not a straight line but shows a downward inflexion when the membrane is depolarized. This shows that the membrane is more easily crossed by K ions in an inward direction than in an outward one. This inward-going (anomalous) rectification, which indicates a decrease in g_K when the membrane is depolarized (35, 59, 64, 125), appears as soon as the capacity current is over, therefore almost instantaneously. For this reason it is considered as a time-independent phenomenon (185) and may be represented as a channel controlled only by a system labeled III in Figure 1C, a. Cesium ions (20 mM) suppress the inward-going rectification of i_{K_1} giving rise in Purkinje fibers to almost linear I-V curves (128) and K-efflux-voltage relationships (246). The conductance of the channel as a function of E_m is indicated approximately in Figure 6; this system, g_{K_1}, corresponds to the background conductance of the cardiac membrane for K ions (see ref. 168). It has been observed that the curve relating the current that flows through g_{K_1} to the membrane potential ($i_{K_1} - E_m$ relationship) depends on $[K]_o$. For two different values of $[K]_o$ the curves cross each other in a characteristic way, a situation that has been attributed to the fact that the conductance is a function of the driving force, $E_m - E_K$ (see ref. 186). The nature of inward-going rectification is not yet clear though several explanations have been put forward (see refs. 4 and 83). For strong hyperpolarizations an outward-going rectification has been described (69).

Several other repolarizing currents have been described in Purkinje fibers (see ref. 190). The repolarizing current i_{x_1}, which is not a pure K current, is governed by mechanisms I and III, but the rectifier function (not shown in

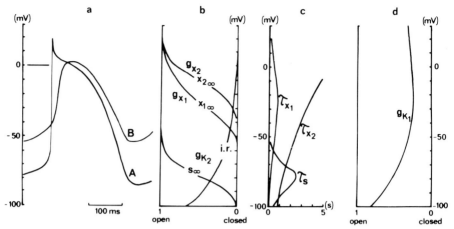

Fig. 6. Schematic composite diagram showing cardiac action potentials and parameters controlling repolarizing currents.Ordinate: membrane potential (mV). In a: same action potentials as in Fig. 5. In b: steady-state conductance variables (s_∞, x_{1_∞}, x_{2_∞}) determining activation of time and voltage-dependent outward currents (corresponding conductances labeled g_{K_2}, g_{x_1}, g_{x_2}). i.r. corresponds approximately to the time-independent, inward-going rectifier controlling g_{K_2}. In c: time constants of activation (τ_3, τ_{x_1}, τ_{x_2}) of i_{K_2}, i_{x_1}, i_{x_2}. In d: approximative background time-independent conductance g_{K_1} exhibiting inward-going rectification. For sake of clarity, the inward-going rectifier controlling g_{x_1} and the positive dynamic current have been omitted in this diagram. [b From Noble and Tsien (188, 189); c adapted from Noble and Tsien (190); d adapted from Noble and Tsien (188).]

Fig. 6) closes the channel only partially when the membrane is depolarized. The gating mechanism that controls i_{x_1} activates in the plateau range (-50 to $+20$ mV). Because of the value of its time constant in this potential range, i_{x_1} plays a major role in the termination of the action potential in Purkinje fibers (168). It corresponds to the process known as delayed rectification. In fact in Purkinje fibers two currents activating in about the same potential range but exhibiting different kinetics have been described, namely i_{x_1} and i_{x_2}: i_{x_2} is much slower than i_{x_1} and does not possess rectifier function; it does not increase sufficiently in the course of an action potential to be very important for repolarization (189). In mammalian myocardium, two currents rather similar to i_{x_1} and i_{x_2} and labeled i_K and i_x have been described (171, 172). The importance of the delayed rectification (essentially due to i_{x_1} or i_K) seems less in myocardium (173) than in Purkinje fibers, although i_K is larger in the cat and guinea pig (140) than in calf and sheep (173). This is also shown by the fact that an electrically induced increase in plateau amplitude, by activating x_1 and switching on more outward current, shortens the action potential in Purkinje fibers (138), whereas in myocardium the action potential is not shortened (62) or even lengthened (16, 211). As i_{x_1}, i_K possesses an inward rectifier function and can be described by Hodgkin-Huxley variables without power relations. Currents very similar to i_{x_1} and i_{x_2} and labeled i_1 and i_2 (33) or I_1 and I_2 (197) have also been described in frog myocardium.

More recently, two other components of repolarizing currents ($i_{x,\text{fast}}$; $i_{x,\text{slow}}$) have been described in frog atrial fibers (27) using a four-electrode arrangement instead of three electrodes, as in a previous study (33), and a different method of analysis. Obviously the analysis is made difficult by current flow (i_{acc}) coming from changes in driving force as a result of potassium accumulation. In frog atrium the current previously labeled i_{x_2} (33) is probably $i_{x,slow}$ combined with i_{acc}, whereas i_{x_1} seems to be a mixing of $i_{x,fast}$ and $i_{x,slow}$. It must be pointed out that the subscript fast refers to the rate of decay of $i_{x,fast}$ at negative membrane potentials; at positive potentials, this current is indeed slower than $i_{x,slow}$ (27). Ignoring the accumulation current can lead in some situations to a threefold overestimation of $\tau_{x,slow}$. The main characteristics of $i_{x,fast}$ is its sigmoid onset. This is in contrast to other outward currents described above, but already observed by De Hemptinne (67). Activation curves for $i_{x,fast}$ and $i_{x,slow}$ have similar positions on the voltage axis (27).

Another repolarizing outward current has been observed in Purkinje fibers in response to strong depolarization. This current, called positive dynamic current (91, 201), initial transient or early outward current (141, 206, 220), or i_{qr} (168, 220) seems to be responsible for at least a part of the initial repolarization of the Purkinje fiber action potential (including the notch often observed at the beginning of the plateau). The first effect of a shortening of the diastolic interval is indeed to reduce or abolish the notch by preventing the complete reactivation of this initial outward current (113). The time course of inactivation and reactivation of this current is clearly voltage dependent (91). The early outward current is inhibited by tetraetylammonium and 4-aminopyridine (141, 142). It is no longer considered as being carried by Cl ions but rather, at least to a large extent, by K ions (141, 220). In calf Purkinje fibers it has been clearly established that the transient outward current is activated by intracellular Ca (219). Cl ions exert a repolarizing effect, however, both in Purkinje fibers where part of the early outward current is chloride sensitive (142) and in ventricular muscle where the action potential duration is markedly increased by Cl removal (16). That such an effect may be indirect is suggested by the observation that chloride substitution induces a reduction in potassium permeability (36). It has been suggested that inactivation of the positive dynamic current results from an accumulation of K in extracellular clefts and/or chloride inside the cell (17).

Cardiac automatism is due to the development during diastole of a slow depolarization that progressively brings the membrane potential to the critical value (threshold potential), at which level a regenerative response is initiated. Since the membrane depolarizes when $i_{inward} > i_{outward}$, the development of slow diastolic depolarization implies 1) the existence of a diastolic inward current and 2) that the inward current becomes progressively greater than the outward current, a requirement that may be satisfied either by an increase in inward current or by a decrease in outward current, or both. Observation that membrane resistance increases during diastolic depolarization (235, 249) has long been considered as a proof that the main phenomenon involved is a decrease in g_K (140, 235, for recent review see 38, 65, 133, 187, 240 and the chapter by Vassalle in this volume). The diastolic decrease in g_K is considered as resulting from the closing (deactivation) of a time- and voltage-dependent channel that has opened during the preceding action potential. In the mammalian sinus node, the resulting current is probably a pure K current, since its reversal potential is around -100 mV for $[K]_o = 2.7$ mM and around -85 mV for $[K]_o = 5.4$ mM. It is activated in the range -50 to $+20$ mV and has been referred to as the pacemaker current component, i_p (194). In the frog sinus venosus two distinct components of outward current (i_{fast} and i_{slow}) have been described (31); they activate in a more negative potential range than the corresponding currents observed in frog atrium. In sinus i_{slow} appears to be a rather selective K current. Both components of K current are deactivating throughout the diastolic depolarization. A progressive activation of the slow inward current also probably participates in the diastolic depolarization of the frog sinus venosus (31) and of the mammalian sinus cell (see ref. 241). Deactivation of i_{x_1} is responsible for the pacemaker mechanism in partially depolarized Purkinje fibers (112), whereas deactivation of a K current of the i_x type is mainly responsible for the diastolic depolarization in frog atrial fibers rendered repetitively active by depolarizing currents (27, 33, 160).

It has recently been observed in spontaneously active preparations of the rabbit sinus node (29, 193) and of frog sinus venosus (31) that hyperpolarizing voltage clamp steps to or beyond the diastolic potential induced the development of an inward current that could hardly be interpreted as resulting from the deactivation of some pure K outward current (82). This new inward current (Fig. 7), which has been called i_f (28, 29) or i_h (264), begins to activate at -50 mV, fully saturates at about -100 mV, and can be expressed by Hodgkin-Huxley kinetics (264). The fully activated current shows no rectifying properties and reverses at $\simeq -25$ mV (264). Such a low reversal potential indicates that the current is carried by several ions, mainly K and Na (32, 82); in addition i_f is depressed or suppressed by cesium and by reducing external Na (82) and is enhanced by epinephrine (28, 29).

An outward K current labeled i_{K_2} was described in Purkinje fibers (188). It was governed by a time- and voltage-dependent variable, s, which increased from 0 to 1 in steady-state conditions when the membrane was depolarized from -100 to -50 mV. In fact i_{K_2} decreased upon depolarization because of the existence of a strong quasi-instantaneous inward-going rectification. The amplitude of i_{K_2} was reduced in Na-free solution (189) and the conductance of its rectifier component appeared to be dependent on $[K]_o$ (188). For more than ten years the diastolic deactivation of i_{K_2} was considered to be responsible for the diastolic depolarization in normally polarized Purkinje fibers (188). Alterations of the parameters controlling i_{K_2} have been reported to occur at high (238) or low pH (34) and can induce during acidosis abnormal repolarization leading to bursts of repetitive activity (54, 60). As it became clear that currents i_{K_2} and i_f exhibit very similar properties (82), it was proposed that the current i_{K_2} is not, as previously thought, a pure K current, deactivating when the membrane is hyperpolarized, but an inward current activated by hyperpolarizations negative to $\simeq -50$ mV (76, 77). As in the sinus node, Na and K both participate in carrying i_f (78). The apparent reversal potential of i_{K_2} near E_K (Fig. 7) previously described may be explained by the strong anomalous rectification of i_{K_1} associated with accumulation-depletion phenomena in intercellular clefts (79, 81).

Control of Potassium Conductance by Internal Calcium

The suggestion (260) that a rise in free Ca inside erythrocytes leads to an increase in membrane permeability to K has been confirmed in neurons, where

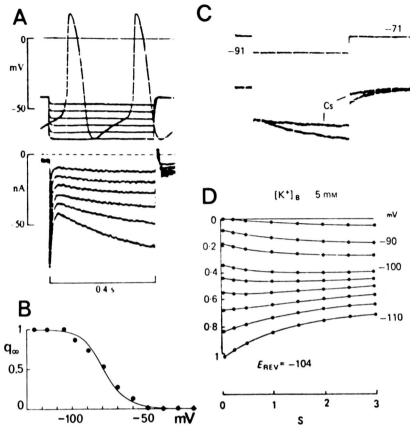

Fig. 7. *A*: spontaneous activity of small pieces (0.3 mm × 0.3 mm) of rabbit sinoatrial node and voltage clamp hyperpolarizing pulses (*upper* records). As a consequence of the pulses, very slow inward (downward) currents, i_f, develop (*lower* records). *B*: steady-state values of the activation parameter (here labeled q) of the pacemaker current i_h (or i_f) as a function of E_m in the rabbit sinoatrial node. *C*: in Purkinje fibers submitted to a hyperpolarizing pulse from −71 to −91 mV, 0.5 mM cesium suppresses the pacemaker current i_{K_2} and shifts the total current to less inward values, as if i_{K_2} were an inward current. *D*: reconstruction of Purkinje fiber pacemaker currents during hyperpolarizing pulses from −70 mV, assuming the existence of the new inward current i_f and progressive K depletion in the clefts during the pulses. In spite of the time-dependent increase in inward i_f current, the progressive increase in outward current due to depletion becomes sufficient at large depolarizations to reverse the kinetics of the current, leading to an apparent reversal potential close to E_K. [*A* from Brown et al. (29); *B* from Yanagihara and Irisawa (264); *C* from DiFrancesco (76); *D* from DiFrancesco and Noble (79).]

injections of Ca ions indeed produce an increase in potassium permeability (see ref. 175). This observation has stimulated the search for a similar mechanism in cardiac fibers, though this mechanism does not exist in every tissue, being absent for example in giant axon (20). A control of g_K by internal Ca concentration [Ca]$_i$ could indeed account for several observations, for example, that an increase in [Ca]$_o$ leads to an elevation and a shortening of the plateau (see ref. 138). If the former effect is easily explained by an increase in i_{Ca} resulting from an enhanced driving force for this ion, the latter might be the consequence of an increased g_K resulting from the elevation of $[Ca]_i$ that follows the increase in inward calcium current.

In Purkinje fiber, intracellular iontophoretic injections of Ca by means of single- or double-barreled microelectrodes produce, at the injection point, an appreciable and transitory hyperpolarization (7–19 mV) and a marked shortening of the action potential ($\simeq$40%) that disappears with a time constant of about 0.5 s (127). When the intracellular Ca concentration is increased by different procedures (repetitive depolarizations, decrease in $[Na]_o$ associated with addition of cyanide) the time-independent outward current in sheep and calf ventricular trabeculae increases. When this increase was due to repetitive depolarizations, it was less marked after Ca current blockers such as verapamil or D600, thus showing that i_{Ca} is partly responsible for the increased $[Ca]_i$ (16). A similar increase in i_{K_1} was also observed in short Purkinje fibers submitted to intracellular injections of Ca (130) or to Ca-rich media, but in this case i_{Ca} seemed not to be very important in the Ca-dependent increase in i_{K_1} (138). This does not exclude the possibility, however, that the factor controlling i_{K_1} is $[Ca]_i$, since Ca can enter the cell via pathways other than the slow channel, for example, the Na-Ca exchange system. (105). Spontaneous electrical fluctuations ($\pm$0.2–0.4 mV) occurring in quiescent atrial fibers of the carp are increased (to $\pm$ 1–2 mV) and become more synchronous in K-free Ringer solution or in the presence of ouabain while they are accompanied by a fine mechanical tremor (6). It has been proposed that they are due to oscillations of a feedback loop including internal Ca release and $[Ca]_i$-induced changes in resting g_K. Periodic and random voltage fluctuations that have been recorded in other cardiac preparations (66, 135, 157) may have a similar origin. It has also been suggested that the outward plateau currents are generated by changes in intracellular free Ca (174) though other observations show that Ca ions do not favor the onset of delayed K current, i_{x_1} (138). The existence of a calcium-activated transient outward current masking partly the slow inward current, i_{si}, has been demonstrated in calf Purkinje fibers (219, 220). Upon removal of $[Ca]_o$ and replacement with strontium (Sr) or barium (Ba), this current is strongly depressed or abolished, thus revealing i_{si}. Chelating Ca_i by intracellular injections of EGTA did not alter i_{si}, but suppressed both the contraction and the transient outward current, thus showing that the latter current results from the i_{si}-induced increase in $[Ca]_i$ (219). Experiments performed in molluscan neurons confirm that Sr_i or Ba_i are poorly effective or ineffective in activating outward currents (107).

Control of Intercellular Coupling by Internal Calcium

Cardiac cells, like many other types of cells, are coupled by gap junctions (214) or nexuses corresponding to membrane areas in close apposition crossed by tubular particles or connexons that show variable degrees of packing into a hexagonal lattice. Each connexon, which has been assumed to be composed of a dimer of hexamers of the gap-junction protein connexin, behaves as a permeable channel from cytoplasm to cytoplasm (for review see ref. 106). In frog atrial fibers, junctional particles are often disposed in rows or circles (167), a structure that also exists in mammalian embryonic cardiac cells (108). It has been proved in noncardiac cells (newt embryo) that the specific conductances of the junctional membrane are much greater than that of the nonjunctional membrane; a value of 10^{-10} S has been obtained for the single channel as a lower limit of conductance, whereas the insulation for the single channel amounts to $>10^{12}$ Ω (164).

In Purkinje fibers it has been known for a long time from cable analysis that the internal longitudinal resistance is so low (2–3 times that of Tyrode solution) that cell-to-cell contacts must be of the low-resistance type [(250) see also ref. 254)]. It has been shown that when half of a bundle of ventricular fibers is charged by radiopotassium and the other half washed by inactive Tyrode solution, the average space constant for the decrease in labeled K concentration corresponds to a distance of about 12 times the length of a cell, indicating a low value for disc resistance. The average resistance of intercalated discs to the movement of K ions has been estimated at 3 $\Omega \cdot cm^2$ (253). However, confirming previous studies (223), it has recently been deduced from transverse to longitudinal resistivity measurements that the intercalated discs are high-resistance membranes (224). This conclusion seems, at the present time, difficult to reconcile with the observation that in ventricular muscle the permeability of the nexal membrane to tetraethylammonium is 21,000 times larger than the permeability of the surface membrane (256) and that in Purkinje fibers, Procion yellow, a dye of a molecular weight near 700, is able to diffuse intracellularly from a freshly cut end of the bundle through about 20 cells in succession by the end of a 4-h period; whereas there is no detectable uptake of the substance through intact peripheral surface membrane (126). From such diffusion studies it may be proposed that the pores within the nexal membrane of the intercalated discs probably have a diameter of somewhat more than 1 nm (256), a value that fits well with the size (about 2 nm in diameter) of the dense, stain-filled core visible in the center of each nexus particle (106). Studies on nexus formation in reaggregated cells revealed a quantal increase in junctional conductance in nascent junctions (165). By bringing in contact spheroidal embryonic chick heart cell aggregates (44) it was shown that transmission of excitation from one aggregate to the other can occur after a short time of contact ($\simeq$ 15 to 30 min in 1.3 and 4.8 mM K, respectively). At this time action potentials in the two aggregates are separated by a rather long delay ($\simeq$50 ms). The delay decreases as time of contact increases (1 ms after 1–2 h) because of the continuous formation of connexons (possibly 100 per cell in 1–2 h). Calculations suggest that only 5 channels per cell are enough to allow liminal transmission between two $\simeq$150 μm diam aggregates in contact by $\simeq$100 cells each (44).

There has been, so far, no indication that ionic permeability of cardiac nexal membranes may be time or voltage dependent, though some other electrical synapses, such as crayfish giant motor synapses, are known to show rectifier properties (95). A firmly established point today is that in several tissues the permeability of the nexal membrane is strongly dependent on the intracellular Ca concentration (72, 164). In cardiac tissues it has been shown that Ca or Sr injections produce a rise in input resistance and, correspondingly, electrical uncoupling (74). Similar increases in longitudinal internal resistance (associated with contracture) have also been obtained by Wojtczak (263) in ventricular trabeculae perfused with glucose-free hypoxic Tyrode solution, whereas elevation of the glucose concentration up to 50 mM prevented contracture and markedly reduced partial electrical uncoupling. Simultaneous measurements of internal longitudinal resistance R_i and intracellular Ca activity, aCa_i, performed in sheep Purkinje fibers (63) have shown that electrical decoupling (measured as relative increase in R_i) begins to develop when aCa_i reaches $\simeq 0.5$ μM, the decoupling being induced either with 10^{-4}M dihydro-ouabain or 10^{-3}M dinitrophenol. In the latter case R_i eventually increases 80 times or more, whereas aCa_i reaches 20 to 55 μM; after decoupling the diameter of junctional particles increases from 8.2 nm at control up to 9.8 nm (the particle diameters grouping around two peaks, 8.1 nm and 10.8 nm), whereas particle height decreases slightly and subunits are more regularly arranged (63). Ca-induced uncoupling of cardiac cells must therefore be considered as a phenomenon able to occur each time cardiac cells are in an energy-depleted state or active transports are inhibited (75). The lower the density of gap junctions between cells, the greater the resulting alterations in conduction will probably be. The lowest density seems to occur in nodal cells (23) where the rate of intracellular diffusion of fluorescein was at least three orders of magnitude lower than in other cardiac tissues.

Electrogenic Active Transports

If a coupled transport of ions extrudes more cations of a given valency than it takes up, an excess of positive charges is ejected from the cells, giving rise to an outward repolarizing current. Conversely, an inward depolarizing current would develop if more cations are taken up than cations extruded. In both cases the transport is electrogenic; on the other hand, if the number of charges of the same sign transported in both directions are equal, the transport is electroneutral.

Electrogenicity of the Na pump (for reviews see refs. 109, 166, 232) has been admitted for a long time since it was established 20 years ago that in erythrocytes 2 K ions are actively transported into the cells for every 3 Na ions extruded (for bibliography, see ref. 203). A similar coupling ratio has been measured in snail neurons (231), making the Na pump a repolarizing system. However, the coupling ratio seems to vary according to the tissue studied and the experimental conditions. In frog atria the data favor the idea of a loose linkage between Na and K transport (109). On the assumption that the pump delivers a current I_p and that R_m is the membrane resistance, the change in membrane potential produced by the electrogenic transport will be $V = I_p \cdot R_m$. Therefore, an electrogenic pump will modify the membrane potential more markedly when the membrane resistance is high. Any electrogenic pump is probably influenced by the membrane potential (see ref. 166) in such a way that strong external positivity (hyperpolarization) tends to prevent or even reverse the outward movement of positive charges carried by the pump, whereas depolarization of the membrane tends to increase it. From this point of view, an electrogenic pump can be characterized by a transport equilibrium potential (109). However, experimental results do not permit a clear conclusion as to whether the functioning of the Na pump is dependent on the membrane potential or not. In skeletal muscle, it has been suggested that membrane depolarization, beyond −70 mV, strongly stimulated Na efflux (124), and the pump current has been considered to reverse (becoming a depolarizing current) when the membrane potential was more negative than −95 mV (227). Similar effects of membrane potential have also been described in neurons (151), but in other cases (120, 233) hyperpolarization did not inhibit the pump. In Purkinje fibers the outward current attributed to the electrogenic Na pump has been shown to be voltage independent (131).

In cardiac tissues (for recent review see ref. 102) the hypothesis of an electrogenic Na pump was first put forward by Deleze (71). Later Page and Storm (198) observed that in cat papillary muscle previously depleted of K and enriched with Na by cooling, rewarming rapidly brought E_m to a value more negative than the calculated E_K value, a phenomenon that was inhibited by ouabain. Glitsch (100) described a similar transient increase of E_m beyond E_K in guinea pig atria during recovery after hypothermia in low-K media. Electrogenic Na pump activity has also been described in Purkinje fibers (117, 239, 261) and in rabbit sinus node (192). According to Vassalle (239), the electrogenicity of the Na pump becomes effective when the driving rate is faster than the intrinsic rate: in the course of a rapid drive the maximum diastolic potential increases slowly as a result of a progressive increase in $[Na]_i$, which stimulates the pump until a steady state is reached after 2 min or more; at this point the pump is supposed to be stimulated at maximal or near maximal level. The hyperpolarization is accelerated by norepinephrine, decreased in magnesium-free Tyrode solution, suppressed when Na is substituted by lithium, or after dinitrophenol or strophanthidin (39, 239). When Purkinje fibers are quiescent or driven at slow rates, the activity of the electrogenic Na pump is reduced (242). Because of this rate-dependent electrogenicity, Purkinje fibers being normally overdriven by the sinus node would be continuously hyperpolarized, and their automaticity would be inhibited (239).

In guinea pig auricles, the active Na efflux approximately doubles when $[Na]_i$ increases from 10 to 40 mmol/liter fiber water and decreases significantly only when $[K]_o$ falls below 0.5 mM (104), a point that is in agreement with measurements in Purkinje fibers (24). As clearly stated by Lüttgau and Glitsch (166), the crucial point in convincingly demonstrating an electrogenic Na transport is to exclude the fact that an electroneutral Na pump causes the hyperpolarization in Na-rich muscle fibers by depletion of K ions just outside the membrane. This possibility seems unlikely in Purkinje fibers because a decrease in $[K]_o$ is known to increase the membrane resistance of these fibers, whereas no change in membrane resistance occurred before, during, or after overdrive; moreover, low-K media increase automaticity of Purkinje fibers, whereas rapid drive is followed by a temporary suppression of automaticity (239). The fact that the effects of temporary stimulation of the Na pump are those expected from a transient increase in outward current, not those expected from K depletion, has been fully confirmed in dog Purkinje fibers (97) in which the pump current has been directly measured (96). The intensity of the outward current generated by the electrogenic Na pump in rewarmed guinea pig atria displays saturation kinetics when measured as a function of $[K]_o$, half saturation occurring at $\simeq 1.5$ mM K (103). The electrogenicity of the Na pump seems to be relatively independent of membrane potential (86). Metabolic inhibitors, especially dihydro-ouabain (131) and cooling (132), have been used to block the sodium pump. Within seconds after cooling or poisoning, the action potential was markedly lengthened and the outward current, mainly the instantaneous background current, was reduced. The latter effect was also observed in Na-free Tyrode solution and was therefore not due to an increase in inward Na current. The reduction in current was also apparently not due to a change in E_K, which suggests that the suppressed outward current was an active electrogenic one. This current could play a role in the decay of normal action potential in Purkinje fibers (132), but apparently it does not contribute as much in ventricular action potential (169). Another type of electrogenic transport, namely an active K uptake giving rise to an inward depolarizing current has been investigated by Haas et al. [(110), see also ref. 51]. Still other possible electrogenic systems may exist in cardiac tissues, for example a Cl-HCO_3 exchange mechanism that is thought to have a coupling ratio (bicarbonate-chloride) varying between about 2:1 and 1:1 (244).

A Na-Ca exchange mechanism has been described in nerve (12, 13) and heart (213, for review see ref. 40). It normally acts as a system for extruding Ca from the cells, but when $[Na]_o$ is reduced or $[Na]_i$ becomes very high, Ca accumulates within the cell through partial reversal of the Na-Ca exchange (105). In nerve fibers this system is electrogenic and therefore is sensitive to changes in membrane potential (14). In such a case, depolarization of the membrane would tend to slow down the exchange or even to reverse it, Na now being extruded and Ca entering the cell. The system will produce an inward depolarizing current when it extrudes Ca from the cell and an outward repolarizing current when it operates in the reverse direction. However, it has also been suggested that three Na ions can exchange for one Ca and one K (13). From experiments performed in the giant axon, a model has been developed where the binding of 4 Na to a carrier induces a Ca binding site on the opposite side of the membrane (179). In cardiac tissues an electroneutral exchange system has been proposed, 2 Na ions and 1 Ca ion competing for 1 transport site (carrier) on both sides of the membrane (21, 209). It has been shown in frog atrial fibers, however, that the transient increase in tonic tension elicited by low-Na media is prevented when the ratios $[Ca]_o/[Na]_o^4$ to $[Ca]_o/[Na]_o^6$ are kept constant. This observation and the fact that Na-free contracture is accompanied by a hyperpolarization and an outward current indicates that tonic tension is regulated by an electrogenic Na-Ca exchange mechanism (121, 123). A similar exchange of 1 Ca for 3 Na ions has also been recently proposed in frog atrial muscle (41). In dog Purkinje fibers a hyperpolarization always accompanies and in fact precedes by a few seconds the Na-free contracture, provided that the Na pump is inhibited by K-free media or ouabain; it has been shown that these associated phenomena, which are insensitive to many conductance inhibitors and are depressed by low pH, are related to electrogenic Na-Ca exchange (56, 57). It has also been suggested that the transient inward current (TI) which occurs during cardiotonic steroid intoxication may be due to electrogenic Na-Ca exchange (139). The generation of electric currents in cardiac fibers by the Na-Ca exchange has been analyzed from a theoretical point of view: assuming an exchange of 1 Ca for 4 Na, the reversal potential of the current is about -40 mV (180).

Concluding Remarks

Cardiac electrical phenomena resemble those in other excitable tissues in that they are essentially due to the operation of voltage- and time-dependent conductances. Extending to heart concepts built up from experiments performed mainly in nerve fibers, it may be considered that cardiac membranes possess specific ionic channels controlled by gates bearing electrical charges. When the gates open because of a change in membrane potential, they allow ions to cross the membrane under their electrochemical gradient. The resulting passive inward and outward currents are rather complex, since two inward (depolarizing) currents, three or four outward (repolarizing) currents, and a special pacemaker current have been described. At least two repolarizing currents are controlled by quasi-instantaneous inward rectifying processes. The existence of two inward currents flowing through the rapid and slow channel explains the difference between cardiac fibers exhibiting action potentials with a rapid upstroke and a rapid conduction (for example Purkinje fibers) and those exhibiting slowly developing and slowly conducted action potentials (essentially nodal cells). However, this difference is not related to the absence of a rapid channel in nodal cell membrane but to its inactivation as a consequence of the low diastolic potential in these cells. For the same reason partially depolarized cells also exhibit only slow action potentials. The slow diastolic depolarization responsible for the spontaneous triggering of a cardiac action potential results from two different mechanisms, depending on the potential range in which the pacemaker potential develops, namely 1) activation of an inward current, i_f, for fibers with a diastolic potential more negative than $\simeq -50$ mV, and 2) deactivation of $i_{x_1}(i_K)$ and concomitant activation of the slow inward current for fibers with a diastolic potential of $\simeq -50$ mV or less. It has been shown that, in addition to passive ionic currents, electrogenic active transports can participate in the development of electrical phenomena. A well-documented example is the electrogenic Na pump. Another one is the electrogenic Na-Ca exchange mechanism. Several other points have recently been developed: one is that the accumulation and depletion of ions (mainly potassium) in extracellular clefts probably change appreciably the extracellular ionic concentrations, even in the course of a single action potential; another point is that membrane conductances for a given ion depend markedly on extracellular and intracellular concentrations of other ions (control of g_K by $[Ca]_i$); the last point concerns the influence of alterations in nexus resistance on cardiac conduction. Clearly increases in $[Ca]_i$ as a result of metabolic alterations appear today as one of the major potential sources of conduction blocks.

Acknowledgments

The author gratefully acknowledges Miss Edith Deroubaix for excellent technical assistance and Drs. Louis De Felice, Jean-Luc Mazet, Denis Noble, Dario DiFrancesco, and Robert DeHaan for discussion and for helpful suggestions.

References

1. ADAMS, D. J., AND P. W. GAGE. Gating currents associated with sodium and calcium currents in an *Aplysia* neuron. *Science* 192: 783–784, 1976.
2. ADAMS, D. J., AND P. W. GAGE. Characteristics of sodium and calcium conductance changes produced by membrane depolarization in an *Aplysia* neurone. *J. Physiol. London* 289: 143–161, 1979.
3. ADAMS, D. J., AND P. W. GAGE. Sodium and calcium gating currents in an *Aplysia* neurone. *J. Physiol. London* 291: 467–481, 1979.
4. ADRIAN, R. H. Rectification in muscle membrane. *Progr. Biophys. Mol. Biol.* 19: 341–369, 1969.

5. ADRIAN, R. H., W. K. CHANDLER, AND A. L. HODGKIN. Voltage clamp experiments in striated muscle fibres. *J. Physiol. London* 208: 607–644, 1970.

6. AKSELROD, S., E. M. LANDAU, AND Y. LASS. Electromechanical noise in atrial muscle cells of carp: a possible ionic feed-back mechanism. *J. Physiol. London,* 290: 387–397, 1979.

7. ARMSTRONG, C. M., AND F. BEZANILLA. Inactivation of the sodium channel. II. Gating current experiments. *J. Gen. Physiol.* 70: 567–590, 1977.

8. ARMSTRONG, C. M., AND W. F. GILLY. Fast and slow steps in the activation of sodium channels. *J. Gen. Physiol.* 74: 691–711, 1979.

9. ATTWELL, D., AND I. COHEN. The voltage clamp of multicellular preparation. *Progr. Biophys. Mol. Biol.* 31: 201–245, 1977.

10. ATTWELL, D., I. COHEN, D. EISNER, M. OHBA, AND C. OJEDA. The steady-state TTX-sensitive ("Window") sodium current in cardiac Purkinje fibres. *Pfluegers Arch.* 379: 137–142, 1979.

11. ATTWELL, D. E., D. A. EISNER, AND I. COHEN. Voltage clamp and tracer flux data: effects of restricted extracellular space. *Q. Rev. Biophys.* 12: 213–261, 1979.

12. BAKER, P. F. Sodium-calcium exchange across the nerve cell membrane. In: *Calcium and Cellular Function,* edited by A. W. Cuthbert, London: Macmillan, 1970, p. 96–107.

13. BAKER, P. F. Transport and metabolism of calcium ions in nerve. *Progr. Biophys. Mol. Biol.* 24: 177–223, 1972.

14. BAKER, P. F., AND P. A. McNAUGTHON. The effect of membrane potential on the calcium transport systems in squid giant axons. *J. Physiol. London* 260: 24P–25P, 1976.

15. BASS, B. G. Restitution of the action potential in cat papillary muscle. *Am. J. Physiol.* 228: 1717–1724, 1975.

16. BASSINGTHWAIGHTE, J. B., C. H. FRY, AND J. A. S. McGUIGAN. Relationship between internal calcium and outward current in mammalian ventricular muscle; a mechanism for control of the action potential duration. *J. Physiol. London* 262: 15–37, 1976.

17. BAUMGARTEN, C. M., G. ISENBERG, T. F. McDONALD, AND R. E. TEN EICK. Depletion and accumulation of potassium in the extracellular clefts of cardiac Purkinje fibers during voltage clamp hyperpolarization and depolarization. *J. Gen. Physiol.* 70: 149–169, 1977.

18. BEELER, G. W., AND J. A. S. McGUIGAN. Voltage clamping of multicellular myocardial preparations: capabilities and limitations of existing methods. *Progr. Biophys. Mol. Biol.* 34: 219–254, 1978.

19. BEGENISICH, T. Magnitude and location of surface charges on *Myxicola* giant axons. *J. Gen. Physiol.* 66: 47–65, 1975.

20. BEGENISICH, T., AND C. LYNCH. Effects of internal divalent cations on voltage-clamped squid axons. *J. Gen. Physiol.* 63: 675–689, 1974.

21. BENNINGER, C., H. M. EINWÄCHTER, H. G. HAAS, AND R. KERN. Calcium-sodium antagonism on the frog's heart: a voltage clamp study. *J. Physiol. London* 259: 617–645, 1976.

22. BEZANILLA, F., AND C. M. ARMSTRONG. Kinetic properties and inactivation of gating currents of sodium channels in squid axons. *Phil. Trans. Roy. Soc. London* 270: 449–458, 1975.

23. BLEEKER, W. K., A. J. C. MACKAAY, M. MASSON-PEVET, L. N. BOUMAN, AND A. E. BECKER. Functional and morphological organization of the rabbit sinus node. *Circ. Res.* 46: 11–22, 1980.

24. BOSTEELS, S., AND E. CARMELIET. The components of the sodium efflux in cardiac Purkinje fibres. *Pfluegers Arch.* 336: 48–59, 1972.

25. BREHM, P., AND R. ECKERT. Calcium entry leads to inactivation of calcium channel in *Paramecium. Science* 202: 1203–1206, 1978.

26. BRENNECKE, R., AND B. LINDEMANN. A chopped-current clamp for current injection and recording of membrane depolarization with single electrodes of changing resistance. *T.I.T. J. Life Sci.* 1: 53–58, 1971.

27. BROWN, B. F., A. CLARK, AND S. J. NOBLE. Analysis of pacemaker and repolarization currents in frog atrial muscle. *J. Physiol London* 258: 547–577, 1976.

28. BROWN, H., AND D. DiFRANCESCO. Voltage clamp investigations of membrane currents underlying pace-maker activity in rabbit sino-atrial node. *J. Physiol. London* 308: 331–351, 1980.

29. BROWN, H. F., AND D. DiFRANCESCO, AND S. J. NOBLE. How does adrenaline accelerate the heart? *Nature London* 280: 235–236, 1979.

30. BROWN, H., D. DiFRANCESCO, D. NOBLE, AND S. J. NOBLE. The contribution of potassium accumulation to outward currents in frog atrium. *J. Physiol. London* 306: 127–149, 1980.

31. BROWN, H. F., W. GILES, AND S. J. NOBLE. Membrane currents underlying activity in frog sinus venosus. *J. Physiol. London* 271: 783–816, 1977.

32. BROWN, H. F., J. KIMURA, AND S. J. NOBLE. Evidence that the current i_f in sino-atrial node has a potassium component. *J. Physiol. London* 308: 33P–34P, 1980.
33. BROWN, H. F., AND S. J. NOBLE. Membrane currents underlying delayed rectification and pace-maker activity in frog atrial muscle. *J. Physiol. London* 204: 717–736, 1969.
34. BROWN, R. H., AND D. NOBLE. Displacement of activation thresholds in cardiac muscle by protons and calcium ions. *J. Physiol. London* 282: 333–343, 1978.
35. CARMELIET, E. *Chloride and Potassium Permeability in Cardiac Purkinje Fibers.* Brussels: Arscia and Presses Académiques Européennes, 1961.
36. CARMELIET, E., AND F. VERDONCK. Reduction of potassium permeability by chloride substitution in cardiac cells. *J. Physiol. London* 265: 193–206, 1977.
37. CARMELIET, E., AND J. VEREECKE. Adrenaline and the plateau phase of cardiac action potential. *Pfluegers Arch.* 313: 300–315, 1969.
38. CARMELIET, E., AND J. VEREECKE. Electrogenesis of the action potential and automaticity. In: *Handbook of Physiology. The Cardiovascular System*, edited by R. M. Berne and N. Sperelakis. Bethesda, MD: Am. Physiol. Soc., 1979, sect. 2, vol. I, chapt. 7, p. 269–334.
39. CARPENTIER, R., AND M. VASSALLE. Enhancement and inhibition of a frequency-activated electrogenic sodium pump in cardiac Purkinje fibers. In: *Research in Physiology: A Liber Memorialis in Honor of Prof. C. McC. Brooks*, edited by F. F. Kao, K. Koizumi, and M. Vassalle. Bologna: Aulo Gaggi, 1971, p 81–98.
40. CHAPMAN, R. A. Excitation-contraction coupling in cardiac muscle. *Progr. Biophys. Mol. Biol.* 35: 1–52, 1979.
41. CHAPMAN, R. A., AND J. TUNSTALL. The interaction of sodium and calcium ions at the cell membrane and the control of contractile strength in frog atrial muscle *J. Physiol. London* 305: 109–123, 1980.
42. CHESNAIS, J. M., E. CORABOEUF, M. P. SAUVIAT, AND J. M. VASSAS. Sensitivity to H, Li and Mg ions of the slow inward sodium current in frog atrial fibres. *J. Mol. Cell. Cardiol.* 7: 627–642, 1975.
43. CHEVAL, J. Enregistrements simultanés du potentiel et du courant transmembranaires à l'aide d'une seule microélectrode intracellulaire. *C. R. Acad. Sci., Paris* 277: 2521–2524, 1973.
44. CLAPHAM, D. E., A. SHRIER, AND R. L. DeHAAN. Junctional resistance and action potential delay between embryonic heart cell aggregates. *J. Gen. Physiol.* 75: 633–654, 1980.
45. CLAY, J., L. J. DeFELICE, AND R. L. DeHAAN. Current noise parameters derived from voltage noise and impedance in embryonic heart cell aggregates. *Biophys. J.* 28: 169–184, 1979.
46. COHEN, I., J. DAUT, AND D. NOBLE. The effects of potassium and temperature on the pace-maker current, I_{K_2}, in Purkinje fibres. *J. Physiol. London* 260: 55–74, 1976.
47. COLATSKY, T. J. Voltage clamp measurements of sodium channel properties in rabbit cardiac Purkinje fibres. *J. Physiol. London,* 305: 215–234, 1980.
48. COLATSKY, T. J., AND R. W. TSIEN. Sodium channels in rabbit cardiac Purkinje fibres. *Nature London* 278: 265–268, 1979.
49. CONTI, F., L. J. DE FELICE, AND E. WANKE. Potassium and sodium ion current noise in the membrane of squid giant axon. *J. Physiol. London* 248: 45–82, 1975.
50. CONTI, F., AND E. NEHER. Single channel recordings of K^+ currents in squid axons. *Nature London* 285: 140–143, 1980.
51. CORABOEUF, E. Aspects cellulaires de l'électrogénèse cardiaque chez les vertébrés. *J. Physiol. Paris* 52: 323–417, 1960.
52. CORABOEUF, E. Voltage clamp studies of the slow inward current. In: *The Slow Inward Current and Cardiac Arrythmias*, edited by D. P. Zipes, J. C. Bailey, and V. Elharrar, the Hague: Nijhoff, 1980, p. 25–95.
53. CORABOEUF, E., E. DEROUBAIX, AND A. COULOMBE. Effect of tetrodotoxin on action potentials of the conducting system in the dog heart. *Am. J. Physiol.* 236 (*Heart Circ. Physiol.* 5): H561–H567, 1979.
54. CORABOEUF, E., E. DEROUBAIX, AND A. COULOMBE. Acidosis-induced abnormal repolarization and repetitive activity in isolated dog Purkinje fibers. *J. Physiol. Paris,* 76: 97–106, 1980.
55. CORABOEUF, E., E. DEROUBAIX, AND J. HOERTER. Control of ionic permeabilities in normal and ischemic heart. *Circ. Res.* 38: 92–98, 1976.
56. CORABOEUF, E., P. GAUTIER, AND P. GUIRAUDOU. Effect of sodium removal on tension and membrane potential after inhibition of the sodium pump in dog Purkinje fibres. *J. Physiol. London* 281: 15P–16P, 1978.

57. CORABOEUF, E., P. GAUTIER, AND P. GUIRAUDOU. Potential and tension changes induced by sodium removal on dog Purkinje fibres: role of an electrogenic sodium-calcium exchange. *J. Physiol. London* 311: 605–622, 1981.

58. CORABOEUF, E., AND G. VASSORT. Effects of some inhibitors of ionic permeabilities on ventricular action potential and contraction of rat and guinea-pig hearts. *J. Electrocardiol.* 1: 19–30, 1969.

59. CORABOEUF, E., F. ZACOUTO, Y. M. GARGOÜIL, AND J. LAPLAUD. Mesure de la résistance membranaire du myocarde ventriculaire de mammifère au cours de l'activité. *C. R. Acad. Sci. Paris* 246: 2934–2937, 1958.

60. COULOMBE, A., E. CORABOEUF, AND E. DEROUBAIX. Computer simulation of acidosis-induced abnormal repolarization and repetitive activity in Purkinje fibers. *J. Physiol. Paris* 76: 107–112, 1980.

61. CRANEFIELD, P. F. *The Conduction of the Cardiac Impulse.* New York: Futura, 1975.

62. CRANEFIELD, P. F., AND B. F. HOFFMAN. Propagated repolarization in heart muscle. *J. Gen. Physiol.* 41: 633–649, 1958.

63. DAHL, G., AND G. ISENBERG. Decoupling of heart muscle cells: correlation with increased cytoplasmic calcium activity and with changes of nexus ultrastructure. *J. Membrane Biol.* 53: 63–75, 1980.

64. DECK, K. A., AND W. TRAUTWEIN. Ionic currents in cardiac excitation. *Pfluegers Arch.* 280: 63–80, 1964.

65. DEHAAN, R. L. Differentiation of excitable membranes *Curr. Top. Dev. Biol.* 16: 117–164, 1980.

66. DEHAAN, R L., AND L. J. DEFELICE. Oscillatory properties and excitability of the heart cell membrane. In: *Theoretical Chemistry: Advances and Perspectives,* edited by H. Eyring, New York: Academic 4: 181–133, 1978.

67. DE HEMPTINNE, A. Properties of the outward currents in frog atrial muscle. *Pfluegers Arch.* 329: 321–331, 1971.

68. DE HEMPTINNE, A. Voltage clamp analysis in isolated cardiac fibers as performed with two different perfusion chambers for double sucrose gap. *Pfluegers Arch.* 363: 87–95, 1976.

69. DE HEMPTINNE, A. Effet du cesium sur l'activité électrique des fibres auriculaires de grenouille (Abstract). *J. Physiol. Paris* 73: 5A, 1977.

70. DELAHAYES, J. F. Depolarization-induced movement of Mn^{2+} across the cell membrane in the guinea pig myocardium: its effect on the mechanical response. *Circ. Res.* 36: 713–718, 1975.

71. DELEZE, J. Possible reasons for the drop of the resting potential of mammalian heart preparations during hypothermia. *Circ. Res.* 8: 553–557, 1960.

72. DELEZE, J., AND W. R. LOEWENSTEIN. Permeability of a cell junction during intracellular injection of divalent cations. *J. Membrane Biol.* 28: 71–86, 1976.

73. DE MELLO, W. C. Some aspects of the interrelationship between ions and electrical activity in specialized tissue of the heart. In: *The Specialized Tissues of the Heart,* edited by A. Paes de Carvalho, W. C. De Mello, and B. F. Hoffman. Amsterdam: Elsevier, 1961, p. 95–107.

74. DE MELLO, W. C. Effect of intracellular injection of calcium and strontium on cell communication in heart. *J. Physiol. London* 250: 231–245, 1975.

75. DE MELLO, W. C. Influence of the sodium pump on intercellular communication in heart fibres: effect of intracellular injection of sodium ion on electrical coupling. *J. Physiol. London* 263: 171–197, 1976.

76. DIFRANCESCO, D. Could i_{K_2} in Purkinje fibres be an inward current activated on hyperpolarization? *J. Physiol. London* 305: 64P–65P, 1980.

77. DIFRANCESCO, D. A new interpretation of the pacemaker current in Purkinje fibres. *J. Physiol. London* 314: 359–376, 1981.

78. DIFRANCESCO, D. A study of the ionic nature of the pace-maker current in Purkinje fibres. *J. Physiol. London* 314: 377–393, 1981.

79. DIFRANCESCO, D., AND D. NOBLE. If "i_{K_2}" is an inward current, how does it display potassium specificity? *J. Physiol. London* 305: 14P–15P, 1980.

80. DIFRANCESCO, D., AND D. NOBLE. The time course of potassium current following potassium accumulation in frog atrium: analytical solutions using a linear approximation. *J. Physiol. London* 306: 151–173, 1980.

81. DIFRANCESCO, D., AND D. NOBLE. Implications of the re-interpretation of i_{K_2} for the modeling of the electrical activity of pacemaker tissue in the heart. In: *Cardiac Rate and Rhythm. Physiological, Methodological and Developmental Aspects,* edited by L. N. Bouman and H. J. Jongsma. The Hague, Netherlands: Nijhoff, 1981. In press.

82. DiFRANCESCO, D., AND C. OJEDA. Properties of the current i_f in the sino-atrial node of the rabbit compared with those of the current i_{K_2} in Purkinje fibres. *J. Physiol. London* 308: 353–367, 1980.

83. DUBOIS, J. M., AND C. BERGMAN. Cesium-induced rectifications in frog myelinated fibres. *Pfluegers Arch.* 355: 361–364, 1975.

84. DUCOURET, P. The effect of quinidine on membrane electrical activity in frog auricular fibres studied by current and voltage clamp. *Br. J. Pharmacol.* 57: 163–184, 1976.

85. EBIHARA, L., N. SHIGETO, M. LIEBERMAN, AND E. A. JOHNSON. The initial inward current in spherical clusters of embryonic heart cells. *J. Gen. Physiol.* 75: 437–456, 1980.

86. EISNER, D. A., AND W. J. LEDERER. Characterization of the electrogenic sodium pump in cardiac Purkinje fibres. *J. Physiol. London* 303: 441–474, 1980.

87. ELLIS, D. The effects of external cations and ouabain on the intracellular sodium activity of sheep heart Purkinje fibres. *J. Physiol. London* 273: 211–240, 1977.

88. ELLIS, E., AND J. W. DEITMER. The relationship between the intra- and extracellular sodium activity of sheep heart Purkinje fibres during inhibition of the Na-K pump. *Pfluegers Arch.* 377: 209–215, 1978.

89. FERRIER, G. R., AND G. K. MOE. Effect of calcium on acetylstrophanthidin induced transient depolarizations in canine Purkinje tissue. *Circ. Res.* 33: 508–515, 1973.

90. FOZZARD, H. A., AND G. W. BEELER. The voltage clamp and cardiac electrophysiology. *Circ. Res.* 37: 403–413, 1975.

91. FOZZARD, H. A., AND M. HIRAOKA. The positive dynamic current and its inactivation properties in cardiac Purkinje fibres. *J. Physiol. London* 234: 569–586, 1973.

92. FOZZARD, H. A., AND C. O. LEE. Influence of changes in external potassium and chloride ions on membrane potential and intracellular potassium ion activity in rabbit ventricular muscle. *J. Physiol. London* 256: 663–689, 1976.

93. FRANKENHAEUSER, B, AND A. L. HODGKIN. The action of calcium on the electrical properties of squid axons. *J. Physiol. London* 137: 218–244, 1957.

94. FUKUDA, J. Voltage clamp study on inward chloride currents of spherical muscle cells in tissue culture. *Nature London* 257: 408–410, 1975.

95. FURSHPAN, E. J., AND D. D. POTTER. Transmission at the giant motor synapses of the crayfish. *J. Physiol. London* 145: 289–325, 1959.

96. GADSBY, D. C., AND P. F. CRANEFIELD. Direct measurement of changes in sodium pump current in canine cardiac Purkinje fibers. *Proc. Natl. Acad. Sci. USA* 76: 1783–1787, 1979.

97. GADSBY, D. C., AND P. F. CRANEFIELD. Electrogenic sodium extrusion in cardiac Purkinje fibers. *J. Gen. Physiol.* 73: 819–837, 1979.

98. GARNIER, D., O. ROUGIER, Y. M. GARGOUÏL, AND E. CORABOEUF. Analyse electrophysiologique du plateau des reponses myocardiques; mise en évidence d'un courant lent entrant en absence d'ions bivalents. *Pfluegers Arch.* 313: 321–342, 1969.

99. GETTES, L. S., AND H. REUTER. Slow recovery from inactivation of inward currents in mammalian myocardial fibres. *J. Physiol. London* 240: 703–724, 1974.

100. GLITSCH, H. G. Über das Membranpotential des Meerchweinchenvorhofes nach Hypothermie. *Pfluegers Arch.* 307: 29–46, 1969.

101. GLITSCH, H. G. An effect of the electrogenic sodium pump on the membrane potential in beating guinea-pig atria. *Pfluegers Arch.* 344: 169–180, 1973.

102. GLITSCH, H. G. Characteristics of active Na transport in intact cardiac cells. *Am. J. Physiol.* 236 (*Heart Circ. Physiol.* 5): H189–H199, 1979.

103. GLITSCH, H. G., W. GRABOWSKI, AND J. THIELEN. Activation of the electrogenic sodium pump in guinea-pig atria by external potassium ions. *J. Physiol. London* 276: 515–524, 1978.

104. GLITSCH, H. G., H. PUSCH, AND K. VENETZ. Effects of Na and K ions on the active Na transport in guinea-pig auricles. *Pfluegers Arch.* 365: 29–36, 1976.

105. GLITSCH, H. G., H. REUTER, AND H. SCHOLZ. The effect of the internal sodium concentration on calcium fluxes in isolated guinea-pig auricles. *J. Physiol. London* 209: 25–43, 1970.

106. GOODENOUGH, D. A. The structure and permeability of isolated hepatocyte gap junctions. *Cold Spring Harbor Symp. Quant. Biol.* 40: 37–44, 1976.

107. GORMAN, A. L. F., AND A. HERMANN. Internal effects of divalent cations on potassium permeability in molluscan neurones. *J. Physiol. London* 296: 393–410, 1979.

108. GROS, D., AND C. E. CHALLICE. Early development of gap junctions between the mouse embryonic myocardial cells. A freeze-etching study. *Experientia* 32: 996–997, 1976.

109. HAAS, H. G. Active ion transport in heart muscle. In: *Electrical Phenomena in the Heart*, edited by W. C. De Mello. New York: Academic, 1972, p. 163–189.

110. HAAS, H. G., R. KERN, AND H. M. EINWÄCHTER. Electrical activity and metabolism in cardiac tissue: an experimental and theoretical study. *J. Membrane Biol.* 3: 180–209, 1970.

111. HAAS, H. G., R. KERN, H. M. EINWÄCHTER, AND M. TARR. Kinetics of Na inactivation in frog atria. *Pfluegers Arch.* 323: 141–157, 1971.

112. HAUSWIRTH, O., D. NOBLE, AND R. W. TSIEN. The mechanism of oscillatory activity at low membrane potentials in cardiac Purkinje fibres. *J. Physiol. London* 200: 255–265, 1969.

113. HAUSWIRTH, O., D. NOBLE, AND R. W. TSIEN. The dependence of plateau currents in cardiac Purkinje fibres on the interval between action potentials. *J. Physiol. London* 222: 27–49, 1972.

114. HELLAM, D. C., AND J. W. STUDT. A core-conductor model of the cardiac Purkinje fibre based on structural analysis. *J. Physiol. London* 243: 637–660, 1974.

115. HILLE, B. Gating in sodium channels of nerve. *Ann. Rev. Physiol.* 38: 139–152, 1976.

116. HILLE, B., A. M. WOODHULL, AND B. I. SHAPIRO. Negative surface charge near sodium channels of nerve: divalent ions, monovalent ions, and pH. *Phil. Trans. R. Soc. London* 270: 301–318, 1975.

117. HIRAOKA, M., AND H. H. HECHT. Recovery from hypothermia in cardiac Purkinje fibers. Considerations for an electrogenic mechanism. *Pfluegers Arch.* 339: 25–36, 1973.

118. HIRAOKA, M., AND T. SANO. Role of slow inward current in the genesis of ventricular arrhythmia. *Jpn. Circ. J.* 40: 1419–1427, 1976.

119. HODGKIN, A. L., AND A. F. HUXLEY. A quantitative description of membrane current and its application to conduction and excitation in nerve. *J. Physiol. London* 117: 500–544, 1952.

120. HODGKIN, A. L., AND R. D. KEYNES. Active transport of cations in giant axons from Sepia and Loligo. *J. Physiol. London* 128: 28–60, 1955.

121. HORACKOVA, M., AND G. VASSORT. Regulation of tonic tension in frog atrial muscle by voltage-dependent Na-Ca exchange (Abstract). *J. Physiol. London* 258: 77P, 1976.

122. HORACKOVA, M., AND G. VASSORT. Calcium conductance in relation to contractility in frog myocardium. *J. Physiol. London* 259: 597–616, 1976.

123. HORACKOVA, M., AND G. VASSORT. Sodium-calcium exchange in regulation of cardiac contractility. *J. Gen. Physiol.* 73: 403–424, 1979.

124. HOROWICZ, P., AND C. J. GERBER. Effects of external potassium and strophanthidin on sodium flux in frog striated muscle. *J. Gen. Physiol.* 48: 489–514, 1965.

125. HUTTER, O. F., AND D. NOBLE. Rectifying properties of cardiac muscle. *Nature London* 188: 495, 1960.

126. IMANAGA, I. Cell-to-cell diffusion of Procion yellow in sheep and calf Purkinje fibers. *J. Membrane Biol.* 16: 381–388, 1974.

127. ISENBERG, G. Is potassium conductance of cardiac Purkinje fibres controlled by $[Ca^{2+}]_i$? *Nature London* 253: 273–274, 1975.

128. ISENBERG, G. Cardiac Purkinje fibers: cesium as a tool to block inward rectifying potassium currents. *Pfluegers Arch.* 365: 99–106, 1976.

129. ISENBERG, G. Cardiac Purkinje fibres. The slow inward current component under the influence of modified $[Ca^{2+}]_i$. *Pfluegers Arch.* 371: 61–69, 1977.

130. ISENBERG, G. Cardiac Purkinje fibres: $[Ca^{2+}]_i$ controls the potassium permeability via the conductance components g_{K_1} and g_{K_2}. *Pfluegers Arch.* 371: 77–85, 1977.

131. ISENBERG, G., AND W. TRAUTWEIN. Effect of dihydro-ouabain and lithium ions on the outward current in cardiac Purkinje fibers. *Pfluegers Arch.* 350: 41–54, 1974.

132. ISENBERG, G., AND W. TRAUTWEIN. Outward current and electrogenic sodium pump. In: *Recent Advances in Studies on Cardiac Structure and Metabolism*, edited by A. Fleckenstein and N. S. Dhalla. Baltimore: University Park Press, 1975, p. 43–49.

133. IRISAWA, H. Comparative physiology of the cardiac pacemaker mechanism. *Physiol. Rev.* 58: 461–498, 1978.

134. JOHNSON, E. A., AND M. LIEBERMAN. Heart: excitation and contraction. *Ann. Rev. Physiol.* 33: 479–532, 1971.

135. KASS, R. S., W. J. LEDERER, R. W. TSIEN, AND R. WEINGART. Role of calcium ions in transient inward currents and aftercontractions induced by strophanthidin in cardiac Purkinje fibres. *J. Physiol. London,* 281: 187–208, 1978.

136. KASS, R. S., S. SIEGELBAUM, AND R. W. TSIEN. Incomplete inactivation of the slow inward current in cardiac Purkinje fibres. *J. Physiol. London* 263: 127P–128P 1976.

137. KASS, R. S., S. A. SIEGELBAUM, AND R. W. TSIEN. Three-micro-electrode voltage clamp experiments in calf cardiac Purkinje fibres: is slow inward current adequately measured? *J. Physiol. London* 290: 201–225, 1979.

138. KASS, R. S., AND R. W. TSIEN. Control of action potential duration by calcium ions in cardiac Purkinje fibers. *J. Gen. Physiol.* 67: 599–617, 1976.
139. KASS, R. S., R. W. TSIEN, AND R. WEINGART. Ionic basis of transient inward current induced by strophanthidin in cardiac Purkinje fibres. *J. Physiol. London* 281: 209–226, 1978.
140. KATZUNG, B. G., AND J. A. MORGENSTERN. Effects of extracellular potassium on ventricular automaticity and evidence for a pacemaker current in mammalian ventricular myocardium. *Circ. Res.* 40: 105–111, 1977.
141. KENYON, J. L., AND W. R. GIBBONS. Influence of chloride, potassium, and tetraethyl-ammonium on the early outward current of sheep cardiac Purkinje fibers. *J. Gen. Physiol.* 73: 117–138, 1979.
142. KENYON, J. L., AND W. R. GIBBONS. 4-aminopyridine and the early outward current of sheep cardiac Purkinje fibres. *J. Gen. Physiol.* 73: 139–157, 1979.
143. KEYNES, R. D., AND E. ROJAS. Kinetics and steady-state properties of the charged system controlling sodium conductance in the squid giant axon. *J. Physiol. London* 239: 393–434, 1974.
144. KLEBER, A. G. Effects of sucrose solution on the longitudinal tissue resistivity of trabecular muscle from mammalian heart. *Pfluegers Arch.* 345: 195–206, 1973.
145. KLINE, R. P., I. COHEN, R. FALK, AND J. KUPERSMITH. Activity-dependent extracellular K⁺ fluctuations in canine Purkinje fibres. *Nature London* 286: 68–71, 1980.
146. KLINE, R., AND M. MORAD. Potassium efflux and accumulation in heart muscle. *Biophys. J.* 16: 367–372, 1976.
147. KOHLHARDT, M., B. BAUER, A. KRAUSE, AND A. FLECKENSTEIN. Selective inhibition of transmembrane Ca conductivity of mammalian myocardial fibres by Ni, Co and Mn ions. *Pfluegers Arch.* 338: 115–123, 1973.
148. KOHLHARDT, M., H. R. FIGULLA, AND O. TRIPATHI. The slow membrane channel as the predominant mediator of the excitation process of the sinoatrial pacemaker cell. *Basic Res. Cardiol.* 71: 17–26, 1976.
149. KOLHARDT, M., H. P. HAASTERT, AND H. KRAUSE. Evidence of nonspecificity of the Ca channel in mammalian myocardial fibre membranes. *Pfluegers Arch.* 342: 125–136, 1973.
150. KOLHARDT, M., H. KRAUSE, M. KÜBLER, AND A. HERDEY. Kinetics of inactivation and recovery of the slow inward current in mammalian ventricular myocardium. *Pfluegers Arch.* 355: 1–17, 1975.
151. KOSTYUK, P. G., O. A. KRISHTAL, AND V. I. PIDOPLICHKO. Potential-dependent membrane current during the active transport of ions in snail neurones. *J. Physiol. London* 226: 373–392, 1972.
152. KOSTYUK, P. G., O. A. KRISHTAL, AND V. I. PIDOPLICHKO. Asymmetrical displacement currents in nerve cell membrane and effect of internal fluoride. *Nature London* 267: 70–72, 1977.
153. KREITNER, D. Evidence for the existence of a rapid sodium channel in the membrane of rabbit sinoatrial cells. *J. Mol. Cell. Cardiol.* 7: 655–662, 1975.
154. KREITNER, D. Effects of polarization and inhibition of ionic conductances on the action potentials of nodal and perinodal fibers in rabbit sinoatrial node. In: *The Sinus Node*, edited by F.I.M. Bonke, The Hague; Nijhoff, p. 270–278, 1977.
155. LANGER, G. A., J. S. FRANK, L. M. NUDD, AND K. SERAYDARIAN. Sialic acid: effect of removal on calcium exchangeability of cultured heart cells. *Science* 193: 1013–1015, 1976.
156. LATORRE, R., AND J. E. HALL. Dipole potential measurements in asymmetric membranes. *Nature London* 264: 361–363, 1976.
157. LEDERER, W. J., AND R. W. TSIEN. Transient inward current underlying arrhythmogenic effects of cardiotonic steroids in Purkinje fibres. *J. Physiol. London* 263: 73–100, 1976.
158. LEE, C. O., AND H. A. FOZZARD. Activities of potassium and sodium ions in rabbit heart muscle. *J. Gen. Physiol.* 65: 695–708, 1975.
159. LEE, K. S., T. A. WEEKS, R. L. KAO, N. AKAIKE, AND A. M. BROWN. Sodium current in single heart muscle cells. *Nature London* 278: 269–271, 1979.
160. LENFANT, J., J. MIRONNEAU, AND J. K. AKA. Activité répétitive de la fibre sino-auriculaire de grenouille: analyse des courants transmembranaires responsables de l'automatisme cardiaque. *J. Physiol. Paris* 64: 5–18, 1972.
161. LENFANT, J., J. MIRONNEAU, Y. M. GARGOUÏL, AND G. GALAND. Analyse de l'activité électrique spontanée du centre de l'automatisme cardiaque de lapin par les inhibiteurs de perméabilités membranaires. *C. R. Acad. Sci. Paris* 266: 901–904, 1968.
162. LIPSIUS, S. L., AND M. VASSALLE. Characterization of a two-component upstroke in the sinus node subsidiary pacemakers. In: *The Sinus Node*, edited by F.I.M. Bonke, The Hague; Nijhoff, p. 233–244, 1977.

163. LIPSIUS, S. L., AND M. VASSALLE. Dual excitatory channels in the sinus node. *J. Mol. Cell. Cardiol.* 10: 753–767, 1978.

164. LOEWENSTEIN, W. R. Permeable junctions. *Cold Spring Harbor Symp. Quant. Biol.* 40: 49–63, 1976.

165. LOEWENSTEIN, W. R., Y. KANNO, AND S. J. SOCOLAR. Quantum jumps of conductance during formation of membrane channels at cell-cell junction. *Nature London* 274: 133–136, 1978.

166. LÜTTGAU, H. C., AND H. GLITSCH. Membrane physiology of nerve and muscle fibres. New York: Fischer, 1976.

167. MAZET, F., AND J. CARTAUD. Freeze-fracture studies of frog atrial fibres. *J. Cell Sci.* 22: 427–434, 1976.

168. MCALLISTER, R. E., D. NOBLE, AND R. W. TSIEN. Reconstruction of the electrical activity of cardiac Purkinje fibres. *J. Physiol. London* 251: 1–59, 1975.

169. MCDONALD, T. F., H. NAWRATH, AND W. TRAUTWEIN. Membrane currents and tension in cat ventricular muscle treated with cardiac glycosides. *Circ. Res.* 37: 674–682, 1975.

170. MCDONALD, T. F., D. PELZER, AND W. TRAUTWEIN. On the mechanism of slow calcium channel block in heart. *Pfluegers Arch.* 385: 175–179, 1980.

171. MCDONALD, T. F., AND W. TRAUTWEIN. Membrane currents in cat myocardium: separation of inward and outward components. *J. Physiol. London* 274: 193–216, 1978.

172. MCDONALD, T. F., AND W. TRAUTWEIN. The potassium current underlying delayed rectification in cat ventricular muscle. *J. Physiol. London* 274: 217–246, 1978.

173. MCGUIGAN, J. A. S. Some limitations of the double sucrose gap and its use in a study of the slow outward current in mammalian ventricular muscle. *J. Physiol. London* 240: 775–806, 1974.

174. MCNAUGHTON, P. Dependence of plateau potassium currents on Ca (Abstract). *Seventh European Congress of Cardiology, Amsterdam.* p. 140, 1976.

175. MEECH, R. W., AND N. B. STANDEN. Potassium activation in *Helix aspersa* neurones under voltage clamp: a component mediated by calcium influx. *J. Physiol London* 249: 211–239, 1975.

176. MEVES, H. The effect of holding potential on the asymmetry currents in squid giant axons. *J. Physiol. London* 243: 847–867, 1974.

177. MIURA, D. S., B. F. HOFFMAN, AND M. R. ROSEN. The effect of extracellular potassium on intracellular potassium ion activity and transmembrane potentials of beating canine Purkinje fibers. *J. Gen. Physiol.* 69: 463–474, 1977.

178. MOBLEY, B. A., AND E. PAGE. The surface area of sheep cardiac Purkinje fibres. *J. Physiol. London* 220: 547–563, 1972.

179. MULLINS, L. J. A mechanism for Na/Ca transport. *J. Gen. Physiol.* 70: 681–695, 1977.

180. MULLINS, L. J. The generation of electric currents in cardiac fibers by Na/Ca exchange. *Am. J. Physiol.* 236 (Cell Physiol. 5): C103–C110, 1979.

181. NARGEOT, J. Current clamp and voltage clamp study of the inhibitory action of DNP on membrane electrical properties of frog auricular heart muscle. *J. Physiol. Paris* 72: 171–180, 1976.

182. NATHAN, R. D., AND R. L. DEHAAN. Voltage clamp analysis of embryonic heart cell aggregates. *J. Gen. Physiol.* 73: 175–198, 1979.

183. NEW, W., AND W. TRAUTWEIN. Inward membrane current in mammalian myocardium. *Pfluegers Arch.* 334: 1–23, 1972.

184. NEW, W., AND W. TRAUTWEIN. The ionic nature of slow inward current and its relation to contraction. *Pfluegers Arch.* 334: 24–38, 1972.

185. NOBLE, D. A modification of the Hodgkin-Huxley equations applicable to Purkinje fibre action and pacemaker potentials. *J. Physiol. London* 160: 317–352, 1962.

186. NOBLE, D. Electrical properties of cardiac muscle attributable to inward going (anomalous) rectification. *J. Cell Comp. Physiol.* 66: 127–136, 1965.

187. NOBLE, D. *The Initiation of the Heartbeat* (2nd ed.). Oxford: Clarendon, 1979.

188. NOBLE, D., AND R. W. TSIEN. The kinetics and rectifier properties of the slow potassium current in cardiac Purkinje fibres. *J. Physiol. London* 195: 185–214, 1968.

189. NOBLE, D., AND R. W. TSIEN. Outward membrane currents activated in the plateau range of potentials in cardiac Purkinje fibres. *J. Physiol. London* 200: 205–231, 1969.

190. NOBLE, D., AND R. W. TSIEN. Reconstruction of the repolarization process in cardiac Purkinje fibres based on voltage-clamp measurements of membrane current. *J. Physiol. London* 200: 233–254, 1969.

191. NOBLE, S. J. Potassium accumulation and depletion in frog atrial muscle. *J. Physiol. London* 258: 579–613, 1976.

192. NOMA, A., AND H. IRISAWA. Electrogenic sodium pump in rabbit sinoatrial node cell. *Pfluegers Arch.* 351: 177–182, 1974.
193. NOMA, A., AND H. IRISAWA. Membrane currents in the rabbit sinoatrial node cell as studied by double microelectrode method. *Pfluegers Arch.* 364: 45–52, 1976.
194. NOMA, A., AND H. IRISAWA. A time- and voltage-dependent potassium current in the rabbit sinoatrial node cell. *Pfluegers Arch.* 366: 251–258, 1976.
195. OCHI, R. The slow inward current and the action of manganese ions in guinea-pig's myocardium. *Pfluegers Arch.* 316: 81–94, 1970.
196. OCHI, R. Manganese-dependent propagated action potentials and their depression by electrical stimulation in guinea-pig myocardium perfused by sodium-free media. *J. Physiol. London* 263: 139–156, 1976.
197. OJEDA, C., AND O. ROUGIER. Kinetic analysis of the delayed outward currents in frog atrium. Existence of two types of preparation. *J. Physiol. London* 239: 51–73, 1974.
198. PAGE, E., AND S. R. STORM. Cat heart muscle in vitro. VIII. Active transport of sodium in papillary muscles. *J. Gen. Physiol.* 48: 957–972, 1965.
199. PAYET, D. Effet de l'anoxie, du calcium et de quelques inhibiteurs sur le courant lent du myocarde de rat. (Ph.D. thesis). University of Sherbrooke, Canada, July, 1977.
200. PAYET M. D., O. F. SCHANNE, E. RUIZ-CERETTI, AND J. M. DEMERS. Inhibitory activity of blockers of the slow inward current in rat myocardium, a study in steady-state and of rate of action. *J. Mol. Cell. Cardiol.* 12: 187–200, 1980.
201. PEPER, K., AND W. TRAUTWEIN. A membrane current related to the plateau of the action potential of Purkinje fibers. *Pfluegers Arch.* 303: 108–123, 1968.
202. POINDESSAULT, J. P., A. DUVAL, AND C. LEOTY. Voltage clamp with double sucrose gap technique. External series resistance compensation. *Biophys. J.* 16: 105–120, 1976.
203. POST, R. L., C. D. ALBRIGHT, AND K. DAYANI. Resolution of pump and leak components of sodium and potassium ion transport in human erythrocytes. *J. Gen. Physiol.* 50: 1201–1220, 1967.
204. RAMON, F., N. ANDERSON, R. W. JOYNER, AND J. W. MOORE. Axon voltage-clamp simulations. IV. A multicellular preparation. *Biophys. J.* 15: 55–69, 1975.
205. REUTER, H. The dependence of slow inward current in Purkinje fibres on the extracellular calcium concentration. *J. Physiol. London* 192: 479–492, 1967.
206. REUTER, H. Slow inactivation of currents in cardiac Purkinje fibres. *J. Physiol. London* 197: 233–253, 1968.
207. REUTER, H. Divalent cations as charge carriers in excitable membranes. *Progr. Biophys. Mol. Biol.* 26: 3–43, 1973.
208. REUTER, H. Localization of beta adrenergic receptors, and effects of noradrenaline and cyclic nucleotides on action potentials, ionic currents, and tension in mammalian cardiac muscle. *J. Physiol. London* 242: 429–451, 1974.
209. REUTER, H. Exchange of calcium ions in the mammalian myocardium. *Circ. Res.* 34: 599–605, 1974.
210. REUTER, H. Properties of two inward membrane currents in the heart. *Ann. Rev. Physiol.* 41: 413–424, 1979.
211. REUTER, H., AND H. SCHOLZ. A study of the ion selectivity and the kinetics properties of the calcium dependent slow inward current in mammalian muscle. *J. Physiol. London* 264: 17–47, 1977.
212. REUTER, H., AND H. SCHOLZ. The regulation of the calcium conductance of cardiac muscle by adrenaline. *J. Physiol. London* 264: 49–62, 1977.
213. REUTER, H., AND N. SEITZ. The dependence of calcium efflux from cardiac muscle on temperature and external ion composition. *J. Physiol. London* 195: 451–470, 1968.
214. REVEL, J. P., AND M. J. KARNOVSKY. Hexagonal array of subunits in intercellular junctions of the mouse heart and liver (Abstract). *J. Cell Biol.* 33: C7, 1967.
215. ROUGIER, O., G. VASSORT, D. GARNIER, Y. M. GARGOUÏL, AND E. CORABOEUF. Existence and role of a slow inward current during the frog atrial action potential. *Pfluegers Arch.* 308: 91–110, 1969.
216. SAUVIAT, M. P. Effects of ervatamine chlorhydrate on cardiac membrane currents in frog atrial fibres. *Br. J. Pharmacol.* 71: 41–49, 1980.
217. SCHANNE, O. F., M. D. PAYET., AND E. RUIZ-CERETTI. Voltage control during inward current flow in rat ventricular muscle using a double sucrose gap technique. *Can. J. Physiol. Pharmacol.* 57: 124–127, 1979.

218. SCHNEIDER, J. A., AND N. SPERELAKIS. Slow Ca^{++} and Na^+ responses induced by isoproterenol and methylxanthines in isolated perfused guinea pig hearts exposed to elevated K^+. *J. Mol. Cell. Cardiol.* 7: 249–273, 1975.

219. SIEGELBAUM, S. A., AND R. W. TSIEN. Calcium-activated transient outward current in calf cardiac Purkinje fibres. *J. Physiol. London* 299: 485–506, 1980.

220. SIEGELBAUM, S. A., R. W. TSIEN, AND R. S. KASS. Role of intracellular calcium in the transient outward current of calf Purkinje fibres. *Nature London* 269: 611–613, 1977.

221. SIGWORTH, F. J., AND E. NEHER. Single Na^+ channel currents observed in cultured rat muscle cells. *Nature London* 287: 447–449, 1980.

222. SPERELAKIS, W. Origin of the cardiac resting potential. In: *Handbook of Physiology. The Cardiovascular System*, edited by R.M. Berne and N. Sperelakis. Bethesda, MD: Am. Physiol. Soc., 1979, sect. 2, vol. I, chapt. 6, p. 187–267.

223. SPERELAKIS, N., T. HOSHIKO, AND R. M. BERNE. Nonsyncytial nature of cardiac muscle: membrane resistance of single cells. *Am. J. Physiol.* 198: 531–536, 1960.

224. SPERELAKIS, N., AND R. L. MACDONALD. Ratio of transverse to longitudinal resistivities of isolated cardiac muscle fiber bundles. *J. Electrocardiol.* 7: 301–314, 1974.

225. SPERELAKIS, N., AND J. A. SCHNEIDER. A metabolic control mechanism for calcium ion influx that may protect the ventricular myocardial cell. *Am. J. Cardiol.* 37: 1079–1085, 1976.

226. STEVENS, C. F. Principles and applications of fluctuation analysis: a nonmathematical introduction. *Federation Proc.* 34: 1364–1369, 1975.

227. TAHARA, T., H. KIMIZUKA, AND K. KOKETSU. An analysis of the membrane hyperpolarization during action of the sodium pump in frog's skeletal muscles. *Jpn. J. Physiol.* 23: 165–182, 1973.

228. TARR, M. Two inward currents in frog atrial muscle. *J. Gen. Physiol.* 58: 523–543, 1971.

229. TARR, M., E. F. LUCKSTEAD, P. A. JURENWICZ, AND H. G. HAAS. Effect of propranolol on the fast sodium current in frog atrial muscle. *J. Pharmacol. Exptl. Therap.* 184: 599–610, 1973.

230. TARR, M., AND J. TRANK. Equivalent circuit of frog atrial tissue as determined by voltage clamp-unclamp experiments. *J. Gen. Physiol.* 58: 511–522, 1971.

231. THOMAS, R. C. Membrane current and intracellular sodium changes in a snail neurone during extrusion of injected sodium. *J. Physiol. London* 201: 495–514, 1969.

232. THOMAS, R. C. Electrogenic sodium pump in nerve and muscle cells. *Physiol. Rev.* 52: 563–594, 1972.

233. THOMAS, R. C. Intracellular sodium activity and the sodium pump in snail neurones. *J. Physiol. London* 220: 55–71, 1972.

234. TRAUTWEIN, W. Membrane currents in cardiac muscle fibers. *Physiol. Rev.* 53: 793–835, 1973.

235. TRAUTWEIN, W., AND D. G. KASSEBAUM. On the mechanism of spontaneous impulse generation in the pacemaker of the heart. *J. Gen. Physiol.* 45: 317–330, 1961.

236. TRAUTWEIN, W., T. F. MCDONALD, AND O. TRIPATHI. Calcium conductance and tension in mammalian ventricular muscle. *Pfluegers Arch.* 354: 55–74, 1975.

237. TSIEN, R. W., W. R. GILES, AND P. GREENGARD. Cyclic AMP mediates the action of adrenaline on the potential plateau of cardiac Purkinje fibres. *Nature London New Biol.* 240: 181–183, 1972.

238. VAN BOGAERT, P. P., J. S. VEREECKE, AND E. CARMELIET. The effect of raised pH on pacemaker activity and ionic currents in cardiac Purkinje fibers. *Pfluegers Arch.* 375: 45–52, 1978.

239. VASSALLE, M. Electrogenic suppression of automaticity in sheep and dog Purkinje fibers. *Circ. Res.* 27: 361–377, 1970.

240. VASSALLE, M. Cardiac automaticity and its control. *Am. J. Physiol.* 233 (*Heart Circ. Physiol.* 2): H625–H634, 1977.

241. VASSALLE, M. Effect of pharmacological interventions on atrial pacemakers. In: *Physiology of Atrial Pacemakers and Conductive Tissues*, edited by R. C. Little, New York: Futura, p. 315–337, 1980.

242. VASSALLE, M., AND R. CARPENTIER. Hyperpolarizing and depolarizing effects of norepinephrine in cardiac Purkinje fibers. In: *Research in Physiology. A Liber Memorialis in Honor of Prof. C. McC. Brooks*, edited by F. F. Kao, K. Koizumi, and M. Vassalle. Bologna: Aulo Gaggi, 1971, p. 373–388.

243. VASSORT, G., O. ROUGIER, D. GARNIER, M. P. SAUVIAT, E. CORABOEUF, AND Y. M. GARGOUÏL. Effects of adrenaline on membrane inward currents during the cardiac action potential. *Pfluegers Arch.* 309: 70–81, 1969.

244. VAUGHAN-JONES, R. D. Regulation of chloride in quiescent sheep-heart Purkinje fibres studies using intracellular chloride and pH-sensitive microelectrodes. *J. Physiol. London* 295: 111–137, 1979.

245. VEREECKE, J., AND E. CARMELIET. Sr action potentials in cardiac Purkinje fibres. I, II. *Pfluegers Arch.* 322: 60–82, 1971.
246. VEREECKE, J., G. ISENBERG, AND E. CARMELIET. K efflux through inward rectifying K channels in voltage clamped Purkinje fibers. *Pfluegers Arch.* 384: 207–217, 1980.
247. WALKER, J. L., AND R. O. LADLE. Frog heart intracellular potassium activities measured with potassium microelectrodes. *Am. J. Physiol.* 225: 263–267, 1973.
248. WATANABE, A. M., AND H. R. BESCH, JR. Cyclic adenosine monophosphate modulation of slow calcium influx channels in guinea-pig hearts. *Circulation Res.* 35: 316–324, 1974.
249. WEIDMANN, S. Effect of current flow on the membrane potential of cardiac muscle. *J. Physiol. London* 115: 227–236, 1951.
250. WEIDMANN, S. The electrical constants of Purkinje fibers. *J. Physiol. London* 118: 348–360, 1952.
251. WEIDMANN, S. The effect of the cardiac membrane potential on the rapid availability of the sodium carrying system. *J. Physiol. London* 127: 213–224, 1955.
252. WEIDMANN, S. *Elektrophysiologie der Herzmuskelfaser.* Stuttgart: Huber, 1956.
253. WEIDMANN, S. The diffusion of radiopotassium across intercalated disks of mammalian cardiac muscle. *J. Physiol. London* 187: 323–342, 1966.
254. WEIDMANN, S. Electrical coupling between myocardial cells. *Progr. Brain Res.* 31: 274–281, 1969.
255. WEIDMANN, S. Heart: electrophysiology. *Ann. Rev. Physiol.* 36: 155–169, 1974.
256. WEINGART, R. The permeability to tetraethylammonium ions of the surface membrane and the intercalated disks of sheep and calf myocardium. *J. Physiol. London* 240: 741–762, 1974.
257. WEINGART, R. The actions of ouabain on intercellular coupling and conduction velocity in mammalian ventricular muscle. *J. Physiol. London* 264: 341–365, 1977.
258. WEINGART, R., R. S. KASS, AND R. W. TSIEN. Is digitalis inotropy associated with enhanced slow inward calcium current? *Nature London* 273: 389–391, 1978.
259. WEISS, R., H. TRITTHART, AND B. WALTER. Correlation of Na-withdrawal effects on Ca-mediated action potentials and contractile activity in cat myocardium. *Pfluegers Arch.* 350: 299–307, 1974.
260. WHITTAM, R. Control of membrane permeability to potassium in red blood cells. *Nature London* 219: 610, 1968.
261. WIGGINS, J. R., AND P. F. CRANEFIELD. Effect on membrane potential and electrical activity of adding sodium to sodium-depleted cardiac Purkinje fibers. *J. Gen. Physiol.* 64: 473–493, 1974.
262. WILSON, W. A., AND M. M. GOLDNER. Voltage clamping with single microelectrode. *J. Neurobiol.* 6: 411–422, 1975.
263. WOJTCZAK, J. Contractures and increase in internal longitudinal resistance of cow ventricular muscle induced by hypoxia. *Circ. Res.* 44: 88–95, 1979.
264. YANAGIHARA, K. AND H. IRISAWA. Inward current activated during hyperpolarization in the rabbit sinoatrial node cell. *Pfluegers Arch.* 385: 11–19, 1980.
265. ZIPES, D. P., J. C. BAILEY, AND V. ELHARRAR. *The Slow Inward Current and Cardiac Arrhythmias.* The Hague: Nijhoff, 1980.

CHAPTER 2

Characteristics of Active Sodium Transport in Intact Cardiac Cells

H. G. GLITSCH
Department of Cell Physiology, Ruhr University Bochum,
Federal Republic of Germany

Methods for Studies on Active
Na Transport

Dependence of Active Na
Transport on [Na]$_i$

Dependence of Active Na
Transport on [K]$_o$

Effect of Other Ions on Active
Na Transport

Ionic Dependence of Cardiac
Transport ATPase Compared to

Active Na Transport

Linkage Between Active Na and
K Transport

Electrogenic Na Pumping

Effect of Active Na Transport
on Cardiac Contraction

Effect of Active Na Transport on
Intercellular Coupling in Heart

Role of Active Na Transport in
Prevention of Cardiac Arrhythmias

Like other animal cells, heart muscle cells have an intracellular sodium concentration ([Na]$_i$) much lower than the Na concentration of the extracellular fluid ([Na]$_o$). [Na]$_i$ is about 15–30 mmol/liter fiber water in cardiac preparations bathed in physiological media (for references see refs. 44 and 79). The internal K concentration ([K]$_i$) is 30 times higher than the external ([K]$_o$) and amounts to about 150 mmol/liter fiber water. Thus, the Na concentration gradient across the cardiac cell membrane is directed inward, whereas the K gradient is directed outward.

In addition to the Na and K concentration gradients, the membrane potential governs the passive Na and K fluxes across the heart cell membrane. The resting potential (about −75 mV in atrial and ventricular cells) promotes the influx of cations into the cell and inhibits the efflux. The potential at which the passive influx of an ion species equals the passive efflux is called the equilibrium potential of this ion. However, the potassium equilibrium potential (E_K) and the sodium equilibrium potential (E_{Na}) differ from the resting potential in heart muscle cells. Assuming the same K activity coefficient inside and outside the cardiac fiber, a calculation of E_K by means of the Nernst equation yields ~ −90 mV. E_K is more negative than the resting potential. Therefore the passive K efflux exceeds the passive K influx at resting potential (*white arrows* in Fig. 1).

37

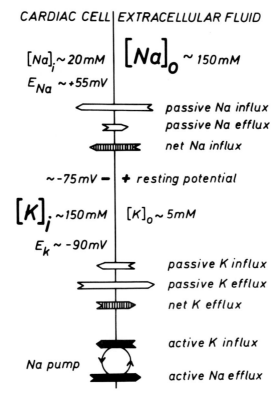

CARDIAC CELL | EXTRACELLULAR FLUID

$[Na]_i \sim 20mM$ $[Na]_o \sim 150mM$

$E_{Na} \sim +55mV$

passive Na influx
passive Na efflux
net Na influx

$\sim -75mV$ — + resting potential

$[K]_i \sim 150mM$ $[K]_o \sim 5mM$

$E_k \sim -90mV$

passive K influx
passive K efflux
net K efflux

active K influx

Na pump

active Na efflux

Fig. 1. Passive and active Na and K fluxes through the cell membrane of a quiescent cardiac cell. *White arrows*: passive unidirectional Na and K fluxes; *hatched arrows*: net passive Na and K fluxes; *black arrows*: active unidirectional Na and K fluxes. Na and K equilibrium potentials were calculated assuming equal Na and K activity coefficients inside and outside the cell from the Na and K concentration shown. [Measurements by cation-sensitive microelectrodes reveal an intracellular Na activity of about 6–8 mM in Purkinje fibers (20, 21, 35) and in ventricular (82) and atrial cells (52). The internal K activity is measured to be about 120–130 mM in Purkinje fibers (83, 88) and about 80–120 mM in the working myocardium (14, 112, 116). From these values a slightly less negative E_K and a more positive E_{Na} ($\sim +70$ mV) are calculated.]

A net K efflux (*hatched arrow* in Fig. 1) should occur until a new K concentration gradient is established and E_K coincides with the resting potential. E_{Na} is calculated to be $\sim +55$ mV (inside positive). That means the passive Na influx into a quiescent cardiac cell is larger than the passive Na efflux (*white arrows* in Fig. 1). Consequently one would expect a net Na influx into the fiber (*hatched arrow*).

In contrast to the expected continuous Na uptake and K loss from the cell, the intracellular Na and K concentrations remain fairly constant over a long period of time in vivo and in vitro. It may be argued that the constancy of the internal Na and K concentrations is due to an impermeability of the cardiac cell membrane toward Na and K ions. Studies with radioactive Na and K isotopes, however, revealed a permeability to both cations (for references see ref. 79). Thus a mechanism must exist that extrudes Na from the cell and takes K up into the fiber against their respective electrochemical gradients (*black arrows* in Fig. 1). Such a transport is called *active* if the underlying mechanism is *directly* coupled to a metabolic reaction that produces the energy required for the transport. Na extrusion from cardiac cells is mainly an active transport. The transport mechanism is often called the "Na pump" in order to visualize that the Na transport is effected against an energy gradient.

During the last 40 years the characteristics of the Na pump have been studied in detail in a variety of cells. An enzyme system within the cell membrane delivers the energy for the uphill transport of Na and K ions by ATP splitting (107). This Na^+-, K^+-activated, Mg^{2+}-dependent ATPase shares many characteristics with the Na pump and is most probably an essential part of the transport mechanism (for references see ref. 7). Obviously, active Na transport is of fundamental significance for the function of electrically excitable cells, because the maintenance of the physiological Na and K gradients across the cell membrane is a prerequisite for the production of action potentials. The Na pump is a most important transport mechanism in cells such as heart muscle fibers that display frequent electrical discharges. In contrast to our knowledge about passive ionic movements, however, little is known about active Na transport in the heart. The present review deals with some characteristics of Na pumping in *intact* cardiac cells. Biochemical aspects of cardiac active Na transport are not within the scope of this review. They have already been described in detail by Schwartz and co-workers (103). For further information the reader is referred to the reviews by Glitsch (44), Haas (57), and Lüttgau and Glitsch (87).

Methods for Studies on Active Na Transport

Until now the active Na transport of heart muscle cells has been studied mainly by means of radioisotopes, flame photometry, Na^+-sensitive microelectrodes, and electrophysiological techniques. The use of Na radioisotopes allows the direct measurement of the unidirectional cardiac Na efflux. Unfortunately, it proves to be extremely difficult to identify kinetically the cellular component of the total Na efflux (e.g., refs. 8, 60, 69; for further references see ref. 79). Furthermore it is necessary to define the active component of the Na efflux by means of specific inhibitors (e.g., cardiac glycosides), because passive components contribute appreciably to the total Na efflux from heart muscle cells (e.g., refs. 9, 58, 61). The cellular Na is more easily derived from measurements of the total cardiac Na concentration by means of flame photometry and determinations of the extracellular space. Of course these measurements only permit studies on the *net* Na movements across cardiac cell membranes. One unidirectional Na flux can be estimated by these methods if the other is absent or negligibly small. The most direct method for measuring intracellular Na in the heart is to record the intracellular Na activity by means of a Na^+-sensitive microelectrode (20, 35, 82). Again, this technique is suitable for measurements of net Na fluxes. Unidirectional fluxes can be estimated only under special conditions. Finally, electrophysiological methods proved to be useful for studies on the kinetics of the cardiac Na pump and on its interaction with the membrane potential or with the mechanism of contraction (e.g., refs. 30, 31, 41, 42).

Dependence of Active Na Transport on $[Na]_i$

If the Na pump maintains the low intracellular Na concentration, one would expect an effect of $[Na]_i$ on the magnitude of the cellular active Na efflux. In fact, it is known from many cells that an increase in $[Na]_i$ stimulates the active Na transport. For example, Brinley and Mullins (11) and Baker et al. (5) measured an increased active Na efflux following an increase in $[Na]_i$ in squid giant axons. Comparable results were reported from skeletal muscle fibers (76, 77, 105). Page and Storm (95) inferred from experiments on cat papillary muscles that active Na transport is enhanced also in the heart by an increased $[Na]_i$. The same conclusion was drawn from observations on active Na pumping in guinea pig atria (43, 45), sheep Purkinje fibers (62), and rabbit sinoatrial cells (90, 91). Glitsch et al. (53) measured net Na movements during rewarming of guinea pig atria with different $[Na]_i$. They also estimated the Na influx into the auricular cells under similar conditions. The magnitude of active Na efflux as a function of $[Na]_i$ was then derived from the data.

As can be seen in Figure 2, the active Na efflux increases with increasing $[Na]_i$ over a wide range of concentrations. For example, the efflux from auricles with an intracellular Na concentration of about 10 mmol/liter fiber water is significantly smaller than the active Na efflux from atria with an $[Na]_i$ of about 40 or 80 mmol/liter fiber water. Due to the scatter in the results no distinct kinetic behavior can be demonstrated. If Michaelis-Menten kinetics are assumed to apply, however, a maximal active Na efflux of about 4 mmol/liter fiber water per min, or about 30 pmol/cm$^2 \cdot$s can be extrapolated. The efflux is half maximal at an $[Na]_i$ of about 22 mmol/liter fiber water. From Figure 2 the active Na efflux

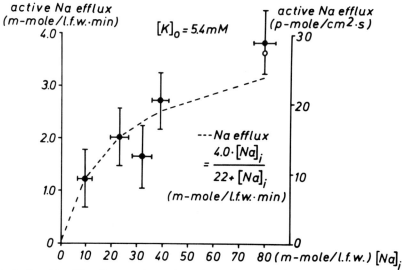

Fig. 2. Active Na efflux from guinea pig atria as function of intracellular Na concentration. Left ordinate: active Na efflux (expressed as mmol/liter fiber water per min). Right ordinate: active Na efflux (expressed as pmol/cm$^2 \cdot$s). Abscissa: exchangeable $[Na]_i$ (mmol/liter fiber water). Extracellular K concentration, 5.4 mM. *Open circle:* mean value of different set of experiments. Curve drawn obeys equation given in the figure. Equation was derived assuming simple saturation kinetics. [From Glitsch et al. (53).]

turns out to be 10 pmol/cm^2·s under physiological conditions (i.e., *exchangeable* [Na]$_i$ ~ 11 mmol/liter fiber water). The active Na efflux from Purkinje fibers is measured to be 2.8 to 7.5 pmol/cm^2·s (9, 21). Its magnitude is linearly related to the internal Na activity in the range from 7.5 to 31 mM (21). According to Haas et al. (60, 61) the active component of Na efflux from frog atria amounts to approximately 3 pmol/cm^2·s.

Dependence of Active Na Transport on [K]$_o$

The intracellular Na concentration of skeletal muscle fibers increases when the extracellular K concentration is diminished (e.g., ref. 37). On the other hand an increase in [K]$_o$ enhances the net Na outward movement from fibers with a high [Na]$_i$ (15). Similar changes in [Na]$_i$ as a function of [K]$_o$ are known to occur in the mammalian (94) and the amphibian (59, 72) myocardium. As in skeletal muscle, an increased [Na]$_i$ after hypothermia is normalized at a faster rate at higher [K]$_o$ in rewarmed cat papillary muscles (95) and guinea pig atria (43). An augmented [Na]$_i$ does not diminish in K-free media (43). These findings demonstrate a dependence of [Na]$_i$ in skeletal and heart muscle cells on [K]$_o$.

If the changes in [Na]$_i$ are due to variations of the active Na transport with [K]$_o$, alterations of [K]$_o$ should affect the magnitude of active Na efflux. In 1955 Hodgkin and Keynes (63) reported that indeed Na efflux from the squid giant axon is diminished in K-free seawater. The Na influx remained unchanged. Similarly, the Na efflux from skeletal muscle cells is increased with increasing [K]$_o$ (e.g., refs. 29, 75). Again, the Na influx is little affected by changes in [K]$_o$ (37, 64). The Na efflux from Purkinje fibers of cow and calf hearts is higher in K-containing than in K-free solutions. Only small changes in Na efflux are observed upon alteration of [K]$_o$ between 5.4 and 16.2 mM (9), however. This is shown in Figure 3, where mean rate coefficients of Na efflux in media with or without K are plotted against time. The Na efflux is enhanced by a factor of 1.43 when [K]$_o$ is increased from 0 to 5.4 mM (at *left* in figure). On raising [K]$_o$ from 0 to 16.2 mM the Na efflux is augmented by a factor of 1.57 (at *right* in figure). Comparable results have been reported from Na efflux measurements in frog atria (61). Active Na efflux from this preparation is little affected by an increase in [K]$_o$ from 1.35 to 27 mM. A possible reason for the relative ineffectiveness of high [K]$_o$ to stimulate further active Na efflux may be that active Na transport in the heart is already strongly activated at low [K]$_o$. Preliminary estimations of the magnitude of active Na efflux from guinea pig atria as a function of [K]$_o$ reveal that half-maximal activation occurs at a [K]$_o$ between 0.2 and 1.5 mM (47, 53). Similarly, active Na efflux from Purkinje fibers is half maximal at 1–2.6 mM K (38, 49). According to these results the cardiac Na pump is nearly fully activated at physiological [K]$_o$. It follows that under physiological conditions active Na efflux is mainly regulated by changes in [Na]$_i$.

Effect of Other Ions on Active Na Transport

So far no cation species is known that can replace intracellular Na ions in the activation of the Na pump in myocardium. Carmeliet (12) concluded from experiments on calf Purkinje fibers and cat ventricular muscles that internal Li ions are probably not pumped outward. Similarly, intracellular choline ions are

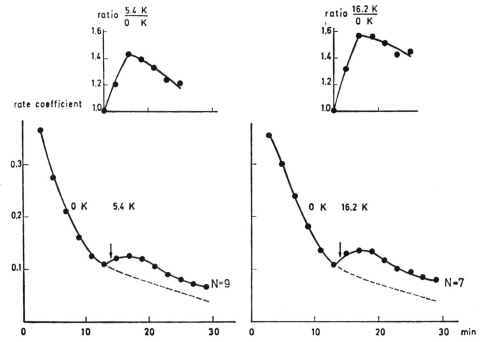

Fig. 3. Influence of extracellular K on ^{22}Na efflux from Purkinje fibers. $[K]_o$ was increased from 0 to 5.4 mM or to 16.2 mM. Mean rate coefficients (ordinate) of indicated number of experiments plotted as a function of time (abscissa). In *insets*, ratios of rate coefficients are shown. Time units of abscissas in *insets* are same as in *lower* graphs. *Dotted line* represents mean rate coefficients obtained in a control series of 15 preparations in which fibers remained in K-free solution. [From Bosteels and Carmeliet (9).]

not actively extruded from cat cardiac cells (10). Several authors have demonstrated that extracellular Rb ions similar to K^+ stimulate an active transport mechanism in heart muscle cells, which is likely to be the Na pump (e.g., refs. 78, 89). The ouabain-sensitive component of Rb influx into guinea pig atrial cells is shown in Figure 4 as a function of $[Rb]_o$. Maximal active Rb influx amounts to about 2 mmol/liter fiber water per min. The active Rb influx is half maximal at a $[Rb]_o$ of 2–3 mM (46). Half-maximal activation of the Na pump in sheep Purkinje fibers by external Rb is observed at 2.6 to 6.3 mM (32, 49). Rb and K ions are equipotent in the activation of active Na efflux from this preparation (31, 48). External Tl^+ is a stronger activator and extracellular NH_4^+, Cs^+, and Li^+ are weaker activator cations than K^+ (31, 32). Contradictory observations have been published on the effect of external Na on active Na pumping in the heart. Bosteels and Carmeliet (9) described a Na_o-activated component of active Na efflux in Purkinje fibers bathed in media containing a physiological K concentration. On the other hand Glitsch et al. (53) did not detect a primary, direct effect of external Na ions on the magnitude of the active Na efflux from guinea pig atrial cells at a $[K]_o$ of 0.4 and 1.35 mM. External Na is not a prerequisite for Na pumping in cardiac cells, however, as pointed out first by Page and Storm (95). Further experiments are required to clarify the role of extracellular Na in active Na transport in heart muscle.

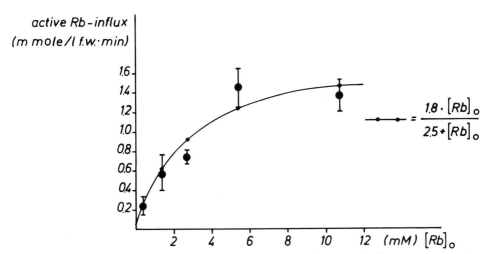

Fig. 4. Active Rb influx into guinea pig atria as function of external Rb concentration (at physiological $[Na]_i$). Ordinate: component of Rb influx inhibited by 10^{-5} M ouabain (expressed as mmol/liter fiber water per min); abscissa: extracellular Rb concentration (mM). Curve drawn obeys equation given in figure. Equation was derived assuming simple saturation kinetics.

Ionic Dependence of Cardiac Transport ATPase Compared to Active Na Transport

It might be instructive to compare the effect of some cations on active Na transport in intact cardiac cells to the ionic dependence of the transport ATPase prepared from heart muscle cells. Portius and Repke (96) described half-maximal ATP splitting by an enzyme preparation isolated from the ventricle of the guinea pig heart at about 30 mM Na (K 130 mM; pH 7.4; 37°C) and at about 2 mM K (Na 135 mM; pH 7.4; 37°C). These values agree well with the data published by Erdmann et al. (36). These authors found half-maximal ATP splitting activity in a ventricular preparation of the same species at about 18 mM Na (K 10 mM; pH 7.4; 37°C) or at about 2 mM K (Na 100 mM; pH 7.4; 37°C). Neither Li nor choline ions can be substituted for Na ions in the activation of the cardiac transport ATPase. Rb, NH_4, and Cs, but not choline activate ATP splitting in Na-containing media without K (96). In Na- and K-containing solutions, however, Rb tends to inhibit ATP splitting by the cardiac transport ATPase (78). Summarizing, the data suggest that active Na transport from heart muscle cells and ATP splitting by the myocardial transport ATPase display a similar ionic dependency.

Linkage Between Active Na and K Transport

The dependence of active Na efflux on $[K]_o$ raises the question whether active K influx is coupled to active Na efflux. Hodgkin and Keynes (63) concluded from experiments on active Na and K transport in giant axons from *Sepia* and *Loligo* that active Na and K fluxes are partially coupled but not rigidly linked. This conclusion was corroborated by the finding that active Na efflux often exceeds active K influx in giant axons (e.g., refs. 5, 104). The sensitivity of active

Na efflux towards external K is determined by the ratio of ATP to ADP inside the axon (27, 28). In skeletal muscle, Na extrusion and active K uptake display similar kinetic characteristics (e.g., ref. 106). Of course, similar kinetics do not prove the existence of a common transport mechanism, but they suggest at least some kind of coupling between active Na efflux and active K influx. As in squid giant axon the coupling ratio (active Na efflux/active K influx linked to active Na efflux) varies widely, corresponding to the experimental conditions (e.g., refs. 105, 106).

Page et al. (94) concluded that the only observation in favor of a chemical linkage between active Na and K transport in the heart is the K dependence of cardiac active Na extrusion. Furthermore, Page (93) reported that both the magnitude and the temperature dependence of cardiac K influx are much the same in media with and without NaCl between 37.5°C and 27.5°C. That is to say, the K influx seems to be independent of variations in $[Na]_i$ and active Na efflux over a certain range. This observation suggests that a rigidly coupled active Na-K exchange does not exist in cardiac cells. As the author pointed out himself, however, this concept is difficult to reconcile with the fact that the membrane potential of heart muscle cells in physiological solutions is less negative than the potassium equilibrium potential. Under these conditions the electrochemical gradient causes a net K efflux from the cardiac cells (see Fig. 1). Thus, some kind of "chemical coupling" between active Na efflux and active K influx must be present if the Na pump is responsible for the K uptake into the fibers.

According to Haas et al. (61) the coupling ratio between active Na and K transport in frog atria varies as a function of the relation between extracellular Na and K concentration. If the sum of $[Na]_o$ and $[K]_o$ is kept constant, active Na efflux is eight times larger than active K influx at a $[K]_o$ of 1.35 mM. The Na efflux is only 0.1 of the active K influx if $[K]_o$ is increased to 108 mM. Similarly, Den Hertog (26) reported an increase in the coupling ratio at low $[K]_o$ in mammalian nonmyelinated nerve fibers. However, changes in $[K]_o$ have only a questionable effect on the coupling ratio in guinea pig atrial cells (47). The coupling ratio in Purkinje fibers is not affected by variations in $[K]_o$ (38), $[Rb]_o$, or $[Cs]_o$ (32). Furthermore, the coupling ratio in this preparation is independent of changes in $[Na]_i$ (32, 41, 51). The latter finding is in line with observations on nonmyelinated nerve fibers (97) and frog sartorius muscle (106). In summary our knowledge of the linkage between active Na and K transport in cardiac cells is still sparse. The factors that govern the coupling ratio are widely unknown. It seems to be a promising field of future research to tackle these problems.

Electrogenic Na Pumping

During the last two decades it has become apparent that active Na transport in many cells probably contributes *directly* to the membrane potential by generating a current. Such a Na pump is termed *electrogenic* as opposed to an *electroneutral* pump, which affects the membrane potential only *indirectly* by maintaining the Na and K concentration gradients across the cell membrane. An electroneutral Na pump does not separate charges. The main finding in favor of an electrogenic Na pump is that membrane potential changes are observed that are not due to alterations in the membrane conductance or in the passive Na or K distribution across the cell membrane. These potential changes share several characteristics in common with active Na transport. Figure 5 shows how electrogenic Na pumping might occur. The scheme displays an anionic carrier Y^- at the inside of the cell membrane that has a much higher affinity toward Na than toward K ions. Consequently Y^- combines with Na^+ to form NaY. The compound NaY permeates the cell membrane. At the outer surface Na^+ dissociates from the carrier. The carrier is now changed to a different form X^- with a high affinity towards K^+. A new compound KX is formed and shuttles back through the cell membrane. Inside the cell, K^+ is released from the carrier, which is again converted to the form Y^- by an energy-requiring reaction. This carrier cycle implies a rigid linkage between an active Na efflux and an active K influx of the same magnitude. The active transport is electroneutral. An electroneutral pump results also if it is supposed that the NaY and KX transport display an influx and an efflux component (shown) or that X^- and Y^- do not exclusively bind either K^+ or Na^+ (see refs. 6, 76). If one assumes additionally that not only the compounds NaY and KX permeate the cell membrane but also the anionic carriers X^- (shown) and Y^-, any rigid linkage between active Na efflux and active K influx is abolished. The pump can now separate charges, and the scheme in Figure 5 visualizes an electrogenic Na pump. The mechanism of charge separation is not yet understood. Experiments that suggest electrogenic Na pumping in heart muscle cells are reviewed in the following paragraph.

In 1960 Délèze (22) reported that the temperature sensitivity of the cardiac resting potential exceeds that predicted for a K electrode. Rewarming the bathing medium causes an almost instantaneous increase in membrane potential. The author demonstrated that this increase in potential requires metabolic energy and is not due to K depletion at the outside of the cell membrane or to an increase in the P_K to P_{Na} ratio of the membrane. He concluded that an electrogenic Na pump contributes to the observed increase in membrane potential of rewarmed cardiac fibers. Similarly, Page and Storm (95) found a rapid

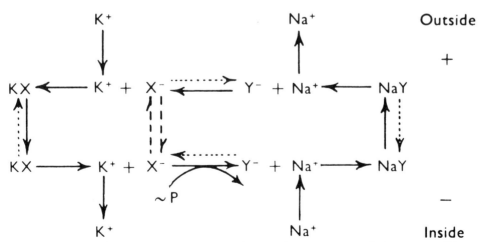

Fig. 5. Scheme of hypothetical carrier mechanism for active Na and K transport. [From Cross et al. (19).]

recovery of the membrane potential in cat papillary muscles rewarmed at 27°–28°C after hypothermia for 2 h at 4°–6°C. As can be seen from Figure 6, the restoration of the membrane potential upon rewarming is accomplished in a very short time. The authors observed complete restoration of the potential in two fibers in less than 15 s. They pointed out that the recovery of the intracellular Na and K concentrations was much slower. After 10 min of rewarming, the membrane potential reached values 6 mV more negative than the K equilibrium potential. Corresponding measurements in guinea pig atria (43) and cultured chicken heart cells (65, 86) confirmed these results.

Further experiments revealed that the membrane potential of rewarmed heart muscle cells usually hyperpolarizes transiently beyond the resting potential level (43, 62). The magnitude of hyperpolarization is significantly correlated with the active Na efflux from the rewarmed cardiac cells (45). This is shown in Figure 7, where the hyperpolarization beyond the steady-state level of the membrane potential in rewarmed guinea pig atria is plotted as a function of

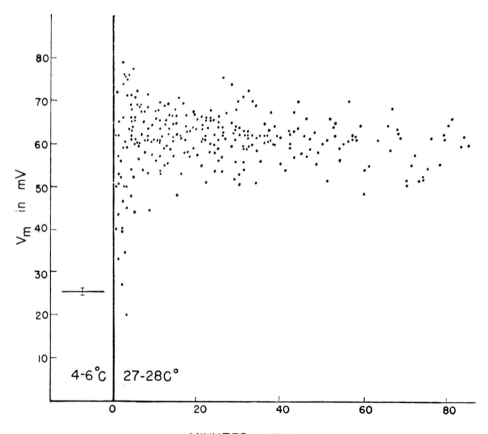

MINUTES AFTER RAISING TEMPERATURE

Fig. 6. Effect of sudden rewarming on membrane potential (V_m) of cat papillary muscles in phosphate-buffered solution with 10 mM K. Ordinate: voltage V_m (mV); abscissa: time from beginning of rewarming in minutes. Period of rewarming is indicated by *vertical bar*. Mean value ± SE in cold was 25 ± 0.7 mV and was 61.3 ± 0.5 mV between 10 and 85 min after rewarming. [From Page and Storm (95).]

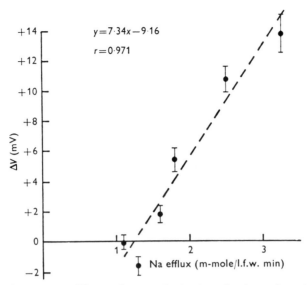

$y = 7.34x - 9.16$

$r = 0.971$

ΔV (mV)

Na efflux (m-mole/l.f.w. min)

Fig. 7. Effect of active Na efflux on hyperpolarization of guinea pig atria after 10 min of rewarming in Tyrode's solution (35°C). Ordinate: hyperpolarization beyond steady-state level of membrane potential at end of rewarming (mV); abscissa: active Na efflux (mmol/liter fiber water per min); correlation coefficient: 0.971 ($P < 0.01$). Equation of regression line is indicated in the figure. [From Glitsch (45).]

active Na efflux. A hyperpolarization of the cardiac cell membrane during rewarming is observed only if the intracellular Na concentration has been increased during hypothermia (45, 62). Li ions cannot replace intracellular Na ions in this mechanism (62, 91, 117). The hyperpolarization is not observed in cardiac cells rewarmed in K-free media (43, 47), where the active Na efflux is reduced. If the extracellular K concentration is diminished from 5.4 to 2.7 mM during rewarming, the membrane potential of sheep Purkinje fibers decreases toward more positive values. This effect is not observed under steady-state conditions (62). The hyperpolarization of the cell membrane during rewarming is blocked by cardiac glycosides (e.g., refs. 62, 95), which are known to be specific inhibitors of active Na transport (102).

The observations described above strongly suggest that the hyperpolarization of the cardiac cell membrane following hypothermia is due to an enhanced active Na extrusion from the cells at the beginning of rewarming. Especially the finding that the membrane potential of rewarmed fibers reaches values beyond the calculated K equilibrium potential favors the idea of an electrogenic Na pump. However, an electroneutral Na pump might produce a similar effect by causing a K depletion at the outside of the cell membrane and thereby a shift of the K equilibrium potential (and the membrane potential) toward more negative values. This question was investigated by Glitsch et al. (47), who measured the membrane potential of quiescent guinea pig auricles as a function of [K]ₒ before and after hypothemia. Figure 8 illustrates the result. Before cooling, the highest value of the membrane potential is obtained in K-free solution and amounts to −92 mV (*filled circles*). The *open circles* represent the membrane potential during the first 20 min of rewarming after 4 h of hypothermia. Unlike the

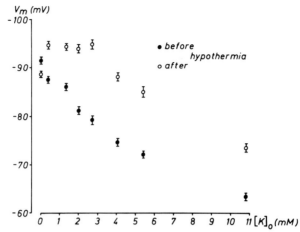

Fig. 8. Membrane potential (V_m) of guinea pig atria at various external K concentrations before (●) and after (○) hypothermia. Means of 6–10 atria. Ordinate: membrane potential (mV); abscissa: external K concentration (mM); temperature 35°C. Unlike Purkinje fibers, guinea pig atrial cells rarely depolarize in K-free solution under steady-state conditions. [From Glitsch et al. (47).]

potential before cooling, the membrane potential of rewarmed atria passes through a maximum at 2.7 mM K and reaches −95 mV. It is more negative than the membrane potential of auricles rewarmed in K-free media (−89 mV). Similarly, the membrane potential of atria rewarmed in solutions with 0.4, 1.35, and 2 mM K is significantly more negative than the potential of the auricles that are rewarmed in a medium without K. This phenomenon is never observed under steady-state conditions. It shows that the hyperpolarization of the cardiac cell membrane during rewarming in K-containing solutions is *not* caused by a K depletion at the outside of the cell membrane due to activation of an electroneutral Na-K exchange pump. Such a pump cannot account for membrane potentials more negative than the potential measured in a K-free medium during rewarming. The same conclusion was reached by Gadsby and Cranefield (42). These authors studied the effects of temporary stimulation of the active Na transport on membrane potential in canine Purkinje fibers. They found that the effects are opposite to those expected from extracellular K depletion. Thus the measurements strongly suggest that the Na pump contributes *directly* to the membrane potential of cardiac cells by an electrogenic mechanism. As mentioned earlier, the nature of this mechanism is not yet understood. Active Na transport in many cells extrudes more Na ions than K ions are actively taken up (cf. refs. 87, 110). Thus an outward current carried by Na ions might produce the electrogenic effect. The current displays saturation kinetics. Half saturation occurs at about 1.0 to 1.5 mM K in canine Purkinje fibers (38) and guinea pig atria (47).

Experiments also pointing to the existence of an electrogenic Na pump in heart were reported by Vassalle (111). The spontaneous activity of cardiac Purkinje fibers is transiently inhibited by a hyperpolarization following stimulation with an unphysiologically high stimulus rate. Substitution of Li for Na_o or exposure to 2,4-dinitrophenol abolishes the hyperpolarization. The author concluded that activation of an electrogenic active Na transport is the major cause for this "overdrive suppression." Wiggins and Cranefield (117) observed a hyperpolarization of the cell membrane when a Na-containing solution is reapplied to Na-depleted canine Purkinje fibers. This hyperpolarization is inhibited by cardiac glycosides. It does not occur at low temperature, in K-free media, or in a bathing fluid containing Li instead of Na. The hyperpolarization is most probably brought about by an electrogenic Na pump. A similar phenomenon exists in guinea pig atria (50). Readmission of K-containing solution after a relatively short period in a K-free medium causes a transient hyperpolarization of the cell membrane in the rabbit sinus node (90, 91) and in Purkinje fibers (39, 40). According to Ito and Surawicz (68), both the active Na transport and a transient increase in K permeability are involved in this hyperpolarization.

A detailed analysis of this phenomenon in canine Purkinje fibers by Gadsby and Cranefield (41, 42) revealed that the hyperpolarization is caused by an outward current. The current is increased under conditions that increase $[Na]_i$. It is inhibited by cardiac glycosides and is activated by an increase in $[K]_o$. This activation displays saturation kinetics (38). The authors concluded that the outward current is probably due to electrogenic Na pumping. A hyperpolarization of the cell membrane following a period in a K(Rb, Cs)-free bathing medium is also observed in a Rb(Cs)-containing solution. This Rb(Cs)-activated response was studied by Eisner and Lederer (32) in voltage clamped sheep Purkinje fibers. In agreement with the findings by Gadsby and Cranefield (41, 42), the authors reported that the hyperpolarization is due to an outward current that is probably carried by Na ions. Cardiac glycosides inhibit the current, whereas an increase in $[Rb]_o$ ($[Cs]_o$) activates the current according to simple saturation kinetics. Glitsch and Pusch (51) and Glitsch et al. (48) investigated the correlation between changes in membrane potential and internal Na activity during the K(Rb)-activated response in the same preparation. The time constant of the decline in membrane potential and in the intracellular Na activity during the response is much the same, regardless of whether K or Rb is used as an extracellular activator cation. This is true within a concentration range from 1.1 to 10.8 mM K (Rb). Similarly Eisner et al. (34) reported that pump current and internal Na activity decline with the same time constant during a Rb-activated response in voltage clamped sheep Purkinje fibers.

Isenberg and Trautwein (66, 67) investigated the possible contribution of Na pumping to the membrane potential of sheep Purkinje fibers under steady-state conditions by means of a voltage clamp method. They described a voltage- and time-independent outward current that was immediately inhibited by cardiac glycosides. The current affected the plateau phase of the action potential. It was attributed to the activity of an electrogenic active Na transport.

The reader will find more detailed information on electrogenic Na pumping in the review articles by Kerkut and York (73), Kernan (74), Lüttgau and Glitsch (87), and Thomas (110).

It is beyond the scope of this review to discuss in detail the particular significance of active Na transport for cellular metabolism and function. The reader is referred to a review by Baker (2), who outlined the effect of active Na pumping on various cellular activities, including control of cell volume, synthesis of macromolecules, and transport of amino acids and sugars. The fundamental significance of the Na pump for cardiac excitability has already been described above. Three further examples for the interaction between active Na transport and various cellular functions are presented here. They are quite specific for cardiac cells and rest mainly on the effect of the Na pump on Ca metabolism: the significance of the Na pump for contraction, intercellular coupling, and for the prevention of cardiac arrhythmias.

Effect of Active Na Transport on Cardiac Contraction

The concentration of free ionized Ca in resting heart muscle cells is about 10^{-7} M. Thus Ca ions have to move from the myoplasm to the outside of the cell membrane against a steep electrochemical gradient. As pointed out by Reuter and Seitz (101), the energy required for Ca efflux from guinea pig atria is *not directly* provided by the cellular metabolism. Both the Q_{10} and the activation energy of the Ca efflux are low. Metabolic poisons such as 2,4-dinitrophenol do not diminish but increase the Ca efflux. The efflux is reduced by 30% in a Ca-free bathing fluid and by 80% in a Na- and Ca-free solution. The flux is largely governed by the ratio $[Ca]_o:[Na]_o^2$. The authors proposed a carrier transport system that effects Ca outward transport by exchange diffusion. Two Na ions (or one Ca ion) are taken up into the cell for each Ca ion extruded. Figure 9 illustrates the suggested carrier system. As can be seen from the figure, net Ca efflux is driven energetically by the Na gradient across the cardiac cell membrane. This gradient in turn is maintained by the Na pump. A similar Ca transport system exists in the squid giant axon (3, 4). From Figure 9 one would expect an inhibition of Ca efflux and an increase in $[Ca]_i$ if $[Na]_i$ is augmented.

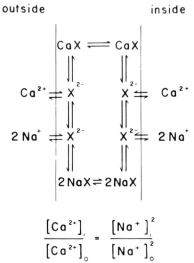

Fig. 9. Carrier scheme for Na-Ca exchange across a membrane; 2 Na ions and 1 Ca ion compete for carrier (X^{2-}) at inside and outside surfaces of membrane. Carrier can move as Na_2X or as CaX across membrane, but unloaded carrier cannot move. In such a transport scheme, a Ca gradient can be established across membrane as a consequence of an existing Na gradient. Respective distribution ratios at equilibrium are given by equation below the scheme. [From Reuter (100).]

Furthermore Glitsch et al. (54) described an enhanced Ca influx into heart muscle cells with a high $[Na]_i$. The Ca influx normalizes with the same rate constant as does the intracellular Na concentration. Due to the increased $[Ca]_i$, the force of contraction is transiently strengthened.

Several authors (4, 54, 79, 80) have suggested that the positive inotropic effect of cardiac glycosides may also be caused by a Na_i-induced increase in $[Ca]_i$ in the following way: According to Schatzmann (102), cardiac glycosides inhibit active Na transport in red blood cells without effects on the cellular metabolism. The compounds affect the Mg^{2+}-dependent, Na^+-, K^+-activated adenosinetriphosphatase, an integral part of the Na pump (108). This enzyme system can be considered to be a cardiac glycosides receptor (for references see ref. 1). Binding of cardiac glycosides to the enzyme and their positive inotropic effect have several characteristics in common. Repke (98, 99) was the first to stress the striking parallelism between inhibition of the adenosinetriphosphatase and certain cardiotonic effects caused by cardiac glycosides. "Therapeutic" doses of these compounds inhibit the transport ATPase (for references see refs. 1, 103). The inhibition of the Na pump produces an increase in $[Na]_i$ in the intact cardiac cell, which in turn may cause the positive inotropic effect via an enhanced Ca influx and a reduced Ca efflux. Some reports are at variance with this hypothesis, however. For example, Okita (92) maintains that inhibition of the Na-K-ATPase is perhaps responsible for certain cardiotoxic effects but not for the positive inotropic effect of cardiac glycosides. Furthermore it has been known for some time that therapeutic concentrations of these compounds (10^{-8}–10^{-9} M) can *stimulate* active Na transport (for references see ref. 85). This point was recently reemphasized by Cohen et al. (13) in their work on glycoside effects on voltage clamped Purkinje fibers. The situation became even more complicated by the observation that various cardiac glycosides exert their positive inotropic effect by different mechanisms. According to Lee et al. (84) and Godfraind and Ghysel-Burton (56), dihydro-ouabain causes its positive inotropic effect via a Na pump inhibition. In the positive inotropic effect of ouabain, however, an additional factor is involved (56). This finding is perhaps related to the existence of two different binding sites for cardiac glycosides in mammalian myocardium (55). The recent observation of an enhanced slow (Ca) inward current in strophanthidin-treated Purkinje fibers (115) does not exclude an important role of the Na pump in the mechanism of cardiac glycoside action. Future research is needed to settle these problems. Whatever the mechanism of glycoside action, the significance of active Na transport for cardiac contraction is well established. The intimate relation between Na pump activity and cardiac contraction has recently been stressed by Eisner and Lederer (33). They observed that inhibition or stimulation of active Na transport in voltage clamped Purkinje fibers induce an increase or decrease in twitch tension, respectively. The authors concluded that the variation in $[Na]_i$ is the rate-limiting step linking Na pump activity and twitch tension in this preparation.

Effect of Active Na Transport on Intercellular Coupling in Heart

An intact intercellular communication is a prerequisite of normal cardiac function. During the past few years the role of the Na pump in cardiac intercellular coupling became apparent. De Mello (23, 25) described a reversible electrical uncoupling between adjacent cells following intracellular Na injection into rabbit atrial strips and dog Purkinje fibers. Uncoupling was achieved within 500 s from the beginning of the injection, as judged from the concomitant increase in the input resistance of the injected cell. Inhibition of active Na transport by cardiac glycosides likewise produced uncoupling. Furthermore, the compounds enhanced uncoupling caused by intracellular Na injection. Intercellular communication was only slightly reduced within 500 s after Na injection into Purkinje fibers bathed in a Ca-poor solution. If afterward the extracellular Ca concentration was raised, electrical uncoupling occurred quickly. Similarly, soaking the fibers in a Na-poor medium accelerated uncoupling evoked by internal Na injection. Intracellular Ca injection also diminished electrical coupling between cardiac cells (24).

From these and other results the author concluded that active Na transport is essential for intercellular coupling between heart muscle cells. Inhibition of the Na pump increases $[Na]_i$, as does internal Na injection. In consequence the Ca influx via the Na-Ca exchange system is augmented, whereas the Ca efflux is inhibited. The increased Ca influx (and the simultaneously reduced Ca efflux) raises the internal Ca concentration. The augmented $[Ca]_i$ finally causes the uncoupling.

Comparable conclusions were reached by Weingart (113, 114). He investigated the mechanism underlying the known glycoside effect on conduction velocity in mammalian heart. Figure 10 illustrates one of his results. Ventricular preparations from calf and cow hearts were bathed in solutions containing ouabain (2×10^{-6} M) and various Ca concentrations. In the figure, normalized changes of the specific longitudinal resistance (R_i) in these fibers are plotted against time. An increase in R_i reflects uncoupling between the ventricular cells. As can be seen from the figure, R_i increases in cardiac fibers poisoned with cardiac glycosides. This phenomenon is probably an important factor in the observed diminution of the conduction velocity. Both onset and steepness of the R_i increase depend on the extracellular Ca concentration. The increase is late and small in a Ca-poor medium. This result is in line with the hypothesis that an augmented $[Ca]_i$ causes uncoupling of adjacent cardiac cells. The high level of $[Ca]_i$ is secondary to an inhibition of the Na pump by cardiac glycosides. The inhibition first yields an increased $[Na]_i$, which then enhances Ca influx and inhibits Ca efflux via the Na-Ca exchange mechanism.

Role of Active Na Transport in Prevention of Cardiac Arrhythmias

Inhibition of the cardiac Na pump can cause various kinds of arrhythmias. For example, digitalis-like drugs that are specific inhibitors of active Na transport may induce automatic or triggered arrhythmias (for references see refs. 16–18). Both a loss in resting potential and the appearance of so-called delayed afterdepolarizations (transient depolarizations, oscillatory afterpotentials) are important in the mechanism of triggered arrhythmias. Obviously the Na pump inhibition results in a depolarization of the cardiac cell membrane by suppression of the pump current and by diminution of the K and Na concentration gradients. The role of active Na transport in the production of delayed after-

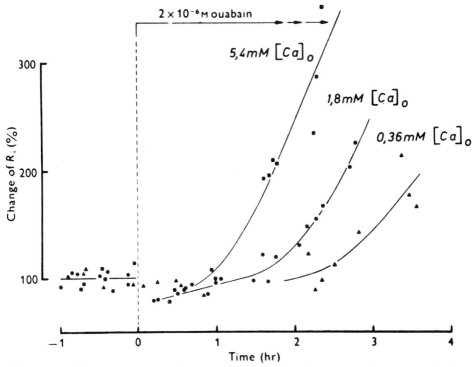

Fig. 10. Modification of ouabain-induced uncoupling by external Ca concentration ([Ca]$_o$). Normalized changes in R_i (ordinate) are plotted against time (abscissa). *Symbols* represent single determinations from three different experiments, carried out at control [Ca]$_o$ of 1.8 mM (●), at 0.36 mM [Ca]$_o$ (▲), and at 5.4 mM [Ca]$_o$ (■), as standard ouabain concentration, 2×10^{-6} M, was administered. Stimulation frequency 0.2 Hz. Time $t = 0$ h marks beginning of exposure to drug; *horizontal arrows on top* indicate its end (the lower the [Ca]$_o$, the longer the ouabain exposure). After an initial decrease, which was absent in the experiment at 0.36 mM [Ca]$_o$, R_i progressively increased in a dose-dependent fashion. [From Weingart (114).]

depolarizations is more complicated. Lederer and Tsien (81) and Kass et al. (70, 71) have studied the characteristics of the current underlying the delayed afterdepolarizations in voltage clamped Purkinje fibers. This transient inward current is largely carried by Na ions. It is most probably controlled by a phasic release of Ca from intracellular stores. The phasic Ca release is a consequence of the Ca overload of the cells, which in turn is evoked via the Na-Ca exchange following an increase in [Na]$_i$. The elevated [Na]$_i$ is due to the Na pump inhibition by the glycosides (70, 109). Inhibition of cardiac active Na transport by lowering the physiological [K]$_o$ likewise induces a transient inward current that causes delayed afterdepolarizations (30).

The examples described above illustrate the role of the Na pump in cardiac excitation spread and contraction and demonstrate again the fundamental functional significance of active Na transport in myocardium.

Acknowledgments

This investigation was supported by SFB 114 "Bionach."

References

1. AKERA, T. Membrane adenosinetriphosphatase: a digitalis receptor? *Science* 198: 569–574, 1977.
2. BAKER, P. F. The sodium pump in animal tissues and its role in the control of cellular metabolism and function. In: *Metabolic Pathways* (3rd ed.). New York: Academic, 1972, vol. VI, *Metabolic Transport*, p. 243–268.
3. BAKER, P. F., M. P. BLAUSTEIN, A. L. HODGKIN, AND R. A. STEINHARDT. The effect of sodium concentration on calcium movements in giant axons of *Loligo forbesi* (Abstract). *J. Physiol. London* 192: 43P–44P, 1967.
4. BAKER, P. F., M. P. BLAUSTEIN, A. L. HODGKIN, AND R. A. STEINHARDT. The influence of calcium on sodium efflux in squid axons. *J. Physiol. London* 200: 431–458, 1969.
5. BAKER, P. F., M. P. BLAUSTEIN, R. D. KEYNES, J. MANIL, T. I. SHAW, AND R. A. STEINHARDT. The ouabain-sensitive fluxes of sodium and potassium in squid giant axons. *J. Physiol. London* 200: 459–496, 1969.
6. BAKER, P. F., AND C. M. CONNELLY. Some properties of the external activation site of the sodium pump in crab nerve. *J. Physiol. London* 185: 270–297, 1966.
7. BONTING, S. L. Sodium-potassium activated adenosinetriphosphatase and cation transport. In: *Membranes and Ion Transport*, edited by E. E. Bittar. London: Wiley, 1970, vol. 1, p. 257–363.
8. BOSTEELS, S., AND E. CARMELIET. Estimation of intracellular Na concentration and transmembrane Na flux in cardiac Purkyně fibres. *Pfluegers Arch.* 336: 35–47, 1972.
9. BOSTEELS, S., AND E. CARMELIET. The components of the sodium efflux in cardiac Purkyně fibres. *Pfluegers Arch.* 336: 48–59, 1972.
10. BOSTEELS, S., A. VLEUGELS, AND E. CARMELIET. Choline permeability in cardiac muscle cells of the cat. *J. Gen. Physiol.* 55: 602–619, 1970.
11. BRINLEY, F. J., JR., AND L. J. MULLINS. Sodium fluxes in internally dialyzed squid axons. *J. Gen. Physiol.* 52: 181–211, 1968.
12. CARMELIET, E. E. Influence of lithium ions on the transmembrane potential and cation content of cardiac cells. *J. Gen. Physiol.* 47: 501–530, 1964.
13. COHEN, I., J. DAUT, AND D. NOBLE. An analysis of the actions of low concentrations of ouabain on membrane currents in Purkinje fibres. *J. Physiol. London* 260: 75–103, 1976.
14. COHEN, S. J., AND H. A. FOZZARD. Intracellular K and Na activities in papillary muscle during inotropic interventions (Abstract). *Biophys. J.* 25: 144a, 1979.
15. CONWAY, E. J., R. P. KERNAN, AND J. A. ZADUNAISKY. The sodium pump in skeletal muscle in relation to energy barriers. *J. Physiol. London* 155: 263–279, 1961.
16. CRANEFIELD, P. F. The conduction of cardiac impulse. Mount Kisco, NY: Futura, 1975, p. 1–404.
17. CRANEFIELD, P. F. Action potentials, afterpotentials, and arrhythmias. *Circ. Res.* 41: 415–423, 1977.
18. CRANEFIELD, P. F., AND A. L. WIT. Cardiac arrhythmias. *Ann. Rev. Physiol.* 41: 459–472, 1979.
19. CROSS, S. B., R. D. KEYNES, AND R. RYBOVÀ. The coupling of sodium efflux and potassium influx in frog muscles. *J. Physiol. London* 181: 865–880, 1965.
20. DEITMER, J. W., AND D. ELLIS. Changes in the intracellular sodium activity of sheep heart Purkinje fibres produced by calcium and other divalent cations. *J. Physiol. London* 277: 437–453, 1978.
21. DEITMER, J. W., AND D. ELLIS. The intracellular sodium activity of cardiac Purkinje fibres during inhibition and reactivation of the Na-K pump. *J. Physiol. London* 284: 241–259, 1978.
22. DÉLÈZE, J. Possible reasons for drop of resting potential of mammalian heart preparations during hypothermia. *Circ. Res.* 8: 553–557, 1960.
23. DE MELLO, W. C. Uncoupling of heart cells produced by intracellular sodium injection. *Experientia* 31: 460–462, 1975.
24. DE MELLO, W. C. Effect of intracellular injection of calcium and strontium on cell communication in heart. *J. Physiol. London* 250: 231–245, 1975.
25. DE MELLO, W. C. Influence of the sodium pump on intercellular communication in heart fibres: effect of intracellular injection of sodium ion on electrical coupling. *J. Physiol. London* 263: 171–197, 1976.
26. DEN HERTOG, A. Some further observations on the electrogenic sodium pump in nonmyelinated nerve fibres. *J. Physiol. London* 231: 493–509, 1973.
27. DE WEER, P. Restoration of a potassium-requiring sodium pump in squid giant axons poisoned with CN and depleted of arginine. *Nature London* 219: 730–731, 1968.

28. DE WEER, P. Effects of intracellular adenosine-5'-diphosphate and orthophosphate on the sensitivity of sodium efflux from squid axon to external sodium and potassium. *J. Gen. Physiol.* 56: 583–620, 1970.

29. EDWARDS, C., AND E. J. HARRIS. Factors influencing the sodium movement in frog muscle with a discussion of the mechanism of sodium movement. *J. Physiol. London* 135: 567–580, 1957.

30. EISNER, D. A., AND W. J. LEDERER. Inotropic and arrhythmogenic effects of potassium-depleted solutions on mammalian cardiac muscle. *J. Physiol. London* 294: 255–277, 1979.

31. EISNER, D. A., AND W. J. LEDERER. The role of the sodium pump in the effects of potassium-depleted solutions on mammalian cardiac muscle. *J. Physiol. London* 294: 279–301, 1979.

32. EISNER, D. A., AND W. J. LEDERER. Characterization of the electrogenic sodium pump in cardiac Purkinje fibres. *J. Physiol. London* 303: 441–474, 1980.

33. EISNER, D. A., AND W. J. LEDERER. The relationship between sodium pump activity and twitch tension in cardiac Purkinje fibres. *J. Physiol. London* 303: 475–494, 1980.

34. EISNER, D. A., W. J. LEDERER, AND R. D. VAUGHAN-JONES. Electrogenic Na pumping in cardiac muscle: simultaneous measurement of intracellular Na activity, membrane current and tension (Abstract). *J. Physiol. London* 300: 42P, 1980.

35. ELLIS, D. The effects of external cations and ouabain on the intracellular sodium activity of sheep heart Purkinje fibres. *J. Physiol. London* 273: 211–240, 1977.

36. ERDMANN, E., H.-D. BOLTE, AND B. LÜDERITZ. The (Na^+-K^+)-ATPase activity of guinea pig heart muscle in potassium deficiency. *Arch. Biochem. Biophys.* 145: 121–125, 1971.

37. FOZZARD, H. A., AND D. M. KIPNIS. Regulation of intracellular sodium concentrations in rat diaphragm muscle. *Science* 156: 1257–1260, 1967.

38. GADSBY, D. C. Activation of electrogenic Na^+/K^+ exchange by extracellular K^+ in canine cardiac Purkinje fibers. *Proc. Natl. Acad. Sci. USA* 77: 4035–4039, 1980.

39. GADSBY, D., AND P. F. CRANEFIELD. Rapid sodium-loading of cardiac Purkinje fibers at 36°C and the resultant stimulation of active sodium extrusion (Abstract). *Biophys. J.* 17: 7a, 1977.

40. GADSBY, D. C., AND P. F. CRANEFIELD. Outward membrane current following rapid sodium-loading of cardiac Purkinje fibers (Abstract). *Biophys. J.* 21: 166a, 1978.

41. GADSBY, D. C., AND P. F. CRANEFIELD. Direct measurement of changes in sodium pump current in canine cardiac Purkinje fibers. *Proc. Natl. Acad. Sci. USA* 76: 1783–1787, 1979.

42. GADSBY, D. C., AND P. F. CRANEFIELD. Electrogenic sodium extrusion in cardiac Purkinje fibers. *J. Gen. Physiol.* 73: 819–837, 1979.

43. GLITSCH, H. G. Über das Membranpotential des Meerschweinchenvorhofes nach Hypothermie. *Pfluegers Arch.* 307: 29–46, 1969.

44. GLITSCH, H. G. Über einige Eigenschaften des aktiven Na^+-Transports am Myokard. *Dtsch. Med. Wochenschr.* 95: 963–970, 1970.

45. GLITSCH, H. G. Activation of the electrogenic sodium pump in guinea-pig auricles by internal sodium ions. *J. Physiol. London* 220: 565–582, 1972.

46. GLITSCH, H. G., W. GRABOWSKI, AND H. PUSCH. Membrane characteristics of atrial cells in K-free, Rb-containing solutions (Abstract). *Pfluegers Arch.* 373: Suppl. R3, 1978.

47. GLITSCH, H. G., W. GRABOWSKI, AND J. THIELEN. Activation of the electrogenic sodium pump in guinea-pig atria by external potassium ions. *J. Physiol. London* 276: 515–524, 1978.

48. GLITSCH, H. G., W. KAMPMANN, AND H. PUSCH. Intracellular Na activity during K or Rb activated response in sheep Purkinje fibres. Correlation with changes in membrane potential. In: *Ion-Sensitive Microelectrodes and Their Use in Excitable Tissues*. New York: Plenum, 1981. In press.

49. GLITSCH, H. G., W. KAMPMANN, AND H. PUSCH. Activation of the Na pump in sheep Purkinje fibres by extracellular K or Rb ions (Abstract). *Pfluegers Arch.* 389: Suppl. R8, 1981.

50. GLITSCH, H. G., AND J. KLARE. On the membrane potential of Na depleted guinea-pig auricles after addition of Na to the extracellular solution (Abstract). *Pfluegers Arch.* 368: Suppl. R3, 1977.

51. GLITSCH, H. G., AND H. PUSCH. Correlation between changes in membrane potential and intracellular sodium activity during K activated response in sheep Purkinje fibres. *Pfluegers Arch.* 384: 189–191, 1980.

52. GLITSCH, H. G., H. PUSCH, AND G. VASSORT. An estimation of the intracellular Na activity in guinea-pig atrial cells (Abstract). *Pfluegers Arch.* 379: Suppl. R2, 1979.

53. GLITSCH, H. G., H. PUSCH, AND K. VENETZ. Effects of Na and K ions on the active Na transport in guinea-pig auricles. *Pfluegers Arch.* 365: 29–36, 1976.

54. GLITSCH, H. G., H. REUTER, AND H. SCHOLZ. The effect of the internal sodium concentration on calcium fluxes in isolated guinea-pig auricles. *J. Physiol. London* 209: 25–43, 1970.

55. GODFRAIND, T., AND J. GHYSEL-BURTON. Binding sites related to ouabain-induced stimulation or inhibition of the sodium pump. *Nature London* 265: 165–166, 1977.

56. GODFRAIND, T., AND J. GHYSEL-BURTON. Independence of the positive inotropic effect of ouabain from the inhibition of the heart Na^+/K^+ pump. *Proc. Natl. Acad. Sci. USA* 77: 3067–3069, 1980.

57. HAAS, H. G. Active ion transport in heart muscle. In: *Electrical Phenomena in the Heart*, edited by W. C. de Mello. New York: Academic, 1972, p. 163–183.

58. HAAS, H. G., H. G. GLITSCH, AND R. KERN. Zum Problem der gegenseitigen Beeinflussung der Ionenfluxe am Myokard. *Pfluegers Arch.* 281: 282–299, 1964.

59. HAAS, H. G., H. G. GLITSCH, AND R. KERN. Kalium-Fluxe und Membranpotential am Froschvorhof in Abhängigkeit von der Kalium-Außenkonzentration. *Pfluegers Arch.* 288: 43–64, 1966.

60. HAAS, H. G., H. G. GLITSCH, AND W. TRAUTWEIN. Natrium-Fluxe am Vorhof des Froschherzens. *Pfluegers Arch.* 277: 36–47, 1963.

61. HAAS, H. G., F. HANTSCH, H. P. OTTER, AND G. SIEGEL. Untersuchungen zum Problem des aktiven K- und Na-Transports am Myokard. *Pfluegers Arch.* 294: 144–168, 1967.

62. HIRAOKA, M., AND H. H. HECHT. Recovery from hypothermia in cardiac Purkinje fibers: considerations for an electrogenic mechanism. *Pfluegers Arch.* 339: 25–36, 1973.

63. HODGKIN, A. L., AND R. D. KEYNES. Active transport of cations in giant axons from *Sepia* and *Loligo*. *J. Physiol. London* 128: 28–60, 1955.

64. HOROWICZ, P., AND C. J. GERBER. Effects of external potassium and strophanthidin on sodium fluxes in frog striated muscle. *J. Gen. Physiol.* 48: 489–514, 1965.

65. HORRES, C. R., J. F. AITON, M. LIEBERMAN, AND E. A. JOHNSON. Electrogenic transport in tissue cultured heart cells. *J. Mol. Cell. Cardiol.* 11: 1201–1205, 1979.

66. ISENBERG, G., AND W. TRAUTWEIN. The effect of dihydro-ouabain and lithium-ions on the outward current in cardiac Purkinje fibers. Evidence for electrogenicity of active transport. *Pfluegers Arch.* 350: 41–54, 1974.

67. ISENBERG, G., AND W. TRAUTWEIN. Temperature sensitivity of outward current in cardiac Purkinje fibers. Evidence for electrogenicity of active transport. *Pfluegers Arch.* 358: 225–234, 1975.

68. ITO, S., AND B. SURAWICZ. Transient, 'paradoxical' effects of increasing extracellular K^+ concentration on transmembrane potential in canine cardiac Purkinje fibers. *Circ. Res.* 41: 799–807, 1977.

69. JOHNSON, J. A. Sodium exchange in the frog heart ventricle. *Am. J. Physiol.* 191: 487–492, 1957.

70. KASS, R. S., W. J. LEDERER, R. W. TSIEN, AND R. WEINGART. Role of calcium ions in transient inward currents and aftercontractions induced by strophanthidin in cardiac Purkinje fibres. *J. Physiol. London* 281: 187–208, 1978.

71. KASS, R. S., R. W. TSIEN, AND R. WEINGART. Ionic basis of transient inward current induced by strophanthidin in cardiac Purkinje fibres. *J. Physiol. London* 281: 209–226, 1978.

72. KEENAN, M. J., AND R. NIEDERGERKE. Intracellular sodium concentration and resting sodium fluxes of the frog heart ventricle. *J. Physiol. London* 188: 235–260, 1967.

73. KERKUT, G. A., AND B. YORK. *The Electrogenic Sodium Pump.* Bristol: Scientechnica, 1971, p. 1–182.

74. KERNAN, R. P. Electrogenic or linked transport. In: *Membranes and Ion Transport,* edited by E. E. Bittar. London: Wiley, 1970, vol. 1, p. 395–431.

75. KEYNES, R. D., AND G. W. MAISEL. The energy requirement for sodium extrusion from a frog muscle. *Proc. Roy. Soc. London Ser. B* 142: 383–392, 1954.

76. KEYNES, R. D., AND R. A. STEINHARDT. The components of the sodium efflux in frog muscle. *J. Physiol. London* 198: 581–599, 1968.

77. KEYNES, R. D., AND R. C. SWAN. The permeability of frog muscle fibres to lithium ions. *J. Physiol. London* 147: 626–638, 1959.

78. KU, D., T. AKERA, T. TOKIN, AND T. M. BRODY. Effects of monovalent cations on cardiac Na^+, K^+-ATPase activity and on contractile force. *Naunyn-Schmiedebergs Arch. Pharmacol.* 290: 113–131, 1975.

79. LANGER, G. A. Ion fluxes in cardiac excitation and contraction and their relation to myocardial contractility. *Physiol. Rev.* 48: 708–757, 1968.

80. LANGER, G. A. Effects of digitalis on myocardial ionic exchange. *Circulation* 46: 180–187, 1972.

81. LEDERER, W. J., AND R. W. TSIEN. Transient inward current underlying arrhythmogenic effects of cardiotonic steroids in Purkinje fibres. *J. Physiol. London* 263: 73–100, 1976.
82. LEE, C. O., AND H. A. FOZZARD. Activities of potassium and sodium ions in rabbit heart muscle. *J. Gen. Physiol.* 65: 695–708, 1975.
83. LEE, C. O., AND H. A. FOZZARD. Membrane permeability during low potassium depolarization in sheep cardiac Purkinje fibers. *Am. J. Physiol.* 237: C156–C165, 1979.
84. LEE, C. O., D. H. KANG, J. H. SOKOL, AND K. S. LEE. Relation between intracellular Na ion activity and tension of sheep cardiac Purkinje fibers exposed to dihydro-ouabain. *Biophys. J.* 29: 315–330, 1980.
85. LEE, K. S., AND W. KLAUS. The subcellular basis for the mechanism of inotropic action of cardiac glycosides. *Pharmacol. Rev.* 23: 193–261, 1971.
86. LIEBERMAN, M., C. R. HORRES, J. F. AITON, AND E. A. JOHNSON. Active transport and electrogenicity of cardiac muscle in tissue culture (Abstract). *Proc. Int. Union Physiol. Sci. Paris* 13: 1316, 1977.
87. LÜTTGAU, H. C., AND H. G. GLITSCH. Membrane physiology of nerve and muscle fibres. *Fortschr. Zool.* 24: 1–132, 1976.
88. MIURA, D. S., B. F. HOFFMAN, AND M. R. ROSEN. The effect of extracellular potassium on the intracellular potassium ion activity and transmembrane potentials of beating canine cardiac Purkinje fibers. *J. Gen. Physiol.* 69: 463–474, 1977.
89. MÜLLER, P. Potassium and rubidium exchange across the surface membrane of cardiac Purkinje fibres. *J. Physiol. London* 177: 453–462, 1964.
90. NOMA, A., AND H. IRISAWA. Electrogenic sodium pump in rabbit sinoatrial node cell. *Pfluegers Arch.* 351: 177–182, 1974.
91. NOMA, A., AND H. IRISAWA. Contribution of an electrogenic sodium pump to the membrane potential in rabbit sinoatrial node cells. *Pfluegers Arch.* 358: 289–301, 1975.
92. OKITA, G. T. Dissociation of Na^+, K^+-ATPase inhibition from digitalis inotropy. *Federation Proc.* 36: 2225–2230, 1977.
93. PAGE, E. Cat heart muscle *in vitro*. VII. The temperature-dependence of steady state K exchange in presence and absence of NaCl. *J. Gen. Physiol.* 48: 949–956, 1965.
94. PAGE, E., R. J. GOERKE, AND S. R. STORM. Cat heart muscle *in vitro*. IV. Inhibition of transport in quiescent muscles. *J. Gen. Physiol.* 47: 531–543, 1964.
95. PAGE, E., AND S. R. STORM. Cat heart muscle *in vitro*. VIII. Active transport of sodium in papillary muscles. *J. Gen. Physiol.* 48: 957–972, 1965.
96. PORTIUS, H. J., AND K. R. H. REPKE. Eigenschaften und Funktion des Na^+ + K^+-aktivierten, Mg^{++}-abhängigen Adenosintriphosphat-Phosphohydrolase-Systems des Herzmuskels. *Acta Biol. Med. Ger.* 19: 907–938, 1967.
97. RANG, H. P., AND J. M. RITCHIE. On the electrogenic sodium pump in mammalian nonmyelinated nerve fibres and its activation by various external cations. *J. Physiol. London* 196: 183–221, 1968.
98. REPKE, K. Über den biochemischen Wirkungsmodus von Digitalis. *Klin. Wochenschr.* 42: 157–165, 1964.
99. REPKE, K. Effect of digitalis on membrane adenosine triphosphatase of cardiac muscle. *Proceedings of the 2nd International Pharmacological Meeting, Prague, 1963*, edited by B. B. Brodie and J. R. Gillette. Oxford: Pergamon, 1965, p. 65–87.
100. REUTER, H. Exchange of calcium ions in the mammalian myocardium. Mechanisms and physiological significance. *Circ. Res.* 34: 599–605, 1974.
101. REUTER, H., AND N. SEITZ. The dependence of calcium efflux from cardiac muscle on temperature and external ion composition. *J. Physiol. London* 195: 451–470, 1968.
102. SCHATZMANN, H.-J. Herzglykoside als Hemmstoffe für den aktiven Kalium- und Natriumtransport durch die Erythrocytenmembran. *Helv. Physiol. Pharmacol. Acta* 11: 346–354, 1953.
103. SCHWARTZ, A., G. E. LINDENMAYER, AND J. C. ALLEN. The sodium-potassium adenosine triphosphatase: Pharmacological, physiological and biochemical aspects. *Pharmacol. Rev.* 27: 3–134, 1975.
104. SJODIN, R. A., AND L. A. BEAUGÉ. Coupling and selectivity of sodium and potassium transport in squid giant axons. *J. Gen. Physiol.* 51: 152s–161s, 1968.
105. SJODIN, R. A., AND L. A. BEAUGÉ. Strophanthidin-sensitive components of potassium and sodium movements in skeletal muscle as influenced by the internal sodium concentration. *J. Gen. Physiol.* 52: 389–407, 1968.

106. SJODIN, R. A., AND O. ORTIZ. Resolution of the potassium ion pump in muscle fibers using barium ions. *J. Gen. Physiol.* 66: 269–286, 1975.

107. SKOU, J. C. The influence of some cations on an adenosine triphosphatase from peripheral nerves. *Biochim. Biophys. Acta* 23: 394–401, 1957.

108. SKOU, J. C. Further investigations on a Mg^{++} + Na^+-activated adenosine triphosphatase possibly related to the active, linked transport of Na^+ and K^+ across the nerve membrane. *Biochim. Biophys. Acta* 42: 6–23, 1960.

109. TSIEN, R. W., AND D. O. CARPENTER. Ionic mechanisms of pacemaker activity in cardiac Purkinje fibers. *Federation Proc.* 37: 2127–2131, 1978.

110. THOMAS, R. C. Electrogenic sodium pump in nerve and muscle cells. *Physiol. Rev.* 52: 563–594, 1972.

111. VASSALLE, M. Electrogenic suppression of automaticity in sheep and dog Purkinje fibers. *Circ. Res.* 27: 361–377, 1970.

112. WALKER, J. L., AND R. O. LADLE. Frog heart intracellular potassium activities measured with potassium microelectrodes. *Am. J. Physiol.* 225: 263–267, 1973.

113. WEINGART, R. Electrical uncoupling in mammalian heart muscle induced by cardiac glycosides. *Experientia* 31: 715, 1975.

114. WEINGART, R. The actions of ouabain on intercellular coupling and conduction velocity in mammalian ventricular muscle. *J. Physiol. London* 264: 341–365, 1977.

115. WEINGART, R., R. S. KASS, AND R. W. TSIEN. Is digitalis inotropy associated with enhanced slow inward current? *Nature London* 273: 389–392, 1978.

116. WIER, W. G. Ionic currents and intracellular potassium in hypoxic myocardial cells (Abstract). *Biophys. J.* 21: 166a, 1978.

117. WIGGINS, J. R., AND P. F. CRANEFIELD. Effect on membrane potential and electrical activity of adding sodium to sodium-depleted cardiac Purkinje fibers. *J. Gen. Physiol.* 64: 473–493, 1974.

CHAPTER 3

Cardiac Automaticity and Its Control

MARIO VASSALLE

Department of Physiology, State University of New York,
Downstate Medical Center, Brooklyn, New York

Automaticity is a property of the heart by which excitation is initiated in the absence of external stimuli. Under normal circumstances only a small fraction of the cardiac cells has the property of automaticity. The cells capable of initiating impulse formation belong to the so-called specialized tissues that include the sinus node, the atrioventricular node, and the Purkinje fibers in the atria and ventricles. These cells share a common characteristic: during diastole the membrane potential slowly declines to less negative values. This slow potential decline is called diastolic depolarization, and it has a different slope in different pacemaker tissues. The fastest rate of diastolic depolarization is found in the sinus node cells (Fig. 1).

If each pacemaker tissue is characterized by the presence of diastolic depolarization, not every pacemaker tissue initiates impulses under normal conditions. The pacemaker tissue initiating the impulses that activate the whole heart (usually the sinus node) is called dominant; all the other pacemakers are called subsidiary. Subsidiary pacemakers are discharged by a propagated impulse originating from the dominant pacemaker: the propagated impulse activates the subsidiary pacemaker tissues before the local diastolic depolarization has progressed enough to attain the threshold potential. Thus under normal circumstances a subsidiary pacemaker does not act as a pacemaker at all. Instead, subsidiary pacemakers do modify importantly the velocity at which impulses originating from the sinus node propagate within the heart. The conduction velocity is low in the atrioventricular node and is high in the ventricular Purkinje fiber network. The importance of the diastolic depolarization of subsidiary pacemakers is seen when the dominant pacemaker or the process of conduction fails. Then survival is ensured by some subsidiary pacemakers becoming dominant and initiating conducted impulses. Common myocardial cells in the atria and in the ventricles lack diastolic depolarization (the potential has a constant value during diastole) and therefore remain quiescent in the absence of propagated impulses.

If diastolic depolarization is common to all cardiac pacemakers, the mechanism underlying diastolic depolarization is not necessarily the same. Furthermore special forms of pacemaker activity may arise under abnormal conditions even in myocardial cells, and the control of pacemaker action may be different, depending on the pacemakers involved.

The pacemaker process has been studied in more detail in cardiac Purkinje fibers than in the sinus node. Therefore the mechanisms underlying diastolic depolarization are considered first in the Purkinje fibers and then in the sinus node.

Ionic Events During Action Potential of Purkinje Fibers

Diastolic depolarization follows an action potential and is related to events occurring during excitation. The ionic events underlying the action potential in Purkinje fibers is briefly recalled here to provide a framework for the analysis of diastolic depolarization but is not reviewed in detail, since the subject is covered in recent reviews (60, 67, 74) and in the chapter by Coraboeuf in this volume.

In each cardiac cell the threshold potential must be attained for excitation to occur. In the sinus node, diastolic depolarization is responsible for the attainment of the threshold and therefore the initiation of an action potential (Fig. 1). The sinus node action potentials electrotonically depolarize neighboring cells to the threshold. The depolarization to the threshold of these neighboring cells locally initiates an action potential. The action potential thus elicited depolarizes contiguous cells to the threshold, and these cells, in turn, become activated. This process of propagation of the action potential continues as long as there is excitable tissue ahead. The electrotonic interaction between cells already excited and cells still at rest brings the membrane potential to the threshold in the latter. Once the threshold is attained, the ionic events are determined by the characteristics of the membrane of the cell being excited.

A brief description of the events during the action potential of Purkinje fibers is as follows: As is well known, sodium ions are more concentrated outside than inside the cell and are positively charged. At rest the inside of the cell is electronegative. Therefore both the concentration and the electrical gradients for sodium are directed inward. If the leak of sodium into a resting fiber is small (in spite of the large driving force), it is due to the fact that the resting membrane is sparingly permeable to sodium. On sudden depolarization to the threshold potential the permeability to sodium increases, and therefore the inward movement of sodium begins to increase. The threshold potential can be defined in several ways: one way is to describe it as the potential at which the inward sodium current begins to overpower outward potassium current. As a consequence of the net entry of positive charges carried by sodium ions, the upstroke (phase 0, Purkinje fiber, Fig. 1) is initiated. In turn, the depolarization induced

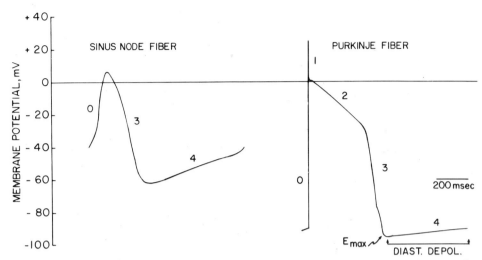

Fig. 1. Action potential of sinus node fibers and of Purkinje fibers. Ordinate shows scale for membrane potential in millivolts.

by sodium entry enhances the sodium conductance. More sodium channels open in the membrane, and therefore more sodium ions enter the cell under the inward-directed driving force represented by the electrochemical gradient. The rate of depolarization during the upstroke increases to attain its peak at around −10 to −20 mV. Regenerative depolarization is the term for the process by which depolarization increases sodium conductance, which then increases sodium influx (which in turn depolarizes the membrane and therefore further increases the sodium conductance, and so on). At zero potential the only inward driving force acting on sodium ions is the concentration gradient. As sodium still enters the cell under this force, the inside of the cell becomes positive. Eventually the internal electropositivity prevents a net influx of sodium: this marks the end of the upstroke.

Thus the upstroke of the action potential of Purkinje fiber is characterized by a brief (less than 1 ms) movement of sodium down its electrochemical gradient, which is permitted by the fact that depolarization to the threshold allows the opening of specific sodium channels. The inward movement of sodium during the upstroke is referred to as the fast sodium current. This current has a voltage dependence: it is turned on (activation) only if the threshold voltage is attained and becomes larger at certain less negative potentials. In addition it has a time dependence: even if the potential were to be kept at a value at which the sodium channels become open, these channels would close (inactivation) as a function of time. Because the process of activation and inactivation occur very rapidly, the channels involved are called fast sodium channels.

The upstroke is followed by an initial phase of rapid repolarization (phase 1, Fig. 1), which is due to the rapid inactivation of the fast sodium channels and to the flow of an outward current (early outward current, i_{qr}) carried mostly by potassium but also by chloride ions (45). The ensuing phase 2 (the plateau, Fig. 1) is characterized by a very slow repolarization; this is due to the near balance of several opposing currents. Some of these currents are time dependent and are slow, because their channels are turned on and off much more slowly than those for the fast sodium current. These currents are triggered by the depolarization induced by the upstroke and the magnitude changes as a function of voltage and time (voltage dependence and time dependence). One of the slow currents is inward and is carried by both calcium and sodium flowing into the cell through a slow channel. The other slow current is outward (i_{x_1}) and is mainly carried by potassium ions. The gradual time-dependent inactivation of the slow inward current and the progressive time-dependent activation of the slow outward current i_{x_1} eventually lead to the termination of the plateau. Phase 3 (late rapid repolarization) follows, and the maximum diastolic potential (E_{max}, Fig. 1) is attained.

As a consequence of the depolarization brought about by the action potential, another potassium channel is fully activated and is responsible for the pacemaker current. The pacemaker current is discussed in detail below.

The picture that emerges from this brief description of the Purkinje fiber action potential is that ions (Na^+, K^+, Cl^-, Ca^{++}), because of their unequal distribution across the cell membrane and because of their charge, are under driving forces that tend to move them down their electrochemical gradients. The magnitude of the movement of these ions across the cell membrane depends on the magnitude of the driving force and on the ease with which the ions can cross the cell membrane (conductance). The magnitude of the conductance is not fixed. It varies to a different extent within a certain range of potentials, and once the conductance is turned on at a given potential, it varies as a function of time.

The currents flowing during the action potential are dependent on the opening of specific channels that have a voltage and a time dependence, as just described. In addition to these currents flowing through specific channels, there are also background or leak currents. An example is represented by the leak of sodium into a resting cell and another by the leak of potassium (i_{K_1}) out of a cell. These currents do not have a time dependence, i.e., they do not vary as a function of time. They have a voltage dependence, however; e.g., the outward potassium current i_{K_1} increases on depolarization. This is not surprising, because an outward leak of potassium would be expected to be larger when the inside of the cell became less negative. The channels for this current have a peculiarity, however; when the potential becomes less negative, the efflux of potassium is hindered (anomalous rectification, inward-going rectification). Because of the anomalous rectification, i_{K_1} increases on depolarization but less than expected. As a consequence, the plateau lasts longer than it otherwise would.

Thus presented, the action potential is brought about by an initial fast inward movement of sodium ions, lasts several hundred milliseconds, in part because of the overlap between a slow calcium-sodium inward current and a slow potassium outward current, and is terminated by a voltage-dependent outward potassium current. During the action potential a potassium channel is fully activated and is responsible for the subsequent diastolic depolarization. The process underlying diastolic depolarization will be considered now.

Diastolic Depolarization in Purkinje Fibers

So far, there have been mentioned four outward potassium currents: 1) the early outward current, i_{qr}, flowing during phase 1; 2) the slow outward current, i_{x_1}, carried predominantly by potassium ions; 3) a background current, i_{K_1}, which is only voltage dependent; and 4) a pacemaker current, i_{K_2}, whose channel is fully activated during the action potential. In this section we will consider only the pacemaker current.

In 1949 Coraboeuf and Weidmann (16) published a short note with the picture of an action potential of a spontaneously active Purkinje fiber showing a pronounced diastolic depolarization that smoothly merged into the upstroke of the following action potential. Draper and Weidmann (26) showed that diastolic depolarization was usually present in Purkinje fibers and was approximately proportional to the extracellular sodium concentration. They suggested that diastolic depolarization could result either from an increase in an inward sodium current or from a decrease in an active extrusion of sodium ions. Either mechanism would have resulted in a net increase in sodium influx and therefore in a depolarizing current. Subsequent measurements of membrane conductance during diastole showed that the conductance decreases during diastolic depolarization. The finding was not compatible with an increased sodium influx, because, if this were the case, the conductance should have increased and not decreased. It was mentioned above, however, that potassium conductance decreases with depolarization (anomalous, inward-going rectification, see refs. 13, 35). Therefore two alternative explanations were possible: either the decrease in conductance was due to the decrease in potential during diastole, or vice versa.

New insight into the problem was acquired with the introduction of voltage clamp technique. The principle of the method consists in displacing the membrane current at a given value and maintaining the voltage constant by passing currents of suitable magnitude and polarity. The membrane current consists of two components: the capacitative current and the ionic current. The capacitative current flows only when the membrane potential varies and, therefore, only at the beginning and the end of the clamp. The current recorded during the clamp consists only of the ionic component as long as the potential does not vary. Thus the voltage clamp method allows the measurement of ionic currents as a function of voltage (by clamping at different potential values) and as a function of time (as the voltage remains unaltered during the clamp). In one method a single strand of Purkinje fibers is ligated twice at a short (1–2 mm) distance (20). The ligated segment is impaled with two microelectrodes: one is used to record the potential and the other to pass current. In another method the sucrose gap technique is employed to pass current through extracellular electrodes and to force it to flow intracellularly in the sucrose compartment (see ref. 44). The advantages and limitations of the voltage clamp method have been discussed recently (32): in general the method is most successful when small, slowly changing currents are involved.

In the course of their analysis of the ionic events underlying the action potential, Deck and Trautwein (21) clamped the membrane potential first at plateau potentials and then back to the resting potential. After the clamp a current flowed that would have caused hyperpolarization, if the fiber had not been clamped back to the resting potential. This would have been expected on the basis that, if a quiescent Purkinje fiber is stimulated, the action potential is followed by a transient hyperpolarization. In other words, the potential is transiently more negative after than before the action potential. The current that flowed on clamping back to the resting potential was outward. If, however, the membrane was clamped to a sufficiently negative potential, the current became inward, thus reversing its direction. Deck and Trautwein (21) concluded from their experiments that a repolarization to about −40 or −50 mV initiates an increase in potassium conductance that decreases during diastole. Therefore, according to this view, when the action potential repolarizes during phase 3 to values more negative than −40 mV, potassium conductance increases. During diastole this potassium conductance decreases, and therefore progressively less potassium leaves the cell; this time-dependent fall in potassium conductance brings about diastolic depolarization.

Independently in 1963–1964 in Berne, Switzerland, the author conducted experiments that were focused on an analysis of diastolic depolarization (72). At the end of an action potential, the membrane potential was clamped at the maximum diastolic potential (E_{max}). In Figure 2, two action potentials are shown in the *upper left corner*. The part of these two action potentials enclosed in the rectangle is shown magnified in the rest of the figure together with the current record (*thick trace*). The voltage record begins wih the last part of phase 3 and continues with diastolic depolarization. With the following repolarization the voltage was clamped at the maximum diastolic potential. As a consequence an inward current flowed, which increased with time (see *thick trace*). In the absence of the clamp, such a current would have caused diastolic depolarization. A tentative explanation for the progressively increasing inward current was that potassium conductance decreased as a function of time: if progressively fewer potassium ions leave the cell and the same amount of sodium leaks into the cell, the net result is an increasing inward current. If the explanation is correct, then clamping to a potential negative to the potassium equilibrium potential should still induce an inward current, but the inward current should decrease as a function of time. This was in fact found (72). The explanation of the finding is as follows. When the membrane potential is more negative than the potassium equilibrium potential, the net driving force acting on potassium

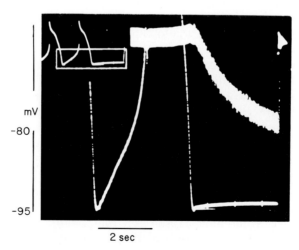

Fig. 2. Effect of clamping the membrane potential at E_{max} in a spontaneously firing Purkinje fiber. Current calibration (2×10^{-8}A) is *vertical bar* at *upper left* of figure. [From Vassalle (72).]

ions is directed inward. Therefore the potassium ions enter the cell. As the potassium conductance decreases with time, however, fewer potassium ions are able to enter the cell: hence there is a progressively decreasing inward current. That potassium conductance decreases as a function of time was demonstrated by another finding. While the potential was kept clamped at the value of the maximum diastolic potential, the membrane conductance was measured: it decreased as a function of time, the voltage remaining constant (72). These findings were compatible with the prevailing view that, during diastole, a potassium conductance that had been turned on during the plateau was deactivated. Other findings, however, made clear that this view required modifications. Thus, in quiescent fibers, clamping the resting potential to a more negative value also resulted in a progressively increasing inward current (72). In other words, in the absence of a preceding action potential, hyperpolarization led to a time-dependent shutting off of a potassium conductance. This meant that the pacemaker current was at least partially activated at the resting potential without a previous depolarization at the plateau. The finding raised the question of the voltage range over which the pacemaker current was activated (see below).

At the end of a hyperpolarizing clamp the membrane potential did not merely return to the original value. Instead, it became transiently less negative than the resting potential. If the clamp pulse had been long enough, the depolarization that followed was large enough to attain the threshold and to initiate an action potential (72). The reason for this behavior was made clear by adopting a different procedure. Instead of just terminating the clamp pulse and allowing the membrane potential to depolarize beyond its original resting value, the membrane was clamped back to the original resting value. In this manner it was possible to measure the membrane current not only during the hyperpolarizing clamp but also after the hyperpolarizing clamp. These records showed that the inward current recorded during the clamp did not disappear instantaneously on returning to the resting potential, but instead slowly declined (72). This is explained as follows: during the hyperpolarizing clamp (and because of the hyperpolarization) the pacemaker current was slowly deactivated; on returning to the resting potential, the pacemaker current was reactivated to the value existing before the hyperpolarizing clamp. The reactivation was not instantaneous, but it was faster than the deactivation during the hyperpolarizing clamp. The inward current that was present after the end of the hyperpolarizing clamp is responsible for the depolarization following the cessation of the hyperpolarization and provides the explanation for the finding of Weidmann (83) that a hyperpolarizing pulse applied during diastole is followed by an acceleration of diastolic depolarization. The fact that the pacemaker current does not require a previous action potential to be at least partially activated is of interest in more than one respect. For example, this partial activation at the resting potential is a prerequisite for the initiation of pacemaker activity by catecholamines (see below).

In another procedure the action potential was terminated at various times during the plateau by clamping the membrane potential back to the resting potential (72). A slowly increasing inward current flowed that would have caused diastolic depolarization in the absence of the clamp. An interesting point was that the time course and the magnitude of this current were the same whether the clamp was applied early or late during the plateau. This finding accounts for the fact that action potentials cut short by a brief clamp are followed by the usual diastolic depolarization whether the action potential is terminated early or late during the plateau (72, 83).

Noble and Tsien (62) carried out a detailed analysis of the voltage dependence of the pacemaker current, which they labeled i_{K_2}. They clamped the membrane at potentials positive and negative to the resting potential and found that in the steady state the pacemaker conductance is maximally activated at about −60 mV and is deactivated at −90 mV. They also found that during hyperpolarizing clamps the current was deactivated slowly, as previously reported (72). During depolarizing pulses, however, the pacemaker conductance was activated more quickly, in agreement with the results obtained with clamps applied at various times during the plateau (see above). The channel for the pacemaker current behaved as an inward-going rectifier (see also ref 39); that is to say, the pacemaker current decreases on depolarization to the plateau range of potentials (62). Also, as the pacemaker channel is rapidly activated and undergoes anomalous rectification, the pacemaker current does not contribute to the time-dependent changes that terminate the plateau (62). On repolarization the voltage-dependent inward-going rectification is removed and the K_2 conductance is still high (62). Thus the pacemaker channel, fully activated on depolarization, is still activated when the maximum diastolic potential is attained. In the steady state, however, this potassium conductance should be close to zero at the maximum diastolic potential. Therefore there is a discrepancy between the value of the conductance at E_{max} (fully activated) and the value that the conductance ought to have at such potential (fully deactivated). Because of this discrepancy the conductance decreases toward the steady-state value that is appropriate for E_{max}. The decrease of conductance is slow, as is characteristic at these negative potentials. The change in current with time is caused by a change in the number of channels open. The gating reaction is described by the variable s, which is equal to 1 when all channels are open and is equal to 0 when all channels are closed (62). As indicated above, at the maximum diastolic potential the variable s is close to 1, that is, the K_2 conductance is high. Instead, at that negative potential it should be close to 0 in the steady state. The deactivation of the pacemaker current is brought about by the fall in variable s toward its steady-state value. But, because of diastolic depolarization the steady-state value of s increases during diastole, as it depends solely on voltage (62). Eventually the variable s will decrease to a value identical to the steady-state value, and the potential (in the absence of other factors) will not vary any longer. The fiber will be quiescent and the resting potential will be stable (62). The pacemaker current will be partially activated and there will be no net transfer of charges across the cell membrane. This series of events characterize the potential changes occurring with electrical stimulation of Purkinje fibers that are not spontaneously active. In contrast, in spontaneously active pacemakers diastolic depolarization will attain the threshold potential and automatic discharge will be renewed. The attainment of the threshold must be favored by an increase in sodium conductance (72) and by a voltage-dependence decrease (inward-going rectification) of g_{K_2} (62) and of the background potassium conductance, g_{K_1}. More recently, in a computer reconstruction of the action potential in Purkinje fibers, it has been found necessary to increase the sodium conductance well ahead of the threshold in order to insure pacemaking activity (60).

That there is indeed a component sensitive to tetrodotoxin (TTX) in the late diastolic depolarization has been demonstrated by recent experiments (79). In fibers driven at a slow rate or spontaneously active, the late diastolic depolarization appears to enter a range of potential at which the inward sodium current begins to increase ahead of the threshold. In Figure 3 the panels show the lowest part of the action potential and the twitch curves of a Purkinje fiber driven at 18/min. The *left panel* shows the control tracings, the *middle panel* shows those in the presence of TTX, and the *right panel* illustrates the recovery in Tyrode solution. Because of the slow driving rate, spontaneous beats were intermingled with driven beats in the control panel. As a consequence of TTX exposure, the initial diastolic depolarization was little affected, but the late part of diastolic depolarization was markedly flattened, and the fiber now responded to each stimulus. It is apparent that the TTX-sensitive component appears well ahead of the threshold. It should be noted that Purkinje fibers are far less sensitive than nerve fibers to TTX and therefore respond to electrical stimuli at concentrations that fully block the fast channel in the nerve. If the Purkinje fibers are not electrically stimulated and are spontaneously active, TTX leads to quiescence, because in the absence of the sodium component the diastolic depolarization remains negative to the threshold. When Purkinje fibers are driven by the sinus node at about 70/min, the diastole is too short for the diastolic depolarization to attain the potential range where the TTX-sensitive sodium component is activated (sodium zone). This and the flattening of diastolic depolarization due to overdrive (see below) contribute to keep the subsidiary pacemakers suppressed.

It is clear from the present description that any increase in the net inward diastolic current brought about either by increasing the sodium inward current or by decreasing the potassium outward current will enhance the rate of discharge of Purkinje fibers. Thus both an increase in [Na]$_o$ (by increasing the

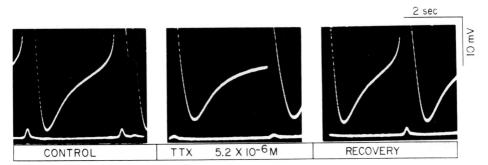

2 sec

10 mV

| CONTROL | TTX   5.2 X 10^{-6}M | RECOVERY |

Fig. 3. A component sensitive to tetrodotoxin (TTX) in diastolic depolarization in Purkinje fibers. In *left panel* there are 2 spontaneous beats. During recovery of 2nd beat, the fiber was activated by electrical stimulus. *Middle panel* was recorded in presence of TTX (5.2×10^{-6}M): spontaneous activity had ceased, and two action potentials were driven. *Right panel* was recorded during recovery in Tyrode's solution. [K]$_o$ = 1.35 mM. [From Vassalle and Scidá (79), adapted from Vassalle (74a).]

electrochemical gradient for this ion) and a decrease in [K]$_o$ (by decreasing K conductance) lead to an increase in rate of discharge of Purkinje fibers (71).

The demonstration that the pacemaker channel is fully activated at about −60 mV (62) provides an explanation for the finding (21) that a repolarization to about −50 mV apparently was required for an increase in the conductance of the pacemaker channel. On repolarization to potentials positive to −50 mV, time-dependent changes attributable to this current would not be apparent, since the pacemaker current would not be deactivated.

Recently the view that the pacemaker current is a K current has been challenged. It has been proposed that the pacemaker current is not a K current deactivating during hyperpolarization negative to −60 mV but rather to an inward current activated over the same range of potentials on the basis of membrane resistance measurements (23, 24). Also, the abolition of the time-dependent inward current on hyperpolarization from the resting potential by cesium has been taken as evidence that the pacemaker current is an inward current (24). The reversal potential of the pacemaker current shifts with [K]$_o$, as would be expected for a potassium current (62). In a computer reconstruction, however, it was shown (25) that the total inward current (resulting from an inward current and changes in potassium current due to depletion) shows a reversal potential at negative values and a shift of such reversal potential with changes in [K]$_o$.

Diastolic Depolarization in Sinus Node

The sinus node shares with Purkinje fibers the presence of a slow diastolic depolarization. In contrast to Purkinje fibers, however, the action potential of sinus node cells has a smaller amplitude, slower upstroke, little or no overshoot, no phase 1, little plateau, smaller E_{max}, and a faster diastolic depolarization. Furthermore, the sinus node action potential has little sensitivity to TTX (53, 86) and to a decreased [Na]$_o$ (28, 65, 66); instead it is sensitive to changes in [Ca]$_o$ (63, 64) and to substances that interfere with the slow channel (53, 57, 85, 87).

The characteristics just described apply to dominant pacemaker cells in the sinus node. It should be added that many of the sinus node cells behave as subsidiary pacemakers. That is to say, they are activated by impulses conducted from the dominant pacemakers rather than from local diastolic depolarization. Because of this the upstroke of the subsidiary pacemakers includes two components: one component is relatively fast and is presumably due to the influx of sodium through a fast channel; the other component is slow and is presumably due to the activation of a slow inward current carried by both Na$^+$ and Ca^{2+} (55). The fast component is abolished by TTX and is substituted by the slow component: in this situation diastolic depolarization proceeds to the threshold, and the whole upstroke is made up of the slow component, as is usually seen in dominant pacemakers. The fast component, conversely, becomes much larger if the take-off potential is more negative (55). It would appear from these findings that part of the differences between dominant and subsidiary pacemakers is due to the fact that in subsidiary pacemakers the upstroke originates at a more negative potential and therefore includes the activation of a relatively fast sodium channel. It may well be that part of the differences between the sinus node and Purkinje fibers are due to the less negative maximum diastolic potential. As a consequence the potential range during diastole is less negative and the fast sodium channel is at least partially inactivated. If the fast sodium channel is substantially inactivated, excitation must rely on the opening of a slow channel. The activation of the slow channel would account for the less negative take-off potential, the slower upstroke, the lack of overshoot, the little sensitivity to TTX (which blocks only the fast channel), the little sensitivity to low $[Na]_o$ (which can be substituted by calcium in the slow channel), the sensitivity to $[Ca]_o$ (as calcium does enter the slow channel), and the inhibitory effect of substances that block the slow channel. The less negative maximum diastolic potential and related consequences may be due to the larger background sodium current in the sinus node (68). It should be noted here that there is a rather important difference between the usually dominant pacemakers and the subsidiary pacemakers: the fast component not only is absent in the former but does not appear even if the take-off potential becomes more negative (56).

As to the current responsible for diastolic depolarization, diastolic depolarization in the sinus node also appears to be caused by the fall in a potassium conductance. Irisawa (42) used a sucrose-gap technique to clamp the membrane potential of the sinus node cells to the maximum diastolic potential. This maneuver resulted in an inward current that increased with time. If the membrane potential was clamped to more negative values, the inward current decreased with time, not unlike the findings in Purkinje fibers. These experiments allowed the conclusion that a potassium current was responsible for diastolic depolarization in the sinus node. The identification of this current was not certain, however. One possibility is that the current underlying diastolic depolarization in the sinus node is the plateau current, i_{x_1}. This current, activated during the action potential, deactivates during diastole. If the maximum diastolic potential is sufficiently low, a decay of i_{x_1} causes diastolic depolarization. In the presence of a time-independent inward leak of sodium and calcium into the cell, a progressive decrease of the outward current, i_{x_1}, would result in a depolarizing inward current and, therefore, in diastolic depolarization. Because i_{x_1} deactivates more rapidly than i_{K_2} (at least in Purkinje fibers, ref. 39), diastolic depolarization would be faster in the sinus node than in Purkinje fibers.

Support for the concept that the decay of i_{x_1} can be responsible for diastolic depolarization is provided by experiments conducted in other cardiac tissues. In partially depolarized Purkinje fibers repetitive activity may be present. The action potentials then are much smaller than usual and resemble in many respects those of the sinus node (4, 5, 18, 38). Voltage clamp analysis has shown that the diastolic depolarization that follows these small action potentials is due to the deactivation of the i_{x_1} current (38). By applying depolarizing pulses to atrial (10, 52) and ventricular muscle tissues (41, 43, 44), spontaneous activity was induced in these fibers. The action potential was shown to be due to the opening of a slow channel, and the subsequent diastolic depolarization was due to the deactivation of the outward current, i_{x_1} (10, 44, 52).

Recently it has been proposed that the pacemaker current in the sinoatrial node is an inward current (59). Thus it was found that K accumulates and depletes during depolarization and hyperpolarization, respectively, and that the total membrane conductance during diastole (measured with short pulses) in the absence or in the presence of voltage clamp either remained constant or increased. Also, no reversal potential was found at various $[K]_o$. On this basis it was concluded that diastolic depolarization in the sinus node is due to a time-dependent increase of a Na^+ or Ca^{2+} conductance.

The background inward current in the sinus node is not carried necessarily by sodium ions as in Purkinje fibers, since lowering $[Na]_o$ has little effect (within limits) on the sinus node rate (28, 65, 66). Possibly this little effect is due to the fact that calcium can substitute for sodium. It is recognized that the rate of discharge of the sinus node is increased by a higher $[Ca]_o$ through a steepening of diastolic depolarization (64). Thus calcium may be carrying inward current not only during the action potential but also during diastole in the sinus node (9).

Control of Pacemaker Activity

Under normal conditions the control of cardiac automaticity seems to be carried out only at the sinus node. Thus the activation of the vagus leads to a slowing of the heart rate, and the stimulation of the sympathetic nerves leads to an acceleration of the heart rate. To insure a coordinated heart function, however, subsidiary pacemakers should also be under some form of control. For example, if the vagus were to suppress not only the sinus but also all subsidiary pacemaker tissues, there would be no ventricular escape. Also, in complete atrioventricular (AV) block, the idioventricular rhythm is usually so slow that the necessity or usefulness of an inhibitory control of the idioventricular pacemakers would not be immediately apparent. The stimulatory action of the sympathetic nerves, on the other hand, should not increase the automaticity of different pacemaker tissues to such an extent that the heart is activated from different sites. These considerations show that the question of the control of cardiac automaticity is confined neither to the sinus node nor to the autonomic system.

Vagus-Dependent Inhibition. The sinus node is under the inhibitory control of the vagus. The mechanism of such an inhibition is a shift of the membrane potential to a more negative level during diastole (40). Such a hyperpolarization is brought about by an increase in potassium conductance by the vagal mediator acetylcholine. This has been demonstrated by the use of radioisotopes (36) and by measurement of membrane resistance (69). Because usually the membrane potential is positive to the potassium equilibrium potential, an increase in potassium conductance leads to a shift of the membrane potential to a more negative value. This results in a longer diastole or, if the threshold for excitation is not reached altogether, in a sinus standstill. The other subsidiary pacemaker tissues in the atria and at the AV junction are also subject to the same type of inhibition by the vagus. On maximal vagal stimulation, however, not only the atria but also the ventricles become quiescent. Furthermore, as vagal stimulation is continued, the ventricles but not the atria begin to be active at a rate that progressively increases to a steady value. The mechanism by which vagal stimulation results in a temporary ventricular standstill has been investigated recently.

Rate-Dependent Inhibition. The effect of vagal stimulation on the ventricles poses two distinct problems. The first problem is the cause of the initial ventricular standstill, and the second is the cause of the initiation of an idioventricular rhythm during a continued vagal stimulation. One possibility is that acetylcholine released by the vagus in the ventricles inhibits Purkinje fiber automaticity. The subsequent escape could be due to a loss of sensitivity to acetylcholine or to a release of catecholamines (either reflexly due to the fall in blood pressure or directly due to stimulation of sympathetic fibers in the vagus). Another possibility is the stretch of Purkinje fibers as the ventricles are progressively distended during the standstill. All these possibilities need to be considered, for there are several reports of vagal innervation of the ventricles and of inhibitory action of acetylcholine on ventricular Purkinje fiber automaticity (6, 30, 46, 70). Also, stretch has been shown to enhance the automatic discharge of isolated Purkinje fibers (27). Recent evidence, however, points to another mechanism for the temporary ventricular standstill during vagal stimulation; namely, the ventricular standstill on vagal stimulation appears to be an instance of a phenomenon called overdrive suppression. That a period of fast drive is followed by a temporary suppression of pacemaker activity was observed in the last century by Gaskell (34) and was studied in some detail by Cushny (19). Erlanger and Hirschfelder (31) also became interested in the phenomenon, and in fact they considered the possibility that it might explain the effect of vagal stimulation on the ventricles.

Overdrive suppression could account for the ventricular standstill during vagal stimulation in the following way. The sinus rate is much faster than that of the idioventricular pacemakers, and therefore it continuously overdrives the idioventricular pacemakers under normal conditions. On the basis of this concept the vagus would inhibit only the atrial and AV junctional pacemakers, and in doing so it would reveal the suppression that the sinus node was exerting on idioventricular pacemakers. This would be the cause of the ventricular standstill. The initiation of the idioventricular rhythm during vagal stimulation would be due to the removal of the inhibition: with the sinus node stopped and the ventricles quiescent, the effects of overdrive would be dissipated. There are several experimental procedures that support this viewpoint. Thus if vagal stimulation is made to last 2 min and the ventricles are driven during the 1st min at a rate similar to that of the sinus node, the ventricular standstill is merely postponed to the end of the ventricular drive (81). This finding shows that 1) the ventricular standstill cannot be due to acetylcholine, because after 1 min of vagal stimulation the ventricles are usually active; 2) the stimulation of sympathetic fibers in the vagus are not the cause of the resumption of the activity, since the ventricular arrest is unaltered after 1 min of stimulation of such fibers; and 3) the ventricular drive is the important parameter for the subsequent suppression. This last conclusion is supported by the fact that if, during the 1st min of a 2-min vagal stimulation, the ventricles are driven at a rate faster than the previous sinus rate, the subsequent ventricular standstill becomes longer (76). And if the driving rate during the 1st min of vagal stimulation is, instead, only slightly higher than that of the idioventricular pacemakers, the ventricular standstill is abolished (76). If the sinus node is slowed by means of graded vagal stimulation and then is stopped by maximal vagal stimulation, the ventricular standstill is much shorter (76). Also, if during a 3-min vagal stimulation the ventricles are driven during the 2nd min of vagal stimulation at a rate similar to that of the sinus node, the pause that follows the drive is fairly similar to that at the beginning of vagal stimulation (48). That the drive itself (in the absence of vagal stimulation) is followed by a period of ventricular standstill can be easily shown by driving the ventricles in the presence of complete atrioventricular block (47, 54, 76). It should be added that ventricular arrest and subsequent resumption of activity occur even when the blood pressure is maintained and stretch is avoided (81); and a similar temporary arrest and resumption of activity occurs also in Purkinje fibers overdriven in vitro (2, 73, 76, 82). These findings show that neither a reflex activation of the sympathetic nerves nor stretch of Purkinje fibers are required for the resumption of the activity after a period of suppression.

The mechanism by which overdrive leads to a subsequent suppression of pacemaker activity has been explored in atrial and ventricular pacemakers. In the atria it has been clearly demonstrated that electrical stimuli liberate both acetylcholine and norepinephrine (3, 51, 58, 84). By the use of pharmacological agents it has been shown that most of the overdrive suppression in the atria is due to a release of acetylcholine and the subsequent acceleration to the release of catecholamines (3, 51, 58, 84). The residual inhibition after the elimination of the acetylcholine component has been attributed to an accumulation of $[K]_o$ (58) presumably lost during the fast activity.

In the ventricles the mechanism appears to be different, and the release of acetylcholine, if any, does not appear to play a role in overdrive suppression. Thus overdrive suppression is not altered by the administration of either neostigmine (76) or atropine (9). These differences are reflected in the time course of the maximum diastolic potential (E_{max}) with drive. In the sinus node fast drive causes an increase in maximum diastolic potential within a few beats (58), as one would expect from a liberation of acetylcholine. In the ventricular Purkinje fibers, instead, with drive there is an initial decrease of the maximum diastolic potential followed by an increase above control values (73). There are reasons to believe that the initial diminution of E_{max} is due to an accumulation of K outside the cell membrane (48, 73, 76). This accumulation of K may contribute to the suppression of short drives, and in fact it is possible to transmit humorally overdrive suppression at the beginning of the drive [when the [K] increases in the coronary sinus; (80)]. Furthermore, the measurement of extracellular K with an ion-sensitive electrode has shown that potassium accumulates when the rate of stimulation increases [(50); see ref. 59], but such increase is not maintained during a prolonged drive (50). This is in agreement with the fact that the increase in [K] in the coronary sinus with the onset of drive of the ventricles subsides during the drive (49, 76).

The late hyperpolarization has been attributed to the activation of an electrogenic sodium pump (73). The pump would be activated initially by the higher $[K]_o$ and by the accumulation of sodium inside the fiber as a consequence of the fast drive. Interfering with the pump or its metabolic supply results in the loss of the hyperpolarization (7, 12, 14, 22, 73). The current generated by the pump has been measured in cardiac tissues after a period of exposure to K^+-free solutions that increases cellular sodium; when potassium is restored an outward current flows that is abolished by ouabain (1, 29, 33). In fact a prolonged outward current has been demonstrated after overdrive in Purkinje fiber (15). Recent evidence shows that even in the sinus node overdrive suppression may be related (at least in part) to the function of an electrogenic pump (17).

It would seem then that the sinus node, by driving the Purkinje fibers at a fast rate, imposes on this tissue a sodium load that results in a decrease in the net inward current during diastole. The electrogenic sodium extrusion causes the increases in E_{max} during the drive and maintains the membrane potential negative to the threshold for a period after the drive. When enough sodium has been extruded during the quiescence and the electrogenic current is suitably reduced, the potential reaches the threshold and spontaneous activity is initiated.

Excitatory Action of Sympathetic Nerves. Stimulation of the sympathetic nerves results in an increase in the rate of discharge of all cardiac pacemakers. The mechanism by which this is accomplished is different in atrial versus ventricular pacemakers.

Direct information about the mechanism by which sympathetic stimulation steepens diastolic depolarization in atrial pacemaker tissues is missing. As mentioned above, however, it is possible to obtain spontaneous activity in atrial fibers by means of depolarizing pulses: diastolic depolarization is brought about by the deactivation of i_{x_1} during diastole (10, 38). Administration of epinephrine results in an enhancement of this diastolic depolarization. Voltage clamp analysis has shown that epinephrine increases both the slow inward current and i_{x_1} (11). The increment in rate of discharge is due presumably to an increment in the slow inward current (61).

Instead, in Purkinje fibers epinephrine steepens diastolic depolarization by shifting the steady-state activation curve for i_{K_2} in a depolarizing direction (37). As a consequence of such a shift, the decline of the pacemaker current during diastole is accelerated. Other features of catecholamine action such as an increase in maximum diastolic potential might be related to the ability of catecholamines to stimulate an electrogenic (77) active sodium extrusion (8, 75, 78).

Concluding Remarks

Recent advances have clarified many aspects of cardiac automaticity. The process of diastolic depolarization and its control have been shown to differ in the atrial and the ventricular pacemakers. The mechanism of the action of catecholamines in modifying cardiac automaticity has been clarified. A mechanism of control, the frequency-dependent inhibition that has been called overdrive suppression, has been shown to be important for the control of subsidiary pacemakers. These acquisitions together with the information obtained about the ionic events during the action potential have enhanced quite substantially the understanding of cardiac automaticity.

Acknowledgments

The original work reported here was supported by grants from the National Heart and Lung Institute, National Institutes of Health, and the New York Heart Association.

References

1. AKASU, T., Y. OHTA, AND K. KOKETSU. The effect of adrenaline on the electrogenic Na⁺ pump in cardiac muscle cells. *Experientia* 34: 488–490, 1978.
2. ALANÍS, J., AND D. BENÍTEZ. The decrease in the automatism of the Purkinje pacemaker fibers provoked by high frequencies of stimulation. *Jpn. J. Physiol.* 17: 556–571, 1967.
3. AMORY, D. W., AND T. C. WEST. Chronotropic response following direct electrical stimulation of the isolated sino-atrial node: a pharmacological evaluation. *J. Pharmacol. Exptl. Therap.* 137: 14–23, 1962.
4. ARONSON, R. S., AND P. F. CRANEFIELD. The electrical activity of canine cardiac Purkinje fibers in sodium-free, calcium-rich solutions. *J. Gen. Physiol.* 61: 786–808, 1973.
5. ARONSON, R. S., AND P. F. CRANEFIELD. The effect of resting potential on the electrical activity of canine cardiac Purkinje fibers exposed to Na-free solution or to ouabain. *Pfluegers Arch.* 347: 101–116, 1974.
6. BAILEY, J. C., K. GREENSPAN, M. V. ELIZARI, G. J. ANDERSON, AND C. FISCH. Effect of acetylcholine on automaticity and conduction in the proximal portion of the His-Purkinje specialized conduction system of the dog. *Circ. Res.* 30: 210–216, 1972.
7. BHATTACHARYYA, M., AND M. VASSALLE. Metabolism-dependence of overdrive-induced hyperpolarization. *Arch. Int. Pharmacodyn. Ther.* 246: 28–37, 1980.
8. BORASIO, P. G., AND M. VASSALLE. Effects of norepinephrine on active K transport and automaticity in cardiac Purkinje fibers. In: *Myocardial Biology, Recent Advances in Studies on Cardiac Structure and Metabolism*, vol. 4, edited by N. S. Dhalla. Baltimore: University Park, 1974, p. 41–57.
9. BROOKS, C. McC., AND H. H. LU. *The Sinoatrial Pacemaker of the Heart.* Springfield, IL: Thomas, 1972.
10. BROWN, H. F., AND S. J. NOBLE. Membrane currents underlying delayed rectification and pacemaker activity in frog atrial muscle. *J. Physiol. London* 204: 717–736, 1969.
11. BROWN, H. F., AND S. J. NOBLE. Effects of adrenaline on membrane currents underlying pacemaker activity in frog atrial muscle. *J. Physiol. London* 238: 51P–53P, 1974.
12. BROWNING, D. J., J. S. TIEDEMAN, A. L. STAGG, D. G. BENDITT, M. M. SCHEINMAN, AND H. C. STRAUSS. Aspects of rate-related hyperpolarization in feline Purkinje fibers. *Circ. Res.* 44: 612–624, 1979.
13. CARMELIET, E. E. *Chloride and Potassium Permeability in Cardiac Purkinje Fibers.* Brussels: Arscia, 1961.
14. CARPENTIER, R., AND M. VASSALLE. Enhancement and inhibition of a frequency-activated electrogenic sodium pump in cardiac Purkinje fibers. In: *Research in Physiology: A Liber Memorialis in Honor of Prof. Chandler McCuskey Brooks*, edited by F. F. Kao, K. Koizumi, and M. Vassalle. Bologna: Aulo Gaggi, 1971, p. 81–98.
15. COHEN, I., R. FALK, AND R. KLINE. Membrane currents following activity in canine Purkinje fibres. *Biophys. J.* 33: 281–288, 1981.
16. CORABOEUF, E., AND S. WEIDMANN. Potentiel de repos et potentiels d'action du muscle cardiaque, mesurés à l'aide d'électrodes intracellulaires. *C. R. Soc. Biol. Paris* 143: 1329–1331, 1949.
17. COURTNEY, K. R., AND P. G. SOKOLOVE. Importance of electrogenic sodium pump in normal and overdriven sinoatrial pacemaker. *J. Mol. Cell Cardiol.* 11: 787–794, 1979.
18. CRANEFIELD, P. F. *The Conduction of the Cardiac Impulse.* Mt. Kisco, NY: Futura, 1975.
19. CUSHNY, A. R. Stimulation of the isolated ventricle, with special reference to the development of spontaneous rhythm. *Heart* 3: 257–278, 1911–1912.
20. DECK, K. A., R. KERN, AND W. TRAUTWEIN. Voltage clamp technique in mammalian cardiac fibres. *Pfluegers Arch.* 280: 50–62, 1964.
21. DECK, K. A., AND W. TRAUTWEIN. Ionic currents in cardiac excitation. *Pfluegers Arch.* 280: 63–80, 1964.
22. DIACONO, J. Suggestive evidence for the activation of an electrogenic sodium pump in stimulated rat atria: apparent discrepancy between the pump inhibition and the positive inotropic response induced by ouabain. *J. Mol. Cell. Cardiol.* 11: 5–30, 1979.
23. DiFRANCESCO, D. Is the current I_{K_2} in cardiac Purkinje fibres an inward current activated on hyperpolarization? (Abstract). *Proc. Int. Union. Physiol. Sci.* 14: 380, 1980.
24. DiFRANCESCO, D. Evidence that 'i_{K_2}' in Purkinje fibres is an inward current activated on hyperpolarization. *J. Physiol. London* 305: 64P–65P, 1980.

25. DiFrancesco, D., and D. Noble. If 'i$_{K_2}$' is an inward current, how does it display potassium specificity? *J. Physiol. London* 305: 14P-15P, 1980.
26. Draper, M. H., and S. Weidmann. Cardiac resting and action potentials recorded with an intracellular electrode. *J. Physiol. London* 115: 74-94, 1951.
27. Dudel, J., and W. Trautwein. Das Aktionspotential und Mechanogramm des Herzmuskels unter dem Einfluss der Dehnung. *Cardiologie* 25: 344-362, 1954.
28. Dudel, J., and W. Trautwein. Der Mechanisms der automatischen rhythmischen Impulsbildung der Herzmuskelfaser. *Pfluegers Arch.* 267: 553-565, 1958.
29. Eisner, D. A., and Lederer, W. J. The role of the sodium pump in the effects of potassium-depleted solutions on mammalian cardiac muscle. *J. Physiol. London* 294: 279-301, 1979.
30. Eliakim, M., S. Bellet, E. Tawil, and O. Muller. Effect of vagal stimulation and acetylcholine on the ventricle. Studies in dogs with complete atrioventricular block. *Circ. Res.* 9: 1372-1379, 1961.
31. Erlanger, J., and A. D. Hirschfelder. Further studies on the physiology of heart-block in mammals. *Am. J. Physiol.* 15: 153-206, 1905-1906.
32. Fozzard, H. A., and G. W. Beeler, Jr. The voltage clamp and cardiac electrophysiology. *Circ. Res.* 37: 403-413, 1975.
33. Gadsby, D. C., and P. F. Cranefield. Direct measurement of changes in sodium pump current in canine Purkinje fibers. *Proc. Natl. Acad. Sci. USA* 76: 1783-1787, 1979.
34. Gaskell, W. H. On the innervation of the heart, with especial reference to the heart of the tortoise. *J. Physiol. London* 4: 43-127, 1884.
35. Hall, A. E., O. F. Hutter, and D. Noble. Current-voltage relations of Purkinje fibers in sodium-deficient solutions. *J. Physiol. London* 166: 225-240, 1963.
36. Harris, E. J., and O. F. Hutter. The action of acetylcholine on the movements of potassium ions in the sinus venosus of the heart. *J. Physiol. London* 133: 58P-59P, 1956.
37. Hauswirth, O., D. Noble, and R. W. Tsien. Adrenaline: mechanism of action on the pacemaker potential in cardiac Purkinje fibers. *Science* 162: 916-917, 1968.
38. Hauswirth, O., D. Noble, and R. W. Tsien. The mechanism of oscillatory activity at low membrane potentials in cardiac Purkinje fibres. *J. Physiol. London* 200: 255-265, 1969.
39. Hauswirth, O., D. Noble, and R. W. Tsien. Separation of the pace-maker and plateau components of delayed rectification in cardiac Purkinje fibres. *J. Physiol. London* 225: 211-235, 1972.
40. Hutter, O. F., and W. Trautwein. Vagal and sympathetic effects on the pacemaker fibers in the sinus venosus of the heart. *J. Gen. Physiol.* 39: 715-733, 1956.
41. Imanishi, S., and B. Surawicz. Automatic activity in depolarized guinea pig ventricular myocardium. Characteristics and mechanisms. *Circ. Res.* 39: 752-759, 1976.
42. Irisawa, H. Electrical activity of rabbit sino-atrial node as studied by a double sucrose gap method. In: *Symposium and Colloquium on the Electrical Field of the Heart*, edited by P. Rijlant. Brussels: Presses Académiques Européennes, 1972.
43. Katzung, B. G. Effects of extracellular calcium and sodium on depolarization-induced automaticity in guinea pig papillary muscle. *Circ. Res.* 37: 118-127, 1975.
44. Katzung, B. G., and J. A. Morgenstern. Effects of extracellular potassium on ventricular automaticity and evidence for a pacemaker current in mammalian ventricular myocardium. *Circ. Res.* 40: 105-111, 1977.
45. Kenyon, J. L., and W. R. Gibbons. 4-Aminopyridine and the early outward current of sheep cardiac Purkinje fibers. *J. Gen. Physiol.* 73: 139-157, 1979.
46. Kent, K. M., S. E. Epstein, T. Cooper, and D. M. Jacobowicz. Cholinergic innervation of the canine and human ventricular conducting system. Anatomic and electrophysiologic correlations. *Circulation* 50: 948-955, 1974.
47. Killip, T., S. Yormak, E. Ettinger, B. Levitt, and J. Roberts. Depression of ventricular automaticity by electrical stimulation. *Pharmacologist* 8: 203, 1966.
48. Krellenstein, D. J., M. B. Pliam, C. McC. Brooks, and M. Vassalle. On the mechanism of idioventricular pacemaker suppression by fast drive. *Circ. Res.* 35: 923-934, 1974.
49. Krellenstein, D. J., M. B. Pliam, C. McC. Brooks, and M. Vassalle. Factors affecting overdrive suppression of idioventricular pacemakers and associated potassium shifts. *J. Electrocardiol. San Diego* 11: 3-10, 1978.
50. Kunze, D. L. Rate-dependent changes in extracellular potassium in the rabbit atrium. *Circ. Res.* 41: 122-127, 1977.

51. LANGE, G. Action of driving stimuli from intrinsic and extrinsic sources on in situ cardiac pacemaker tissues. *Circ. Res.* 17: 449–459, 1965.

52. LENFANT, J., J. MIRONNEAU, AND J. -K. AKA. Activité repetitive de la fibre sino-auriculaire de grenouille: analyse des courants membranaires responsables de l'automatisme cardiaque. *J. Physiol. Paris* 64: 5–18, 1972.

53. LENFANT, J., J. MIRONNEAU, AND G. GALAND. Activité électrique du noeud sino-auriculaire de lapin, analyse au moyen des inhibiteurs de perméabilités membranaires. *J. Physiol. Paris* 60: Suppl. I: 272–273, 1968.

54. LINENTHAL, A. J., P. M. ZOLL, G. H. GARABEDIAN, AND K. HUBERT. Ventricular slowing and standstill after spontaneous or electrically stimulated runs of rapid ventricular beats in atrioventricular block. *Circulation* 22: 781, 1960.

55. LIPSIUS, S. L., AND M. VASSALLE. Dual excitatory channels in the sinus node. *J. Mol. Cell. Cardiol.* 10: 753–767, 1978.

56. LIPSIUS, S. L., AND M. VASSALLE. Characterization of a two-component upstroke in the sinus node subsidiary pacemakers. In: *The Sinus Node, Structure, Function and Clinical Relevance,* edited by F. I. M. Bonke. The Hague: Nijhoff, 1978, sect. 3, p. 233–244.

57. LU, H. -H., AND C. McC. BROOKS. Role of calcium in cardiac pacemaker cell action. *Bull NY Acad. Med.* 45: 100, 1969.

58. LU, H. -H., G. LANGE, AND C. McC. BROOKS. Factors controlling pacemaker action in cells of the sinoatrial node. *Circ. Res.* 17: 460–471, 1965.

59. MAYLIE, J., M. MORAD, AND J. WEISS. A study of pacemaker potential in rabbit sinoatrial node: measurement of potassium activity under voltage clamp conditions. *J. Physiol. London.* 311: 161–178, 1981.

60. McALLISTER, R. E., D. NOBLE, AND R. W. TSIEN. Reconstruction of the electrical activity of cardiac Purkinje fibres. *J. Physiol. London* 251: 1–59, 1975.

61. NOBLE, D. *The Initiation of the Heartbeat.* Oxford: Clarendon, 1975.

62. NOBLE, D., AND R. W. TSIEN. The kinetics and rectifier properties of the slow potassium current in cardiac Purkinje fibres. *J. Physiol. London* 195: 185–214, 1968.

63. REITER, M., AND J. NOÉ. Die Bedeutung von Calcium, Magnesium, Kalium und Natrium für die rhythmische Erregungsbildung im Sinusknoten des Warmblüterherzens. *Pfluegers Arch.* 269: 366–374, 1959.

64. SEIFEN, E., H. SCHAER, AND J. M. MARSHALL. Effect of calcium on the membrane potentials of single pacemaker fibres and atrial fibres in isolated rabbit atria. *Nature* 202: 1223–1224, 1964.

65. TODA, N., AND T. C. WEST. Interactions of K, Na, and vagal stimulation in the S-A node of the rabbit. *Am. J. Physiol.* 212: 416–423, 1967.

66. TODA, N., AND T. C. WEST. Interaction between Na, Ca, Mg and vagal stimulation in the S-A node of the rabbit. *Am. J. Physiol.* 212: 424–430, 1967.

67. TRAUTWEIN, W. Membrane currents in cardiac muscle fibers. *Physiol. Rev.* 53: 793–835, 1973.

68. TRAUTWEIN, W., AND D. G. KASSEBAUM. On the mechanism of spontaneous impulse generation in the pacemaker of the heart. *J. Gen. Physiol.* 45: 317–330, 1961.

69. TRAUTWEIN, W., S. W. KUFFLER, AND C. EDWARDS. Changes in membrane characteristics of heart muscle during inhibition. *J. Gen. Physiol.* 40: 135–145, 1956.

70. TSE, W. W., M. S. YOON, AND J. HAN. Effect of acetylcholine on automaticity of Purkinje fibers (Abstract). *Federation Proc.* 34: 375, 1975.

71. VASSALLE, M. Cardiac pacemaker potentials at different extra- and intracellular K concentrations. *Am. J. Physiol.* 208: 770–775, 1965.

72. VASSALLE, M. Analysis of cardiac pacemaker potential using a "voltage clamp" technique. *Am. J. Physiol.* 210: 1335–1341, 1966.

73. VASSALLE, M. Electrogenic suppression of automaticity in sheep and dog Purkinje fibers. *Circ. Res.* 27: 361–377, 1970.

74. VASSALLE, M. Generation and conduction of impulses in the heart under physiological and pathological conditions. *Pharmacol. Therap. B* 3: 1–39, 1977.

74a. VASSALLE, M. The role of the slow inward current in impulse formation. In: *The Slow Inward Current and Cardiac Arrhythmias,* edited by D. P. Zipes, J. C. Bailey, and V. Elharrar. The Hague: Nijhoff, 1980.

75. VASSALLE, M., AND O. BARNABEI. Norepinephrine and potassium fluxes in cardiac Purkinje fibers. *Pfluegers Arch.* 322: 287–303, 1971.

76. VASSALLE, M., D. L. CARESS, A. J. SLOVIN, AND J. H. STUCKEY. On the cause of ventricular asystole during vagal stimulation. *Circ. Res.* 20: 228–241, 1967.

77. VASSALLE, M., AND R. CARPENTIER. Hyperpolarizing and depolarizing effects of norepinephrine in cardiac Purkinje fibers. In: *Research in Physiology: A Liber Memorialis in Honor of Prof. Chandler McCuskey Brooks*, edited by F. F. Kao, K. Koizumi, and M. Vassalle. Bologna: Aulo Gaggi, 1971, p. 373–388.
78. VASSALLE, M., R. CARPENTIER, AND P. C. CHAN. The effects of norepinephrine and cyclic AMP on ($Na^+ + K^+$)-activated ATPase. *Physiologist* 15: 293, 1972.
79. VASSALLE, M., AND E. E. SCIDÁ. The role of sodium in spontaneous discharge in the absence and in the presence of strophanthidin (Abstract). *Federation Proc.* 38: 880, 1979.
80. VASSALLE, M., D. J. KRELLENSTEIN, M. B. PLIAM, AND C. McC. BROOKS. Potassium-related humoral transmission of overdrive suppression. *J. Mol. Cell. Cardiol.* 9: 921–931, 1977.
81. VASSALLE, M., F. J. VAGNINI, A. GOURIN, AND J. H. STUCKEY. Suppression and initiation of idioventricular automaticity during vagal stimulation. *Am. J. Physiol.* 212: 1–7, 1967.
82. VICK, R. L. Suppression of latent cardiac pacemaker: relation to slow diastolic depolarization. *Am. J. Physiol.* 217: 451–457, 1969.
83. WEIDMANN, S. Effect of current flow on the membrane potential of cardiac muscle. *J. Physiol. London* 115: 227–236, 1951.
84. WEST, T. C. Effects of chronotropic influences on subthreshold oscillations in the sino-atrial node. In: *The Specialized Tissues of the Heart,* edited by A. Paes de Carvalho, W. C. De Mello, and B. F. Hoffman. New York: Elsevier, 1961, p. 81–94.
85. WIT, A. L., AND P. F. CRANEFIELD. Effect of verapamil on the sinoatrial and atrioventricular nodes of the rabbit and the mechanism by which it arrests reentrant atrioventricular nodal tachycardia. *Circ. Res.* 35: 413–425, 1974.
86. YAMAGISHI, S., AND T. SANO. Effect of tetrodotoxin on the pacemaker action potential of the sinus node. *Proc. Jpn. Acad.* 42: 1194–1196, 1966.
87. ZIPES, D. P., AND J. C. FISCHER. Effects of agents which inhibit the slow channel on sinus node automaticity and atrioventricular conduction in the dog. *Circ. Res.* 34: 184–192, 1974.

CHAPTER 4

The Structure and Function of the Myocardial Cell Surface

GLENN A. LANGER

Departments of Medicine and Physiology and the American Heart Association,
Greater Los Angeles Affiliate, Cardiovascular Research Laboratory,
University of California at Los Angeles, Center for the
Health Sciences, Los Angeles, California

One has some hesitation in writing a review article on a subject that has only recently begun to receive significant attention. Surface coat structure and function in multicellular organisms has been investigated with increasing intensity since Bennett in 1963 (4) drew attention to what he termed the glycocalyx ("sweet husk") on the surface of cells. The glycocalyx denoted the coating external to the unit membrane bimolecular lipid leaflet. Bennett proposed that this region might "selectively bind certain ions such as calcium or potassium, thus rendering them available in high concentration close to the plasma membrane for uptake into the cell ..." Despite this proposal the surface structure reviewed by Bennett received virtually no attention from those interested in myocardial function until very recently. Therefore the information is somewhat sparse for the heart, but what information we do have for this tissue and from other tissues indicates that the surface material plays a major role in the control of transmembranous ionic exchange. This review should, then, be considered to be an early progress report on the study of the myocardial cellular surface. A great deal more remains to be done in the investigation of this complex structure, particularly with respect to the specific role or roles played by the molecules that constitute the coat. Where appropriate I will draw on information from other tissues in order to present as comprehensive a picture of this relatively new subject as is possible at this early stage of investigation.

Structure-Function Relationship of the Myocardial Surface

It is fair, I believe, to state that when the term cell membrane is mentioned, most of those not directly involved in membrane studies think of the 7.5-nm double track or trilaminar structure that surrounds the cell. In fact this structure, termed the unit membrane, represents only one component of the total sarco-lemmal membrane complex. Closer examination of even routinely stained electron-microscopic sections discloses a "fuzzy" coat or layer about 50 nm thick external to the unit membrane (Fig. 1; refs. 28, 37). This surface layer has two components: an inner less dense component 20 nm thick and an outer slightly more dense component 30 nm thick. The entire coat has been referred to with a variety of terms: basement membrane, basal lamina, external lamina, surface coat, boundary layer, or Bennett's term, glycocalyx. These terms do not

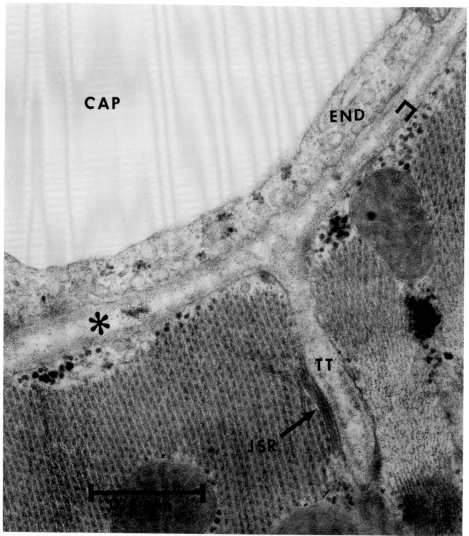

Fig. 1. Electron micrograph of capillary-cell surface region. CAP, capillary; END, endo-thelium; TT, transverse tubule; JSR, junctional sarcoplasmic reticulum. Brackets indicate glycocalyx composed of surface coat and external lamina. Note the close proximity of the capillary wall to cell wall and invagination of the glycocalyx into transverse tubule. Asterisk indicates interstitial space.

indicate that the coat has two structural components and, because each may have a distinct function, they should be referred to separately (17). The inner coat is termed the surface coat and the outer layer the external lamina. When referring to the entire complex, I will use Bennett's original notation, glycocalyx.

The glycocalyx is applied to the external surface of the cell, but in the myocardium it follows the unit membrane as it invaginates to form the transverse tubular (T) system, (Fig. 1; refs. 16, 29). As Fawcett and McNutt (15) point out, the presence of the glycocalyx is used in the identification of T tubules deep within the myocardial cell. Whatever the function of the coats, it is probably safe to assume that it applies to the T tubule system as well as to the external cellular surface. It is of interest that the glycocalyx does not extend into and fill the T tubules of skeletal muscle cells (37) but is restricted to the surface. Cardiac muscle has, then, a much greater amount of the coat material per cell. It is present on the entire electrically active external surface of the cells, including the T tubules.

The glycocalyx on the cellular surface blends with the ground substance of the interstitial space and abuts the basement membrane of closely apposed capillaries (Fig. 1; ref. 16). Frank and Langer (16) showed that as much as 36% of the cellular surface was in virtual direct contact with capillaries. This makes it likely that a significant component of capillary-cellular exchange occurs through a region in which the entire interstitial space is occupied solely by capillary basement membrane and the glycocalyx of the heart cell. Therefore it is likely that there is essentially direct exchange between capillary and cell and that interposed basement membrane—glycocalyx layers play a significant role in the control of this exchange. In other areas where there is more interposed interstitial space, exchange would be influenced by binding and passage through this area, but substances would be subject to the same influences of the glycocalyx before entrance to or exit from the cell.

As is discussed below, the glycocalyx and interstitial space ground substance have a high concentration of fixed, negatively charged sites with an affinity for various cations, including calcium (Ca). This characteristic, in association with the intimate capillary-to-cell relationship described above, seems to provide a reasonable explanation for an interesting experimental observation, and further insight into the excitation-contraction (EC) coupling sequence in the heart (47).

The experimental observation is as follows. When ventricular tissue is vascularly perfused with Ca-free solution, contractile force declines monoexponentially to 10–15% of control with a half time ($t_{1/2}$) of approximately 0.8 min (47, 50). When Ca is reintroduced, force returns to 85–90% of control, again monoexponentially, with a $t_{1/2}$ of approximately 0.2 min. Thus force is regained at about four times the rate at which it declines. The $t_{1/2}$ value of 0.2 min for regain of force is very similar to the $t_{1/2}$ for exchange of substances within the vascular space (47). These observations are reasonably explained by a model (27) in which the Ca involved in EC coupling is bound at surface sites before entry to the cell. Studies (33, 59) of sarcolemmal fractions show binding sites with high Ca affinity, and the study by Philipson and Langer (47) suggests that these sites are those which bind Ca before its release to the myofilaments. Because many of these sites are in close proximity to the capillaries, it is likely that the rate at which they bind Ca is limited by capillary flow. Therefore the recovery of contractile force would occur as rapidly as Ca equilibrates within the capillaries ($t_{1/2} = 0.2$ min). The decline in force upon Ca-free perfusion would be much slower because dissociation of Ca from the sites is rate limiting. This is relatively slow ($t_{1/2} = 0.8$ min). Therefore the regain of force would be limited by the rate of intravascular Ca exchange, and the decline of force would be limited by the rate of Ca dissociation from the binding sites, wherever they may be located on the cellular surface.

The model indicates a major role for cellular surface sites in the control of Ca exchange in the myocardium. Definition of the possible mechanisms operative in this control depends on a detailed knowledge of the molecular structure of these sites in the myocardium. Such knowledge is not available at present, but information gained from other tissues combined with results so far available from the heart indicate an important role for the sarcolemmal-glycocalyx complex in the control of myocardial ionic exchange.

Histochemistry of the Myocardial Surface

Howse and associates (22) were among the first to direct specific attention to the histochemistry of the cardiac surface coats. They compared various species from crustacea to mammal and found that the "external lamina" (surface coat and external lamina were not differentiated in this study) was common to all, though it was a great deal thicker in the more primitive species. The periodic acid-Schiff (PAS), alcian blue, ruthenium red, and colloidal iron reactions used emphasized the acidic character and carbohydrate content of the surface layer. The colloidal iron reaction was carried out at pH 1.2, and at this level of acidity the reaction is specific for sialic acid residues (3), though this was not emphasized in this early study. All species showed significant colloidal iron staining at the surface. The role of sialic acid is discussed below.

Howse and his collaborators (22) proposed that the external lamina might serve as a selective barrier and, in addition, regulate calcium concentration at the cell surface. It was also suggested that as a circulatory system developed in the higher species the function was shared by the basal lamina of the capillaries abutting the cell surface (see above) and therefore the surface coats of the muscle cells thinned as the evolutionary scale was ascended.

A histochemical study of mouse heart (19) made with diverse markers emphasized the similarity of the coatings on the membrane surface and on the T tubules and that the coating was heavily endowed with negatively charged molecules. Staining characteristics indicated that many of the charged groups were glycoproteins. Parsons and Subjeck (44) indicated that the glycoprotein chains extend outward from the unit membrane to contribute to the structure of the surface coat. Studies in the red blood cell show that many of the glycoproteins of the surface coat penetrate into or through the lipid bilayer (54). It is now recognized that the surface coat is not simply a layer of adherent secreted material but represents an integral part of the cell membrane (36). It appears that all of the carbohydrates of the glycoproteins are located at the outer surface of plasma membranes and are not found at the inner side.

As indicated above, the colloidal iron reaction carried out at low pH is specific for sialic acid. In a study of the myocardial cell surface in tissue culture cells, neonatal rat heart cells, and adult rabbit heart cells (17) there was a two-layered pattern of colloidal iron staining (Fig. 2). Colloidal iron particles were bound immediately external to the unit membrane with a second layer found in the outermost region of the external lamina. Application of neuraminidase (an enzyme that specifically cleaves sialic acid from the oligosaccharide chains) prior to staining eliminated most of the iron particles in both layers. This confirms the specificity of the colloidal iron reaction. The pattern indicates that sialic acid molecules with their negatively charged carboxyl groups are present next to the lipid bilayer and on the external lamina at the interstitial or capillary interface. The possible importance of sialic acid in the regulation of cation permeability is discussed below.

In summary, the histochemical studies provide evidence for the polyanionic nature of the glycocalyx. The negative charge is ascribable primarily to the acidic mucopolysaccharide content of the coats. The surface coat is an integral part of the sarcolemma and the external lamina is superficial to this layer. The nine-carbon amino sugar, sialic acid, is distributed in both layers and contributes via its ionized carboxyl group to the negative charge.

Relation of Glycocalyx to the Unit Membrane

It is now generally accepted that the "fluid mosaic model" of the bilayer unit membrane is most realistic (52). The bilayer is composed of phospholipid with the hydrophobic fatty acid chains directed inward from the outer and inner

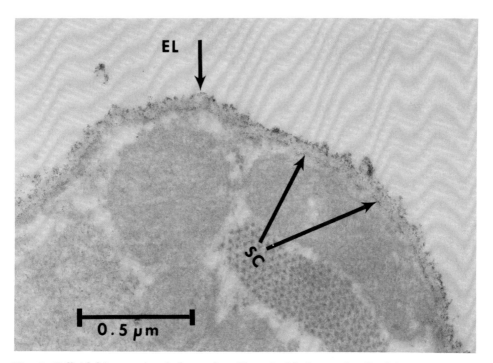

Fig. 2. Colloidal iron stain of glycocalyx. Note double layer of colloidal iron particles, one immediately external to unit membrane on surface coat (SC) and other more superficial on external lamina (EL). Under conditions of staining, colloidal iron represents location of sialic acid.

surfaces. The hydrophilic polar heads of the phospholipids (phosphate groups), sugars, and certain amino acids extend into the aqueous phase at both surfaces. The phosphate group of the phospholipids may be free—in the form of phosphatidic acid—or have various groups such as choline, ethanolamine, or serine attached. An important feature of the fluid mosaic model is the presence of integral proteins embedded in the lipid bilayer. These are amphipathic globular proteins with hydrophilic ends protruding from either surface and the hydrophobic ends extending into the fatty region. Some of these proteins extend through the entire bilayer from outer to inner surface; others extend only part way from the outer surface. The carbohydrates of these glycoproteins extend from the outer hydrophilic ends to form the surface coat (54). The carbohydrates most frequently found in animal cells are the sugars, with galactose usually in the largest amount (60). Smaller amounts of mannose, fucose, and glucose are present. Two acetylamino sugars, N-acetylglucosamine and N-acetylgalactosamine, are usually present, and finally sialic (N-acetylneuraminic) acid is demonstrable. Sialic acid, notably, is always found to be terminal on the oligosaccharide chains (34). As viewed by Singer and Nicolson (52) the integral proteins are able to move laterally in the lipid, though the attachments at outer and inner surfaces would be expected to affect this movement. The integral proteins that pass entirely through the bilayer are presumed to be involved in the formation of transmembrane pores.

The relationship between the unit membrane and the surface coat (SC) and external lamina (EL) is schematized in Figure 3. The scheme is drawn from analysis of diverse tissues but, from what is currently known (30, 37) about the myocardial cell, the general structure almost certainly applies. There are at least two general classes of negatively charged sites: 1) those associated with the polar heads of the phospholipids and 2) those on the oligosaccharides of the SC

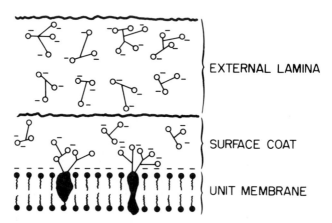

EXTERNAL LAMINA

SURFACE COAT

UNIT MEMBRANE

Fig. 3. Schematic representation of unit membrane and glycocalyx. *Closed circles* represent hydrophilic polar heads at either side of unit membrane. Irregular structures embedded in unit membranes are integral proteins with negatively charged attached oligosaccharide chains (*open circles*) extending into surface coat. Other negatively charged oligosaccharide chains are present in external lamina. In addition, note negatively charged layer at the unit membrane-surface coat interface due, in part, to phospholipids.

and EL. The possible relationship of these regions, which are capable of cation binding, and their respective roles in the control of transmembrane cation flux is not known. As indicated by Nicolson (41), however, there appears to be little tendency for glycoproteins (or glycolipids) to rotate across the membrane. Phospholipids have been seen to undergo flip-flop rotations but at rates that are very slow (over the course of hours). Neither phospholipids nor oligosaccharides seem to play a *direct* role in the beat-to-beat movement of ions via pores or carriers. It seems more likely that these groups function in selective binding of cations before they move across the bilayer. This is discussed further below.

The EL is adherent to the SC. Vandenburgh and associates (56) studied the 3,5-diiodosalicylate-extractable substance (LIS) from isolated sarcolemmal tubes from rat skeletal muscle. This material is stained by ruthenium red, and it is uncertain from the study whether or not it includes the SC, though the authors indicate that LIS specifically extracts external lamina substance (ELS). The ELS was found to be rich in glucuronic acid and hexosamines and to have a greater proportion of acidic amino acids than the whole sarcolemma. It contained 8% acid mucopolysaccharides, which would at least in part account for its ability to bind ruthenium red, lanthanum, and colloidal iron (17), as well as, presumably, physiological cations.

I have indicated that the EL and SC are indeed separate entities, and this is dramatically shown when the heart is perfused with solutions in which the calcium concentration $[Ca]_o$ is essentially zero (17, 40). The study by Frank et al. (17) shows, with the aid of the colloidal iron stain, that zero $[Ca]_o$ separates the two layers of stain and therefore the EL from the SC. The two layers separate by 0.5 μm at the maximum, with formation of a bleb that is anchored at either end as the EL descends into the T tubule. The external lamina remains attached to the cell, since its extension into the confined region of the T system seems to prevent significant separation from the SC within the tubules.

It seems clear that Ca ion is important in the maintenance of connections between the EL and SC. The nature of these connections is not known, but the work of Cook and Bugg (9) raises the possibility of sugar-to-calcium bridges. Specifically they described the situation in which Ca is chelated by a pair of hydroxyl groups from each of two α-fucose molecules. Fucose, like sialic acid, is a common terminal sugar of oligosaccharides or glycoproteins, and Ca has been demonstrated to be the link in hydrated fucose-calcium-fucose bridges. It seems possible that Ca removal might result in the rupture of such carbohydrate couplings and allow separation of the EL and SC.

I have stressed the anionic sites of the glycocalyx because these seem to be of the greatest physiological significance. It should be mentioned, however, that cationic sites are demonstrable with special techniques. If acidic groups are first esterified, then negatively charged markers (colloidal α-stannic acid and negative iron colloid) stain the surface coats of various cells, indicating the presence of positively charged basic sites, many of which are amino groups of the surface proteins (6).

Role of Sialic Acid

There are, as discussed above, many contributors to the negative charges of the surface cells, e.g., polar groups of the phospholipids, sulfated glycoproteins, and glycosaminoglycans. In addition, sialic acid is present at the surface of most animal cells, including the heart, with evidence of distribution in both the SC and EL components of the surface complex (see above). Because of its abundance, because its carboxyl groups would be expected to contribute to the negative charge, and because there is an enzyme (neuraminidase) in very pure form capable of specifically cleaving its O-glycosidic link to the oligosaccharide chain, sialic acid was a likely candidate for initial study of the glycocalyx in the heart (17, 30).

To make the myocardial cell surface readily accessible to neuraminidase, a technique for following ionic flux of tissue culture cells was used (17). ^{45}Ca uptake and washout were compared to controls in cells from which approximately 60% of the sialic acid had been removed by brief exposure to neuraminidase. The effect of this removal on Ca exchange was striking. The rate of Ca uptake and washout increased five to six times and lanthanum ion, normally restricted to the lipid bilayer and glycocalyx, entered the cell and displaced more than 80% of the exchangeable Ca. Despite these marked effects on Ca exchange (and La exchange), potassium exchangeability was not significantly affected. The lack of effect on K exchange or, more specifically, the absence of enhanced K leakage in the cultured heart cells was consistent with the effects reported in leukemic cells (18) and Ehrlich ascites cells (57) where neuraminidase treatment caused, if anything, a decrease in K exchangeability. Because conductance studies (1, 42) indicate that the control of K exchange is located within the lipid bilayer, it can be assumed that sialic acid removal does not disrupt this component of the sarcolemma. These results support the proposal by Kraemer (26) that most of the sialic acid could be removed without destruction of the "permeability barrier" for potassium.

The study demonstrates that a critical component for the control of Ca exchange in cultured cells lies within the glycocalyx and, more specifically, is dependent upon the presence of sialic acid. Sialic acid in the red blood cell (35) and in rat liver cells (51) binds Ca. The fact that neuraminidase treatment reduces surface site binding of La (17) added to the fact that La competes for and displaces Ca from rapidly exchangeable cellular sites (28) indicates that sialic acid may account for a significant number of superficial Ca-binding sites in the cultured cells. It has been assumed that Ca is bound at the ionized carboxyl group of the molecule but a recent nuclear magnetic resonance study indicates that the glycerol side chain may bind Ca by effectively "wrapping around" the ion (24).

Since this report was initially written three years ago, additional information is available with respect to the role of sialic acid in cardiac tissue. Philipson and his collaborators (45) studied Ca binding in purified sarcolemma extracted from rabbit ventricles. It was found that removal of 57% of sialic acid with neuraminidase reduced Ca binding of the membranes by only 13%. Phospholipids extracted from the sarcolemmal vesicles accounted for approximately 80% of the Ca bound. Of the phospholipids present, phosphatidylserine and phosphatidylinositol seem to be most important in binding. Therefore, though sialic acid does account for a component of binding, it is at present undefined as to which sites are most important physiologically and whether these sites vary with preparation, species, or age. In this respect a recent study in adult guinea pig left atria indicates that up to 79% removal of sialic acid has no effect on contractile function, including inotropic and toxic responses to ouabain (20). In this preparation the sialic acid removal was associated with a decrease in La staining on the cellular surface just as occurred in cultured cells (17), but there was no evidence of La penetration intracellularly, which was striking in the neuraminidase-treated cultures. This study indicates a marked difference between the response of cultured monolayers and whole, adult tissue. In contrast to this study is that of Bailey and Fawzi (2) in the same tissue, adult guinea pig heart. They found that treatment of intact hearts with neuraminidase completely prevented the positive inotropic effect of $10^{-7}M$ and $10^{-6}M$ ouabain. In this treatment the enzyme also removed from the same tissue approximately 50% of low-affinity Ca binding sites in isolated cardiac myocytes. In addition the treatment revealed an increased number of high-affinity sites. The toxic response to ouabain was not affected by neuraminidase treatment, in agreement with the study on atria. These studies do indicate that neuraminidase, a large enzyme, can be used in whole-tissue preparations without major problems of diffusion. Further studies will be required to assess the importance of sialic acid in control of Ca movements in the heart, including an examination of different tissues from various species of different ages.

Role of Sialic Acid Compared to Phospholipid

As reported above, sialic acid removal from cultured cells produced a large increase in Ca permeability without alteration in K permeability. This indicated that lipid bilayer structure was unaffected. A recent study (31) compares the effects of neuraminidase and phospholipase treatment of cultured cells on permeabilities and freeze-fracture pattern. Sialic acid removal produced no change in the integral protein pattern in the bilayer, and this, therefore, is consistent with the lack of effect on K permeability. Application of phospholipase C to the cultured cells produced large increases in both Ca and K permeability and was associated with major disruption of the integral protein pattern. This phospholipase catalyzes the cleavage of phosphorylated base from the C-3 position of phospholipid, thereby removing the polar head. Application of phospholipase A_2 affected neither Ca nor K permeability and produced no change in integral protein pattern. Phospholipase A_2 releases a fatty acid from the C-2 position and does not affect the polar groups of the phospholipid molecules. These results in the cultured model indicate the following: 1) specific effect of sialic acid removal on Ca permeability without demonstrable effect on lipid bilayer structure; 2) both Ca and K permeabilities are increased by removal of polar heads from bilayer phospholipid, and this is associated with significant change in bilayer structure; and 3) removal of fatty acid from C-2 of phospholipid alters neither ionic permeability nor bilayer structure. It seems, in this model at least, that there are at least two sites controlling Ca permeability—one in the glycocalyx associated with sialic acid and another in the bilayer associated with polar head of the phospholipids. The latter site is also involved in K permeation.

Role of Sarcolemmal-Glycocalyx Complex in Contractile Control

Though knowledge of the myocardial surface is incomplete, it is fair to state that its ability to bind cations, particularly Ca, almost certainly contributes to the regulation of myocardial function. Studies on isolated sarcolemma from heart (5, 33, 46, 47, 59) indicate that there are two classes of Ca-binding sites within the cell membrane. There are a relatively small number of sites with high affinity ($K_m \sim 20\,\mu M$) and a much larger number of sites with lower affinity ($K_m \sim 1.2$ mM). The high-affinity sites may be involved with maintenance of the structural integrity of the membrane and probably play a limited role in the regulation of contractile force. The low-affinity sites may be of importance in the control of contractility. The curve relating contractile force to $[Ca]_o$ up to 10 mM superimposes on the curve that relates Ca bound on low-affinity sites on sarcolemma to $[Ca]_o$ up to 10 mM (46).

Another series of studies compared the affinity of a series of cations for rapidly exchangeable Ca-binding sites with their ability to uncouple the EC process. In tissue culture studies (32) the order of affinity for the sites was $La^{3+} > Cd^{2+} > Zn^{2+} > Mn^{2+} > Mg^{2+}$, and the sequence for uncoupling was essentially the same. In another study (5), ability of certain cations to displace Ca from sarcolemmal vesicles derived from neonatal rat heart was compared with the ability of these same cations to EC uncouple the intact neonatal rat heart. Again the selectivity sequence ($Y^{3+} > Nd^{3+} > La^{3+} > Cd^{2+} > Co^{2+} > Mg^{2+}$) of sarcolemmal Ca-binding sites was the same as the uncoupling sequence. The potency of an ion as a Ca displacer and Ca uncoupler depended on its charge (trivalent usually more potent than divalent) but most particularly upon the ion's nonhydrated radius. The closer this radius is to that of Ca (0.99 Å) the more effective is its uncoupling ability and the more Ca it displaces. The critical importance of nonhydrated radius was recently confirmed by Cartmill and dos Remedios (8).

A final series of experiments (46) relating sarcolemmal-bound Ca to contractile control used the recent study of Tillisch et al. (55), in which the contractile response of ventricular tissue to perfusion with $[Na]_o$ between 75 and 200 mM was evaluated. An initial force transient (peak at ~ 2 min) following the $[Na]_o$ alteration is most likely based on the response of sarcolemmal bound Ca to competition for surface binding sites with Na. The effect of different $[Na]_o$ on Ca bound to isolated sarcolemma was compared to the effect of different $[Na]_o$ on the force transient in the intact, functional ventricular tissue. The two relationships, i.e., $[Na]_o$ vs. force and $[Na]_o$ vs. Ca bound were superimposable.

These studies indicate that Ca bound on or within the sarcolemmal-glycocalyx complex plays an important role in the control of contraction. A recent study by Isenberg and Klockner (23) demonstrates that one component of the glycocalyx at least has little to do with supply of that Ca manifesting itself as a slow inward current on voltage clamp studies. In rat heart myocytes dissociated from adult tissue by exposure to solution containing low Ca, collagenase, and hyaluronidase, the magnitude of the slow inward Ca current is unchanged. The electron micrographs show a greatly depleted glycocalyx with at least the external lamina removed. More specific staining and ultrastructural analysis is required to assess the status of the surface coat in these cells. The study does show, however, that the component of Ca flux associated with the slow inward current is at least not derived from the EL component of the glycocalyx. As indicated earlier there are a number of other sites, including phospholipids and proteins, from which the Ca might be derived.

Whether the bulk of Ca directed to the myofilaments is derived directly from surface sites or whether surface-bound Ca upon release serves to induce release from the sarcotubular system, which then activates the myofilaments, may vary as to tissue and species. In skinned cardiac fibers Fabiato and Fabiato (12, 13) have conclusively demonstrated that a quantity of Ca substantially less than that necessary to activate force development directly [less than 5 μM/kg wet tissue (53)] is capable of "triggering" significant Ca release from the sarcoplasmic reticulum. There is a wide spectrum among species (as well as atrium versus ventricle) with respect to the quantity of Ca proposed to pass directly to the myofilaments as compared to the quantity released from the sarcotubules (13). Rat ventricle may depend largely on "triggered release," whereas rabbit ventricle requires more direct myofilament activation. Studies recently completed (14) indicate, however, that the sarcotubules of all mammalian species are saturated with Ca even at free Ca concentrations present during diastole (10^{-7}M and less). This indicates that, if triggered release is a mechanism operative in the intact cell, its force development is modulated by gradation of the amount of Ca that crosses the sarcolemma. The latter seems to be derived from cellular surface sites.

Irrespective of the quantity of surface-bound Ca that participates either as a trigger or as direct activator, it seems to enter the cell via two routes. These routes are currently modeled as a "pore" or "channel" system and a "carrier" system (29). The Ca "stored" in the surface sites is visualized as feeding the two systems. The pores are thought to function as selective channels for the electrogenic movement of ions—including Ca. Movement through this system is followed with voltage clamp technique, which indicates that relatively small amounts of Ca (barely enough to reach mechanical threshold) cross the membrane via this route. The carrier system would account for the major component of the transmembrane Ca movement in systems in which triggered sarcotubular release was not predominant. Movement of the carrier couples at least 3 Na ions to 1 Ca ion—thus its operation is electrogenic (48). Under physiological conditions $[Ca]_o$ and $[Na]_o$ are quite stable and $[Ca]_i$ is under control of the sarcotubular pump. Therefore $[Na]_i$ is probably the dominant signal in the determination of the activity of the carrier system. Interventions that give rise to increased $[Na]_i$ would stimulate movement of Na outward and Ca inward (with an associated outward current) and would be expected to give rise to a positive inotropic response. The mechanism by which surface-bound Ca might be transferred to the carrier system for movement across the membrane is unknown.

The importance of surface-bound Ca in heart vis-à-vis skeletal muscle is suggested by a comparison of the cellular surface area and sarcotubular cisternal volume of the two tissues. The total surface area including the T system (per 100 μm^3 fiber volume) for white vastus muscle (fast-twitch from the guinea pig) is about 24 μm^2 (11) and for ventricular muscle is about 97 μm^2 [for a ventricular cell 80 μm long × 12 μm wide with 0.2 μm diam T tubules (15)]. The lateral cisternal volume of frog sartorius is 4.1% of fiber volume (39) and for rat ventricle it is 0.3% of fiber volume (43). Therefore the ratio of surface area to lateral cisternal volume is 6 μm^{-1} for skeletal muscle and 323 μm^{-1} for heart muscle. This means that heart muscle has about 54 times more cellular surface area relative to its cisternal volume than skeletal muscle. Stated another way, if Ca binding and storage are proportional to volume or surface area, then heart muscle has only 7% the cisternal capacity of skeletal muscle but 400% the surface-binding capacity per unit volume of the cell. It should be reemphasized that the T tubules of skeletal muscle do not demonstrate a glycocalyx (15). The point to be noted is that heart muscle has many times the surface binding capacity of skeletal muscle but much less cisternal capacity. This implies but of course does not prove that surface Ca plays a significantly greater role in the heart than in skeletal muscle.

The glycocalyx with its binding sites for Ca may also play an essential role in the prevention of Ca overloading by the cell. As discussed above, perfusion of the heart with zero $[Ca]_o$ peels the EL from the SC. Reperfusion with normal $[Ca]_o$ then produces a mechanical contracture associated with a net gain of Ca by the cells (10). As with removal of sialic acid in cultured cells, the selective permeability of the membrane is disrupted when the glycocalyx is compromised. It has been demonstrated in the normal heart that in the face of even greatly increased concentrations of extracellular Ca ($[Ca]_o$ to 12 mM) intracellular Ca levels are prevented from rising to a value that would induce contraction or contracture (50). It is possible that influx of Ca is prevented from rising in a linear fashion as the negative-binding, and presumably transfer, sites for Ca become saturated.

There is considerable evidence that ischemia and anoxia damage the myocardial surface (21, 25). The resulting contracture seems to be associated, at least in part, with a large influx of Ca in both the ischemic (49) and anoxic (7) conditions (for this study electron-probe analysis of La entry was used as a sensitive indicator of Ca entry). It is quite possible that the determinant of functional reversibility after ischemic or anoxic injury is the sarcolemma and that intracellular organelles can recover if membrane selectivity is maintained.

Acknowledgments

The electron micrographs were kindly supplied by Dr. Joy Frank and are greatly appreciated.

This research was supported by Public Health Service Grant HL-11351-13 and a grant from the Castera Foundation.

References

1. ARMSTRONG, C. M., AND B. HILLE. The inner quarternary ammonium ion receptor in potassium channels of node of Ranvier. J. Gen. Physiol. 59: 388–400, 1972.
2. BAILEY, L. E., AND A. B. FAWZI. Neuraminidase dissociates ouabain inotropy from toxicity. J. Mol. Cell. Cardiol. 12: 527–530, 1980.
3. BENEDETTI, E. L., AND P. EMMELOT. Studies in plasma membranes. IV. The ultrastructural localization and content of sialic acid in plasma membranes isolated from rat liver and hepatoma. J. Cell Sci. 2: 499–511, 1967.
4. BENNETT, H. S. Morphological aspects of extracellular polysaccharides. J. Histochem. Cytochem. 11: 14–23, 1963.
5. BERS, D. M., AND G. A. LANGER. Uncoupling cation effects on cardiac contractility and sarcolemmal Ca^{2+} binding. Am. J. Physiol. 237(Heart Circ. Physiol. 6): H332–H341, 1979.
6. BLANQUET, P. R., AND E. PUVION. Colloidal α-stannic acid and negative iron colloid as differential electron stains for surface proteins. J. Histochem. Cytochem. 23: 174–186, 1975.
7. BURTON, K. P., H. K. HAGLER, G. H. TEMPLETON, J. T. WILLERSON, AND L. M. BUJA. Lanthanum probe studies of cellular pathophysiology induced by hypoxia in isolated cardiac muscle. J. Clin. Invest. 60: 1289–1302, 1977.

8. CARTMILL, J. A., AND C. G. DOS REMEDIOS. Ionic radius specificity of cardiac muscle. *J. Mol. Cell. Cardiol.* 12: 219–223, 1980.

9. COOK, W. J., AND C. E. BUGG. Calcium-carbohydrate bridges composed of uncharged sugars. Structure of a hydrated calcium bromide complex of α-fucose. *Biochim. Biophys. Acta* 389: 428–435, 1975.

10. CREVEY, B. J., G. A. LANGER, AND J. S. FRANK. Role of Ca in maintenance of rabbit myocardial cell membrane structural and functional integrity. *J. Mol. Cell. Cardiol.* 10: 1081–1100, 1978.

11. EISENBERG, B., AND A. M. KUDA. Stereological analysis of mammalian skeletal muscle. II. White vastus muscle of the adult guinea pig. *J. Ultrastruct. Res.* 51: 176–187, 1975.

12. FABIATO, A., AND F. FABIATO. Contractions induced by a calcium-triggered release of calcium from the sarcoplasmic reticulum of single skinned cardiac cells. *J. Physiol. London* 249: 469–495, 1975.

13. FABIATO, A., AND F. FABIATO. Calcium-induced release of calcium from the sarcoplasmic reticulum of skinned cells from adult human, dog, cat, rabbit, rat and frog hearts and from fetal and new-born rat ventricles. *Ann. NY Acad. Sci.* 307: 491–522, 1978.

14. FABIATO, A., AND F. FABIATO. Use of chlortetracycline fluorescence to demonstrate Ca^{2+} induced release of Ca^{2+} from the sarcoplasmic reticulum. *Nature London* 281: 146–148, 1979.

15. FAWCETT, D. W., AND N. S. MCNUTT. The ultrastructure of the cat myocardium. I. Ventricular papillary muscle. *J. Cell Biol.* 42: 1–45, 1969.

16. FRANK, J. S., AND G. A. LANGER. The myocardial interstitium: Its structure and its role in ionic exchange. *J. Cell Biol.* 60: 586–601, 1974.

17. FRANK, J. S., G. A. LANGER, L. M. NUDD, AND K. SERAYDARIAN. The myocardial cell surface, its histochemistry and the effect of sialic acid and calcium removal on its structure and cellular ionic exchange. *Circ. Res.* 41: 702–714, 1977.

18. GLICK, J. L., AND S. GITHENS III. Role of sialic acid in potassium transport of K 1210 leukemia cells. *Nature London* 208: 88, 1965.

19. GROS, D., AND C. E. CHALLICE. The coating of mouse myocardial cell. A cytochemical electron microscopical study. *J. Histochem. Cytochem.* 23: 727–744, 1975.

20. HARDING, S. E., AND J. HALLIDAY. Removal of sialic acid from cardiac sarcolemma does not affect contractile function in electrically stimulated guinea pig left atria. *Nature London* 286: 819–821, 1980.

21. HEARSE, D. J., S. M. HUMPHREY, W. G. NAYLER, A. SLADE, AND D. BORDER. Ultrastructural damage associated with reoxygenation of the anoxic myocardium. *J. Mol. Cell. Cardiol.* 7: 315–324, 1975.

22. HOWSE, H. D., V. J. FERRANS, AND R. G. HIBBS. A comparative histochemical and electron microscopic study of the surface coatings of cardiac muscle cells. *J. Mol. Cell. Cardiol.* 1: 157–168, 1970.

23. ISENBERG, G., AND U. KLOCKNER. Glycocalyx is not required for slow inward calcium current in isolated rat heart myocytes. *Nature London* 284: 358–360, 1980.

24. JAQUES, L. W., E. B. BROWN, J. M. BARRETT, W. S. BREY, JR., AND W. WELTNER, JR. Sialic acid. A calcium-binding carbohydrate. *J. Biol. Chem.* 252: 4533–4538, 1977.

25. JENNINGS, R. B., C. E. GANOTE, AND K. A. REIMER. Ischemic tissue injury. *Am. J. Pathol.* 81: 179–198, 1975.

26. KRAEMER, P. M. Cytotoxic, hemolytic and phospholipase contaminants of commercial neuraminidase. *Biochim. Biophys. Acta* 167: 205–208, 1968.

27. LANGER, G. A. Events at the cardiac sarcolemma: localization and movement of contractile-dependent calcium. *Federation Proc.* 35: 1274–1278, 1976.

28. LANGER, G. A., AND J. S. FRANK. Lanthanum in heart cell culture: effect on calcium exchange correlated with its localization. *J. Cell Biol.* 54: 441–455, 1972.

29. LANGER, G. A., J. S. FRANK, AND A. J. BRADY. The myocardium. In: *Cardiovascular Physiology II.* edited by A. C. Guyton and A. W. Cowley. Baltimore: University Park, 1976, vol. 9, p. 191–237. (Int. Rev. Physiol. Ser.)

30. LANGER, G. A., J. S. FRANK, L. M. NUDD, AND K. SERAYDARIAN. Sialic acid: effect of removal on calcium exchangeability of cultured heart cells. *Science* 193: 1013–1015, 1976.

31. LANGER, G. A., J. S. FRANK, AND K. D. PHILIPSON. Correlation of alteration in cation exchange and sarcolemmal ultrastructure produced by neuraminidase and phospholipases in cardiac cell tissue culture. *Circ. Res.* In press.

32. LANGER, G. A., S. D. SERENA, AND L. M. NUDD. Cation exchange in heart cell culture: correlation with effects on contractile force. *J. Mol. Cell. Cardiol.* 6: 149–161, 1974.

33. LIMAS, C. J. Calcium-binding sites in rat myocardial sarcolemma. *Arch. Biochem. Biophys.* 179: 302–309, 1977.
34. LLOYD, C. W. Sialic acid and the social behavior of cells. *Biol. Rev.* 59: 325–350, 1975.
35. LONG, C., AND B. MOUAT. The binding of calcium ions by erythrocytes and ghost cell membranes. *Biochem. J.* 129: 829–836, 1971.
36. MARTINEZ-PALOMO, A. The surface costs of animal cells. *Int. Rev. Cytol.* 29: 29–75, 1970.
37. MCNUTT, N. S., AND D. W. FAWCETT. Myocardial ultrastructure. In: *The Mammalian Myocardium,* edited by G. A. Langer and A. J. Brady. New York: Wiley, 1974, p. 1–49.
38. MCNUTT, N. S., AND R. S. WEINSTEIN. The ultrastructure of the nexus. A correlated thin-section and freeze-cleave study. *J. Cell Biol.* 47: 666, 1970.
39. MOBLEY, B. A., AND B. R. EISENBERG. Sizes of components in frog skeletal muscle measured by methods of stereology. *J. Gen. Physiol.* 66: 31–45, 1975.
40. MUIR, A. R. The effects of divalent cations on the ultrastructure of the perfused rat heart. *J. Anat.* 101: 239–261, 1967.
41. NICOLSON, G. L. Transmembrane control of the receptors on normal and tumor cells. I. Cytoplasmic influence over cell surface components. *Biochim. Biophys. Acta* 457: 57–108, 1976.
42. OCHI, R., AND H. NISHIJE. Effect of intracellular tetraethylammonium ion on action potential in the guinea pig's myocardium. *Pfluegers Arch.* 348: 305–316, 1974.
43. PAGE, E., L. P. MCCALLISTER, AND B. POWER. Stereological measurements of cardiac ultrastructures implicated in excitation-contraction coupling. *Proc. Natl. Acad. Sci. USA* 68: 1465–1466, 1971.
44. PARSONS, D. F., AND J. R. SUBJECK. The morphology of the polysaccharide coat of mammalian cells. *Biochim. Biophys. Acta* 265: 85–113, 1972.
45. PHILIPSON, K. D., D. M. BERS, AND A. Y. NISHIMOTO. The role of phospholipids in the Ca^{2+} binding of isolated sarcolemma. *J. Mol. Cellular Cardiol.* 12: 1159–1173, 1980.
46. PHILIPSON, K. D., D. M. BERS, A. Y. NISHIMOTO, AND G. A. LANGER. Binding of Ca^{2+} and Na^+ to sarcolemmal membranes: relation to control of myocardial contractility. *Am. J. Physiol.* 238 (*Heart. Circ. Physiol.* 7): H373–H378, 1980.
47. PHILIPSON, K. D., AND G. A. LANGER. Sarcolemmal bound calcium and contractility in the mammalian myocardium. *J. Mol. Cell. Cardiol.* 11: 857–875, 1979.
48. PITTS, B. J. R. Stoichiometry of sodium-calcium exchange in cardiac sarcolemmal vesicles. *J. Biol. Chem.* 254: 6232–6235, 1979.
49. SHEN, A. C., AND R. B. JENNINGS. Kinetics of calcium accumulation in acute myocardial ischemic injury. *Am. J. Pathol.* 67: 441–452, 1972.
50. SHINE, K. I., S. D. SERENA, AND G. A. LANGER. Kinetic localization of contractile calcium in rabbit myocardium. *Am. J. Physiol.* 221: 1408–1417, 1971.
51. SHLATZ, L., AND G. V. MARINETTI. Calcium binding to the rat liver plasma membrane. *Biochim. Biophys. Acta* 290: 70–83, 1972.
52. SINGER, S. J., AND G. L. NICOLSON. The fluid mosaic model of the structure of cell membranes. *Science* 175: 720–731, 1972.
53. SOLARO, R. J., R. M. WISE, J. S. SHINER, AND F. N. BRIGGS. Calcium requirements for cardiac myofibrillar activation. *Circ. Res.* 34: 525–530, 1974.
54. STECK, T. L. The organization of proteins in the human red blood cell membrane. A review. *J. Cell Biol.* 62: 1–19, 1974.
55. TILLISCH, J. H., L. K. FUNG, P. M. HOM, AND G. A. LANGER. Transient and steady-state effects of sodium and calcium on myocardial contractile response. *J. Mol. Cell. Cardiol.* 11: 137–148, 1979.
56. VANDENBURGH, H. H., M. F. SHEFF, AND S. I. ZUCKO. Chemical composition of isolated rat skeletal sarcolemma. *J. Membr. Biol.* 17: 1–12, 1974.
57. WEISS, L., AND C. LEVINSON. Cell electrophoretic mobility and cationic flux. *J. Cell. Physiol.* 73: 31–36, 1969.
58. WENDT, I. R., AND G. A. LANGER. The sodium-calcium relationship in mammalian myocardium: effect of sodium perfusion on calcium fluxes. *J. Mol. Cell. Cardiol.* 9: 551–564, 1977.
59. WILLIAMSON, J. R., M. L. WOODROW, AND A. SCARPA. Calcium binding to cardiac sarcolemma. In: *Recent Advances in Studies in Cardiac Structure and Metabolism. Basic Functions of Cations in Myocardial Activity,* edited by A. Fleckenstein and N. S. Dhalla. Baltimore: University Park, 1975, vol. 5, p. 61–71.
60. WINZLER, R. J. Carbohydrates in cell surfaces. *Int. Rev. Cytol.* 29: 77–125, 1970.

CHAPTER 5

Myocardial Effects of Magnesium

KENNETH I. SHINE
Department of Medicine and the American Heart Association, Greater
Los Angeles Affiliate, Cardiovascular Research Laboratory,
University of California, Los Angeles, School of Medicine,
Los Angeles, California

The routine availability of serum magnesium measurements in man has revealed a wide variety of disorders that involve this cation. Some disorders, such as hypomagnesemia in malabsorption states, may occur in the presence of normal heart function. Others, such as hypomagnesemia induced by diuretic therapy (53, 77) or during the treatment of renal disease (33), characteristically occur in the presence of myocardial disease. Cardiac rhythm disturbances have been documented in patients with hypomagnesemia (14, 24, 43), particularly during digitalis therapy (4, 43). These abnormalities have been treated with parenteral magnesium (62, 77). Rhythm disturbances in hypermagnesemic states are less frequently reported, but elevated magnesium concentrations have been employed to improve mechanical recovery from myocardial ischemia or anoxemia (20, 37). The mechanisms underlying the role of magnesium ions in these phenomena are not clear and have been the subject of considerable speculation (4, 38, 77). Recently the availability of cardiac muscle preparations with altered sarcolemmal membranes (25, 45, 46) has allowed detailed examination of the effects of intracellular magnesium on tension development. These results can be compared to the effect of magnesium in the activity of myofibrillar ATPase (68, 89). Studies with isolated perfused cardiac tissue have delineated the effects of extracellular magnesium on cardiac function (81, 82). The rates of exchange of ^{28}Mg have been described for rat ventricle (65, 66).

I shall review the information made available during the past 10–12 years from studies in isolated cardiac muscle and intact animals and shall formulate a hypothesis regarding the mechanisms by which magnesium affects the heart.

Special emphasis is placed upon species variability of responses to magnesium. Earlier reviews (23, 103, 107) on the pharmacological effects of magnesium included sections on cardiac tissues. Polimeni and Page (66) have written an excellent review dedicated to cardiac muscle. In the subsequent discussion, "free" ionized magnesium is designated Mg^{2+}.

Cellular Content of Magnesium

Page and Polimeni (65, 66) determined a total magnesium content of 17.3 ± 0.2 mmol/kg cell water for the rat ventricle. They estimated approximately 12% of cardiac magnesium to be in the mitochondria (7, 66) and 2–3% in the myofibrils. A large proportion of intracellular magnesium is complexed with adenosine triphosphate (ATP), adenosine monophosphate (AMP), and adenosine diphosphate (ADP). Additional magnesium is bound to enzyme-coenzyme complexes.

The ionic activity for Mg^{2+} in heart muscle was estimated at less than 1.0 mM by Polimeni and Page from levels of enzyme activities and other magnesium-dependent processes. Higher values (3.0–6 mM) have been suggested by ^{31}P nuclear magnetic relaxation (NMR) studies in frog gastrocnemius muscle (17) and by a dye method in barnacle muscle (11). Both methods are limited by the fact that values obtained from the intracellular milieu are compared with those in standard or model solutions that may not recreate the intracellular environment. Studies of ^{31}P NMR in intact frog skeletal muscle gave a value of 0.6 mM, which was unaffected by changes in intracellular Mg^{2+} from 0 to 1.6 mM over 90 min (35). Page and Polimeni's estimate remains the best current value for cardiac tissues. The factors that regulate Mg^{2+} ionic activity remain uncertain. Under special circumstances heart mitochondria can accumulate Mg^{2+} and presumably might regulate cytosolic Mg^{2+} by uptake and release of the cation (10, 90). Alteration of adenine nucleotide contents and distribution could also influence Mg^{2+} activity. On the other hand variation of cytosolic Mg^{2+} as a function of extracellular Mg^{2+} is a relatively slow process. In the rat ventricle Polimeni and Page (65) described a single rate of exchange at 37°C in isolated hearts perfused with an extracellular Mg^{2+} of 0.56 mM. Because 50% or more of serum magnesium is bound to protein, this concentration is a reasonable representation of ionized Mg^{2+} in vivo. They observed a half time ($t_{1/2}$) for [28]Mg exchange of 182 min (0.0038 min^{-1}). Similar results were obtained in my laboratory for [28]Mg efflux in isolated perfused rabbit septa (unpublished results). The rate was not influenced by frequency of contraction or the amount of ventricular work. [28]Mg moved with saturation kinetics, and the phenomenon of countertransport was demonstrated. The data suggested carrier-mediated exchange. With exception of the 2% tightly bound to myofibrils, all of the cellular magnesium in rat ventricle was exchangeable with extracellular Mg^{2+}.

Regulation of Force Development

The actin and myosin filaments of cardiac muscle attach through cross-bridges that contain myofibrillar ATPase activity. Hydrolysis of MgATP by this enzyme yields energy for the contractile process. At low Ca^{2+} concentrations (10^{-8} to 10^{-10} M), MgATP induced tension in skinned crayfish muscle fibers (73).

Reuben et al. (73) suggested that an interaction of MgATP with the globular portion of the myosin molecule allowed cross-bridge interaction to produce this tension. At higher MgATP concentrations, tension development decreased, presumably from substrate inhibition (73). These interrelationships have been further quantitated recently for crayfish skeletal muscle (63). Similar relationships between tension development and MgATP were demonstrated for skinned (mechanically disrupted) cardiac cells by Fabiato and Fabiato (25). An interesting effect of Mg^{2+} was suggested by the observations of Griffiths et al. (34) in fibers of glycerinated insect flight muscles that can be stretch activated. In the absence of Mg^{2+} there was a significant increase in the instantaneous stiffness of the muscle fibers and a much greater ratio of stiffness to tension than observed in the presence of Mg^{2+} and Ca^{2+}. These observations suggested an alteration of the elastic properties of the muscles that may arise from alterations in the nature or location of cross-bridge attachments. Although this phenomenon may be limited to these highly specialized flight muscles, its occurrence in cardiac muscles has not been excluded.

Calcium sensitivity of the cross-bridge requires the troponin and tropomyosin proteins (71). Mg^{2+} may influence the interaction of these proteins by increasing the affinity of F actin for tropomyosin. The interactions between Mg^{2+}, Ca^{2+}, troponin, and the myofibrillar proteins have been studied extensively (9, 68–71, 105). Tropomyosin occupies a position during relaxation that blocks cross-bridge formation between actin and myosin. This inhibition is terminated by an interaction with the troponin complex. This complex includes the calcium-sensitive protein, TN-C, as well as TN-I and TN-T. Potter and Gergely (71) identified six cation binding sites on TN-C. Two sites are Mg^{2+} specific, two are Ca^{2+} specific, and two sites can bind either cation. Potter and Gergely demonstrated that the Ca^{2+}-specific sites had relatively low affinity for Ca^{2+} (binding constant 2×10^5 M^{-1}), compared to a higher affinity (binding constant 2×10^7 M^{-1}) for the Mg-Ca competitive sites. The four sites that bound Mg^{2+} had a binding constant of 4×10^4 M^{-1} for that cation. If the intracellular Mg^{2+} is 0.25 mM, these relationships would cause the high-affinity Ca^{2+} sites to be 91% occupied by Mg^{2+} during relaxation. Since the estimated intracellular Mg^{2+} is greater than 0.25 mM, this argues that Ca^{2+} sensitivity of TN-C occurs at the low-affinity sites. Such a conclusion was supported by studies of Ca^{2+} binding of TN-C using extrinsic fluorescence, spin-label mobility, and sulfhydryl reactivity (72) and by nuclear magnetic resonance (75). In both of these studies conformational changes of TN-C could be attributed to Ca^{2+} binding at the low-affinity noncompetitive sites. In another proton magnetic resonance study, Levine et al. (52) confirmed that Mg^{2+} did not bind to the sites with low Ca^{2+} affinity and agreed that the high-affinity sites were occupied by Mg^{2+} during relaxation. They argue that during contraction there still may be some displacement of Mg^{2+} from high-affinity sites by Ca^{2+} and that detailed differences in protein conformation do occur, depending on which cation is bound at these sites. Because tension development in skinned muscle fibers can be strongly influenced by the ratios of Ca^{2+} to Mg^{2+}, some competition between the two cations for the higher-affinity troponin binding sites would be expected. Potter (74) has emphasized that such competition may take place at the myosin molecule, but the details of the interactions remain to be completely defined.

In disrupted cell preparations, Mg^{2+} inhibits the ability of Ca^{2+} to produce tension and to stimulate myofibrillar ATPase activity (68). In rat cardiac cells whose sarcolemma had been mechanically disrupted, Mg^{2+} decreased the sensitivity of tension development to Ca^{2+} (25). Donaldson et al. (18, 19) showed that the curve relating percentage of maximum tension to P_{Ca} was shifted to the right by increases of Mg^{2+} from 5×10^{-5} M to 10^{-2} M. In addition, the maximal tension induced by either Ca^{2+} or Sr^{2+} could be increased at higher Mg^{2+}. They suggested that Ca^{2+} and Sr^{2+} had inhibitory effects at higher concentrations and that this inhibition was counteracted by high Mg^{2+}.

Solaro and Shiner (89) studied the myofibrillar ATPase activity of rabbit and canine myofibrils from skeletal and cardiac muscles. Increasing Mg^{2+} concentration from 1.0 to 10 mM depressed ATPase activity induced by increasing Ca^{2+} concentrations from 10^{-7} to 10^{-5} M. Surprisingly a decrease of Mg^{2+} from 1.0 mM to 0.04 mM depressed ATPase activity in cardiac muscle, although the anticipated enhancement of enzyme activity was observed in skeletal myofibrils. This latter observation emphasized that protein binding of cations can be different for cardiac and skeletal muscles, an observation also made by Fabiato and Fabiato (25). Solaro and Shiner (89) observed that calcium binding by myofibrils was increased at higher Mg^{2+} levels. This increased binding was attributed to troponin sites and presumably made less Ca^{2+} available to the myofibrillar ATPase activity. Several loci for Ca^{2+}-Mg^{2+} competition exist in myofibrils, including sites on myofibrillar ATPase, myosin, troponin, and tropomyosin. Best et al. (6) argued that $MgATP^{2-}$, ATP^{4-}, and Mg^{2+} all interact to produce maximal tension development in rat cardiac muscle. The complexity of analysis of the in vivo effects of Mg^{2+} are underscored by the difficulties in predicting how a change in Mg^{2+} will influence the ratios of these substances, particularly when their distribution within the cell is uncertain.

In addition to the effects of Mg^{2+} on the contractile proteins themselves, the ion may influence calcium binding and release at other intracellular sites. Ford and Podolsky (29) found that increases of Mg^{2+} from 0.02 to 1.4 mM inhibited the quick contraction induced by Ca^{2+} in skinned fibers. This result was ascribed to a retarded propagation of a regenerative response along the sarcotubules. In skinned skeletal muscle and skinned single cardiac cells (25) elevations in the concentrations of Mg^{2+} increased the capacity and rate of binding for Ca^{2+} by the sarcoplasmic reticulum. Although calcium-triggered calcium release from the sarcoplasmic reticulum was inhibited by high Mg^{2+} in the skeletal muscle fiber, such triggered release could be achieved in cardiac muscle. Fabiato and Fabiato (25) argued, however, that at a higher Ca^{2+} concentration this difference arose from a smaller calcium-binding capacity of cardiac sarcoplasmic reticulum and that calcium-triggered calcium release could be of physiological importance in cardiac tissue at intracellular concentrations of Mg between 10^{-4} and 10^{-3} M, i.e., at levels estimated by Page and Polimeni. In contrast to the results of Fabiato and Fabiato (25), Dunnett and Nayler (21) were not able to overcome the Mg^{2+} inhibition of Ca^{2+} efflux in guinea pig cardiac sarcoplasmic reticulum vesicles by elevating the concentration of Ca^{2+} in the medium. Although striking differences between the response of sarcoplasmic reticulum from cardiac and skeletal muscle to Mg^{2+} were demonstrated in these studies, the precise role of calcium-triggered calcium release in cardiac muscle in vivo remains unresolved

(22). In all of the studies of disrupted cardiac cells, important changes in tension development required manipulation of Mg^{2+} by millimolar amounts in comparison to calcium sensitivity to micromolar changes.

Shine and Douglas (81) examined the effects of 0–20 mM Mg^{2+} on tension development in isolated blood-perfused rat interventricular septa at 28°C. Increased Mg^{2+} produced a decline in tension development at any given Ca^{2+} level in the perfusate. The decline in tension was rapid, with a half time of 36 s (Fig. 1). Recovery of tension development upon reduction of Mg^{2+} occurred with a half time of 25 s. The rapidity with which developed tension changed contrasted sharply with the half time for ^{28}Mg exchange of 182 min previously noted for rat ventricle. Action potential measurements showed that the mechanical results were not attributable to electrophysiological alterations, although conduction velocity was slowed. The antagonist effect of Mg^{2+} and Ca^{2+} in rat ventricle was described by a regression equation in which either developed tension or dT/dt was proportional to $[Ca]_o/([Mg]_o + 0.7)^{1/2}$. This emphasizes the relatively greater potency for tension development of Ca^{2+} in comparison to Mg^{2+}. It indicates the strong dependence of tension development on extracellular calcium, whereas tension development was only modestly influenced at lower concentrations of Mg^{2+}. Shine and Douglas (81) measured the ^{45}Ca uptake by rat ventricle perfused with blood containing 2.5 mM Ca^{2+} and 10 mM Mg^{2+} (Fig. 1). Within 3 min of perfusion with increased Mg^{2+}, the tissue ^{45}Ca counts had decreased significantly, consistent with a net loss of ^{45}Ca from the muscle. Although some rapid displacement of ^{45}Ca from sarcolemmal binding sites might account for this net loss, no displacement could be demonstrated during ^{45}Ca efflux experiments. Shine and Douglas concluded that Mg^{2+} reduced the movement of extracellular calcium into the rat myocardial cell by a competitive process localized at the sarcolemmal membrane. This process could have arisen from a combination of rapid displacement from sarcolemmal binding sites and

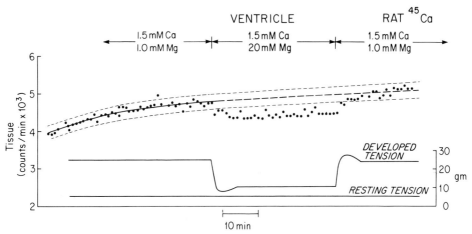

Fig. 1. Response of isolated blood-perfused rat interventricular septum to increase of perfusate Mg^{2+} from 1 to 10 mM. Very rapid decline in tension on elevation of Mg shown with a more rapid recovery of tension upon return to 1.0 mM Mg. Decline in tissue counts of ^{45}Ca measured by Geiger-Müller probe during 20 mM Mg perfusion follows similar pattern with stabilization at new lower level until return to 1.0 mM Mg.

reduction of ^{45}Ca movement into the cell. The effect was completely reversible.

These results were confirmed by Kovács and O'Donnell (49) in isolated trabecular muscle of rat ventricle, although these investigators described resistance to high Mg^{2+} when extracellular Ca^{2+} was 2.5 mM. They also demonstrated that elevated extracellular Mg^{2+} decreased ^{45}Ca exchange in a calcium pool exchangeable with $t_{1/2} < 7$ min, which they attributed to calcium bound to a superficial site on the sarcolemmal membrane.

In studies by Vierling et al. (99, 100) guinea pig papillary muscles also demonstrated competitive effects of Ca^{2+} and Mg^{2+} on tension development. When extracellular Ca^{2+} was 2.15 mM, the addition of 10 mM Mg^{2+} was antagonized by the addition of 0.83 mM Ca^{2+}, the effect of Mg^{2+} largely resulted from reduction of the velocity of contraction. Increases of extracellular Mg^{2+} produced only minimal effects on action potentials. These included a slight increase in action potential duration and a small decrease in the velocity of depolarization. Changes in extracellular Na^+ influenced tension development but did not alter the competition between Ca^{2+} and Mg^{2+}. The authors (99) concluded that Na^+, Ca^{2+}, and Mg^{2+} competed for a common receptor at the cellular surface. The apparent dissociation constant for the complex between Mg^{2+} and receptor was estimated at 2.5 mM in the presence of an extracellular Na^+ concentration of 140 mM.

Several laboratories (5, 60, 93) have reported the presence of a Ca^{2+} ATPase activity in purified preparations of isolated sarcolemma vesicles. This calcium-sensitive enzyme, which has a requirement for Mg^{2+}, could contribute to net Ca^{2+} efflux from the myocardial cell. If such a mechanism were important for regulation of contractility, it might be susceptible to changes in Mg^{2+}. Fabiato and Fabiato (26) point out that contamination of these preparations by ATPase from sarcoplasmic reticulum has not yet been adequately excluded, so that the importance of these mechanisms remain to be defined in more detail.

Langer et al. (51) studied the displacement of ^{45}Ca by lanthanum, cadmium, zinc, manganese, and magnesium in neonatal rat heart tissue culture. The relative affinity of these cations for calcium binding sites, determined by their displacement of ^{45}Ca, was in the same sequence as their ability to inhibit development tension in perfused neonatal rat hearts. Mg^{2+}, 1.0 mM, displaced 20% of the ^{45}Ca counts displaced by lanthanum and by calcium itself, but decreased tension by only 6% of the amount produced by lanthanum. Lanthanum prevented all subsequent ^{45}Ca exchange in the tissue culture, whereas ^{45}Ca uptake continued in the presence of 1.0 mM Mg^{2+} after the initial displacement. These results indicate an effect of Mg^{2+} upon rapidly exchangeable, superficially located sarcolemmal sites that are important to tension development. As shown by Shine and Douglas (81) in perfused rat hearts, new steady-state levels of tension development and ^{45}Ca content were established with a few minutes of changing extracellular Mg^{2+}.

The importance of species variation in response to Mg^{2+} was emphasized by the resistance of the tension developed by isolated blood-perfused rabbit septa to extracellular magnesium concentrations up to 40 mM (82). Although Solaro and Shiner (89) showed that elevation of Mg^{2+} above 10 mM inhibited myofibrillar ATPase obtained from rabbit heart, 40 mM Mg^{2+} in blood perfusate did not decrease tension development in intact perfused rabbit ventricle (82).

Mg^{2+} depression of tension development was described for intact dog ventricles by Bristow et al. (12, 13) and by Stanbury and Farah (94). Their data do not allow precise dose-response analyses but suggest a response of dog ventricle to variations in extracellular Mg^{2+} that was intermediate between rat and rabbit at the relatively more resistant end of the spectrum. Mg^{2+} markedly depressed tension development in frog ventricle, but this resulted from electrophysiological effects and was not reversed by increased extracellular Ca^{2+} (1, 2).

In summary Mg^{2+} could alter tension development in cardiac tissue by effects on the myofibrillar cross-bridge, by alterating Ca^{2+} release or sequestration within the cell, or by an action at the sarcolemmal membrane. The available data from intact cardiac cells argue for a sarcolemmal site of action that varies in different species.

Effects of Mg^{2+} on Na-K-ATPase

Mg^{2+} plays a key role as cofactor in the Na-K-ATPase activity of cardiac muscle (85, 86). The successful administration of magnesium salts for treatment of arrhythmias produced by digitalis in hypomagnesemic states (8, 36, 77, 78) led to the speculation (77) that Mg^{2+} increased the activity of Na-K-ATPase in the presence of digitalis glycosides. On the contrary, Skou et al. (85, 86) showed that Mg^{2+} was required for glycoside inhibition of the ATPase. Increase in Mg^{2+} from 10^{-8} M to 10^{-3} M progressively decreased the concentration of g-strophanthidin required for 50% inhibition of Na-K-ATPase prepared from beef brain. In the intact cell, Skou argues that Mg^{2+} interacts with Na-K-ATPase at the inner aspect of the cell membrane, where it alters the state of the enzyme for glycoside binding at the outer surface.

Shine and Douglas (82) demonstrated that arrhythmias induced by acetylstrophanthidin in isolated perfused rabbit interventricular septa could be suppressed by increased extracellular Mg^{2+}. They measured ^{42}K uptake by Geiger-Müller probe closely apposed to the muscle during glycoside administration. Although net ^{42}K uptake by the muscle was strikingly suppressed by acetylstrophanthidin, 10 mM Mg^{2+} did not decrease this inhibition (Fig. 2). In ^{42}K efflux experiments, however, 10 mM Mg^{2+} consistently reduced or obliterated the net ^{42}K loss induced by glycosides (Fig. 3). These results indicated that Mg^{2+} could inhibit the effluent loss of ^{42}K induced by digitalis glycoside without any measurable change in the magnitude of Na-K-ATPase inhibition. The data suggested that the effects of magnesium upon tissue responses to digitalis glycoside were largely mediated by direct effects of Mg^{2+} on K^+ permeability.

Mg-K Interactions

Increased extracellular Mg^{2+} abruptly decreased ^{42}K efflux from isolated blood perfused rat interventricular septa (81). This effect was independent of the Mg-Ca antagonism previously described. If extracellular Ca^{2+} was increased at the same time that Mg^{2+} was elevated, depression of tension development was prevented, but ^{42}K efflux still decreased (Fig. 4). In contrast increased extracellular Mg^{2+} did not affect ^{42}K exchange in perfused rabbit septa unless a perturbation of potassium kinetics was first produced. As previously noted (82), 10 mM Mg^{2+} diminished the ^{42}K efflux changes produced by acetylstrophanthi-

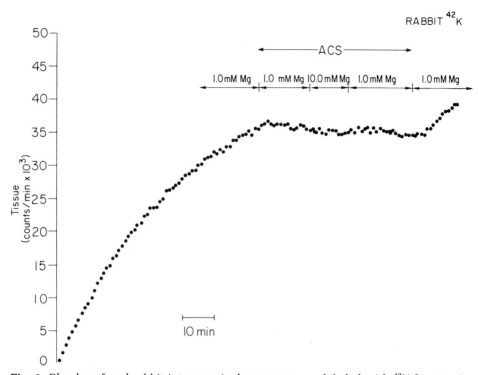

Fig. 2. Blood-perfused rabbit interventricular septum was labeled with ^{42}K for 70 min. Perfusion with 2.5×10^{-6} M acetylstrophanthidin abruptly interrupted ^{42}K uptake. Mg, 10 mM, did not reverse this inhibition of ^{42}K uptake. ^{42}K uptake resumed promptly upon removal of acetylstrophanthidin.

din. Elevated Mg^{2+} also inhibited the increase in ^{42}K exchange produced in rabbit septa by elevation of extracellular K^+ to 16 mM.

The role of Mg^{2+} in digitalis-induced potassium losses had two important implications. Increased tension development from glycoside was well maintained in rabbit septa despite complete obliteration of the increased net ^{42}K losses. This argued against the hypothesis that digitalis inotropy arose from increased calcium influx that was coupled to digitalis-induced potassium losses. The dissociation between glycoside-induced inotropy and ^{42}K loss has also been demonstrated during respiratory acidosis (67). Equally important, the loss of potassium from glycoside-treated cardiac muscle contributes to toxic rhythm disturbances. Mg^{2+} and other substances may provide approaches that could improve the therapeutic-to-toxic ratio of glycosides.

Chronic Mg Deficiency States

The relationships between Mg^{2+}, K^+, and digitalis glycoside are particularly important during prolonged magnesium deficiency. Elucidation of these interrelationships are complicated by the profound systemic effects that chronic deficiency produces. Puppies fed a magnesium-deficient diet for 1–6 mo showed retarded growth, neuromuscular instability, convulsions, and histopathological changes in the aorta, coronary arteries, and peripheral arteries (101). Fibrotic plaques were observed in the arteries and within the myocardium itself. Calci-

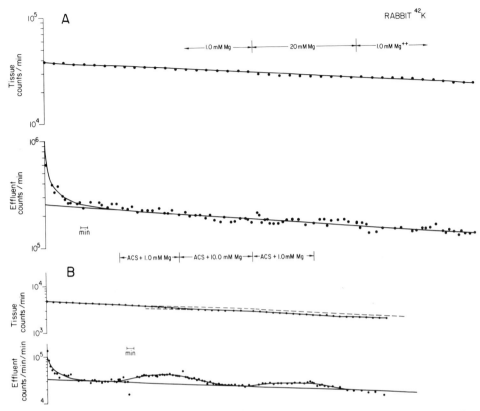

Fig. 3. Tissue counts of ^{42}K measured by Geiger-Müller probe and effluent counts of ^{42}K from blood collected as it left perfused rabbit interventricular septa. In *A* there is no effect of 20 mM Mg on tissue or effluent curves. In *B*, infusion of 2.5 × 10^{-6} M acetylstrophanthidin caused increased rate of ^{42}K loss from the septum, demonstrated by an increase of counts in effluent and steeper decline of tissue probe counts. Both these effects were inhibited by increase of perfusate Mg^{2+} to 10 mM. This inhibition was rapidly reversible upon return to 1.0 mM Mg^{2+} perfusate.

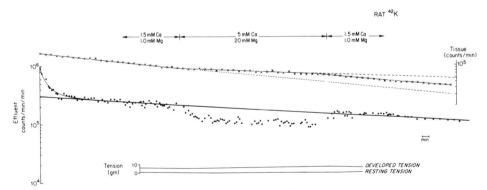

Fig. 4. Tissue probe counts and effluent counts of ^{42}K for blood-perfused rat interventricular septum. Mg, 20 mM, produced marked decrease in appearance of ^{42}K counts in effluent and slowed the rate of decline in tissue counts. This effect occurred in the face of well-maintained tension and was accomplished simultaneously increasing the perfusate Ca^{2+} content from 1.5 to 5.0 mM while Mg^{2+} was increased from 1 to 20 mM.

fication of myocardium and kidneys developed. Losses of 33–42% of tissue magnesium were demonstrated in rats fed a magnesium-deficient diet for weeks to months (106). Rats developed nephrocalcinosis with renal impairment despite supplementation of the deficient diet with potassium (55, 88). Rats consistently developed tissue potassium depletion while maintained on Mg-deficient diets, although serum potassium levels showed considerable variation in this species, in the dog (80, 101), and in the Cebus monkey (102).

Whang and Welt (106) showed that potassium losses from rat diaphragm maintained in a low-magnesium bath were prevented by addition of Mg^{2+} to the bath. Potassium supplementation did not prevent tissue losses of the cation from magnesium-deficient animals. Whang and Welt also drew attention to the absence of a significant increase of muscle sodium in the magnesium-deficient rat. In some experimental animals, muscle calcium rose to partially account for the potassium loss, but accurate extracellular chloride measurements were not available to determine if this anion was lost with the potassium. The only cells investigated that did accumulate sodium were the erythrocytes, but these showed no change in potassium content.

Magnesium-deficient monkeys and dogs demonstrated increased susceptibility to toxic arrhythmias from acetylstrophanthidin (47). Hypercalcemia and hypocalcemia can be induced by magnesium deficiency in some species (55), but this finding did not correlate with toxic arrhythmias. Although some of the antiarrhythmic effects of Mg^{2+} may result directly from the electrical or peripheral effects of the cation, the increased sensitivity of animals to digitalis glycosides was most consistently related to changes in tissue content of potassium. Alterations of [^{3}H]digoxin binding to myocardium in magnesium-deficient animals have been described (32), but no alteration of Na-K-ATPase activity was measured in dogs with digitalis-induced arrhythmias after Mg^{2+} administration (92).

Electrophysiological Effects of Mg^{2+}

Alterations of Mg^{2+} have little effect upon the cardiac action potential unless calcium has been reduced or eliminated from the bathing solution (40, 41, 48). In this situation Mg^{2+} shortened the action potential's plateau and its total duration. Chesnais et al. (15) indicated that this effect resulted from the ability of Mg^{2+} to suppress sodium through "slow channels." This slow-channel sodium current was enhanced in the absence of Ca^{2+} and Mg^{2+}. They demonstrated in frog heart that Mg^{2+} itself could not carry slow inward current at physiological Mg^{2+} concentrations, although Mg^{2+} may penetrate slow channels at extracellular concentrations of 5.5–11.0 mM. Chesnais et al. (15) and Antoni et al. (1, 2) demonstrated that Mg^{2+} could increase the sodium current through fast channels in the frog heart, thereby increasing the initial rate of depolarization. Surawicz and associates (96) showed that Mg^{2+} shortened the Q-T duration and S-T segments of the electrogram in hypocalcemic isolated perfused rabbit heart. This result was consistent with a shortening of action potentials. Watanabe and Dreifus (104) manipulated the K^+ concentration of perfused rabbit hearts while altering Mg^{2+}. They demonstrated increased action potential amplitude and maximal rates of depolarization that could be explained by an enhanced sodium current as described for the frog. They also described an increase in the resting membrane potential and prolongation of the action potential duration by 6.3 mM Mg^{2+} in 7.5 mM K^+. The result is less anomalous if it is recalled that Mg^{2+} significantly decreased ^{42}K efflux in perfused rabbit ventricle when extracellular K was elevated (81). Such an action would be expected to prolong action potential duration in elevated K^+. Watanabe and Dreifus (104) concluded that high Mg^{2+} primarily produces a Ca^{2+} effect with the exception of the effects on action potential duration in high K^+.

Spah and Fleckenstein (91) have proposed the existence of a "new perferentially Mg-carrying, transport system besides the fast Na and the slow Ca channels in the excited myocardial sarcolemma membrane." They demonstrated that $MgCl_2$ (14–34 mM) restored electrical activity to guinea pig papillary muscles that were depolarized by addition of 19 mM K in a low-Ca^{2+} (0.4 mM) Tyrode's solution. Because the Mg^{2+}-dependent action potential was not blocked by tetrodotoxin, lidocaine, or Ca antagonists such as verapamil, D600, nifedipine, or diltiazim, Spah and Fleckenstein argued for a third transport system that perferentially carried Mg current. Unfortunately, they did not include experiments with reduction or removal of extracellular sodium. Because a slow channel exists for sodium in addition to the tetrodotoxin-sensitive fast channels, the possibility that high Mg^{2+} restored Na flow through this slow channel or that Mg traversed this channel has not yet been excluded. Considering the experimental conditions and the very high Mg^{2+} concentrations required, the physiological significance of these observations are problematical.

According to Chesnais et al. (15) the increased rapid sodium current produced by Mg^{2+} might be attributed to a change in equilibrium potential for sodium, conductance, or both. Magnesium-induced E_{Na} changes, measured as changes in the reversal potential for rapid sodium current, were too small to explain the increased sodium current. It was more likely that a direct effect to increase sodium conductance was produced. Enhanced sodium pump activity would be expected to hyperpolarize the muscle membrane, whereas hyperpolarization was very small in the frog heart (1) and slight depolarization was observed in rabbit hearts exposed to high Mg^{2+} (82).

In the frog heart (1, 2) 5–20 mM Mg produced progressive shortening of the action potential duration, and normal excitability was eventually lost. But the muscle was able to develop tension approaching control values during potassium contracture, rapid electrical stimulation, and upon exposure to epinephrine and procaine amide. These results emphasized that intracellular contractile mechanisms remained intact despite failure of excitation in very high extracellular Mg^{2+}. Rat ventricles show similar responses to epinephrine in high extracellular Mg^{2+} (64).

The surface electrocardiogram obtained from magnesium-deficient animals (47, 101, 102) most commonly show peaking of the T wave with either elevated or depressed S-T segments. Hypomagnesemic dogs characteristically showed atrial and junctional dysrhythmias (94) as early manifestations of digitalis toxicity rather than ventricular ectopy.

Infusion of magnesium produces effects similar to hyperkalemia (43, 59, 74, 87, 97), including decreased AV conduction, decreased intraventricular conductance and suppression of sinus node function, probably as a result of a sinus node exit block. Cardiac arrest in the dog requires 27–44 mM Mg, by which time the curarelike peripheral effects of high Mg and the central nervous system effects suppress respiration and depress blood pressure (58). Sinus tachycardia and sinus bradycardia occur with changes in Mg^{2+} levels, but the responses are variable primarily because of the multiple reflex responses to the peripheral effects of the cation (23, 24, 57, 61).

Magnesium loss from myocardium has been demonstrated after ischemic myocardial injury, and it has been suggested that a relationship between this loss and sudden death may exist (16). In view of the other profound alterations in electrolyte content in ischemic reperfused tissue, this association remains speculative at present.

Epidemiological associations between low magnesium content in drinking water and increased incidence of death from ischemic heart disease have been reported (76). It is most likely that such a relationship results from an effect upon vascular tone or the propensity to atherosclerosis rather than as the result of direct myocardial mineral deficiency. However, the precise mechanisms remain to be established.

Mg^{2+} Administration for Arrhythmias

The sulfate and chloride salts of Mg have been used for treatment of atrial, junctional, and ventricular tachyarrhythmias occurring in humans during myocardial ischemia (36, 42), alcoholism (28, 39, 44, 56, 95), digitalis poisoning (62, 77, 78, 98), diuretic therapy (54), valvular and coronary heart disease (3, 8, 14, 24, 30, 43), and in delirium tremens (27, 28). The early studies did not include serum magnesium levels before or after treatment, and many of the later studies showed variable results without adequate control of potentially important variables. For example, the noncardiac effects of Mg^{2+} include a decrease of peripheral resistance (57, 61) that lowers ventricular afterload. In ischemic heart disease or congestive heart failure, arrhythmias may decrease as a consequence of the improvement in myocardial oxygen supply/demand relationship produced by afterload reduction. Stanbury and Farah (94) described an increase of coronary artery blood flow in the dog after Mg^{2+} infusion. This change in coronary flow after Mg^{2+} infusion may also have arisen from a nonvascular mechanism, since heart rate decreased during the infusion. This observation should be clarified because Mg^{2+} has been used in cardioplegic solutions during cardiac surgery. A direct effect on coronary vessels would significantly influence interpretation of the results. Central nervous system effects of Mg^{2+} (23, 58) might also influence arrhythmias.

In digitalis toxicity the evidence for a direct cardiac effect of Mg^{2+} is more convincing (4). Acute hypomagnesemia induced by dialysis of dogs against a low-magnesium bath lowered the cardiac threshold for atrial, junctional, and ventricular arrhythmias during acetylstrophanthidin administration (33, 78). This sensitivity to digitalis glycoside was corrected by infusion of $MgSO_4$. In these animals, acetylstrophanthidin produced a net potassium loss from the heart as determined by arteriovenous differences. The net loss was prevented by administration of $MgSO_4$ (62, 78). The result is consistent with the observations of Shine and Douglas (82) in the rabbit ventricle, although the relationships between arrhythmia production and potassium losses in the dog were less clearly defined than in the paced isolated rabbit septum. It is noteworthy that administration of $MgSO_4$ to dogs prior to any digitalis exposure caused a decrease in arterial and coronary sinus potassium concentrations (78). This also suggests an effect of Mg^{2+} upon potassium movements in noncardiac tissue, but relevant data for canine skeletal muscle are not available for comparison to Whang and Welt's observations in the rat (106).

Mg^{2+} and Digitalis Toxicity

Seller (77) proposed that reversal of digitalis toxicity by administration of Mg^{2+} arose from stimulation by the cation of Na-K-ATPase activity. Several observations argue against this interpretation. Shine and Douglas (82) showed that Mg^{2+} administration could decrease the net ^{42}K loss induced by acetylstrophanthidin in rabbit septa without altering the depressed influx of ^{42}K produced by the glycoside. Magnesium did not change digitalis-inhibited Na-K-ATPase when the enzyme complex was isolated from dog hearts treated with or without β-adrenergic blockade (92). Skou et al. (86) demonstrated that Mg^{2+} actually enhanced g-strophanthidin binding by Na-K-ATPase. Moreover Skou's argument that Mg^{2+} affects Na-K-ATPase at the inner surface of the sarcolemmal membrane makes it difficult to explain the very rapid (1–4 min) onset of the Mg^{2+} effects upon potassium losses produced by digitalis glycoside in the rabbit (82) and the dog (62, 79). The rapidity of this response was particularly striking in Seller's experiments (78), since simultaneous administration of Mg^{2+} and acetylstrophanthidin prevented a net potassium loss. As previously noted, Whang and Welt (106) did not note any change in intracellular sodium content during hypomagnesemia in rats. Finally, if the inotropic effects of digitalis glycoside result from an inhibition of Na-K-ATPase (50), the well-maintained mechanical response of muscles treated with both Mg and acetylstrophanthidin (82) argue against a change in the Na-K-ATPase-digitalis interaction. It is much more likely that the effects of Mg^{2+} on potassium losses during digitalis administration arise from a direct effect of Mg^{2+} on potassium channels in the sarcolemmal membrane. Although Mg^{2+} might have this effect by competition with Ca^{2+}, the well-sustained inotropy suggests that calcium movements continue unaffected during Mg^{2+} administration in digitalis-sensitive cardiac muscle.

Myocardial Preservation

Several laboratories (20, 37, 38) have employed solutions containing 10–80 mM Mg^{2+}, usually as Mg aspartate, to improve cardiac recovery from ischemia or anoxemia. These solutions often included other substances such as procaine amide, elevated K^+, glucose, insulin, or replacement of NaCl with sorbitol or mannitol. Computer simulations in isolated perfused rat hearts (31) indicated that Mg^{2+} may play an important regulatory role in intermediary metabolism. Hearse and Humphrey (37) demonstrated that elevated Mg^{2+} decreased enzyme release and improved mechanical function in isolated perfused rat heart subjected to ischemia. This result in rat heart can be attributed, at least in part, to decreased calcium availability to rat heart, as demonstrated by Shine and Douglas (82). This interpretation is consistent with the protective effect of reduced extracellular Ca^{2+} during reperfusion of ischemic rabbit septa (84). However, the resistance of mechanical function and calcium uptake of rabbit heart to increased Mg^{2+} emphasizes the species variation of Mg-Ca interactions and reiterates the potential danger of extrapolation of results obtained in rat heart to other species. Moreover, aspartate itself can reduce ischemic injury (83). If Mg^{2+} itself is useful in cardioplegic solutions, the mechanisms in species other than the rat remain to be demonstrated.

Physiological Consequences of Changes in Extracellular Mg^{2+}

Chronic magnesium deficiency can deplete tissue of magnesium and potassium and produce various other effects. The data reviewed in this report, however, suggest that acute variations of extracellular Mg^{2+} have little effect on intracellular Mg^{2+} activity and rarely acutely alter intracellular metabolic processes. This hypothesis is supported by the evidence for Mg-Ca interactions in the rat, which occur at the sarcolemmal membrane, and the observation that frog ventricle can develop essentially normal amounts of tension if the contractile mechanism is activated by epinephrine, potassium depolarization, and other interventions despite very high extracellular Mg^{2+}. Although Na-K-ATPase activity requires Mg^{2+}, the information reviewed here fails to show any evidence for a change in this enzyme's activity by an alteration of extracellular Mg^{2+}. On

Table 1. Interactions of Magnesium in Several Species

Species	Magnitude of Interaction	Ref. No.
Mg-Ca		
Neonatal rat	+	51
Adult rat	++	81
Guinea pig	++	99, 100
Rabbit	−	82
Dog	±	12, 13, 94
Mg-electrophysiological effects		
Frog	++	1, 2, 15
Rat	±	81
Guinea pig	+	99, 100
Dog	+*	40, 41
Rabbit	+*	82, 96
Mg-K		
Rat, ventricle	++	81
Rat, diaphragm	+	106
Rabbit	−	82
Rabbit, digitalis	+	82
Rabbit, high K*	+	82
Dog, digitalis	+	77, 78
Mg-Na		
Frog	++	1, 2, 15
Mg-Na-Ca		
Guinea Pig	++	99, 100

++, Marked interaction; +, some interaction; −, no demonstrated interaction.   * Strongly dependent on extracellular Ca^{2+}.

the contrary, all the mechanical and electrical effects of extracellular Mg^{2+} can be adequately explained by a series of interactions, Mg-K, Mg-Ca, and Mg-Na, which occur at the cell surface (Table 1). The species variability of response to Mg^{2+} most likely arises from differences in the sarcolemmal sites of these interactions. The relatively slow exchange of Mg^{2+} across the sarcolemmal membrane, contrasted with the large pool of complexed magnesium within the cell, suggested that intracellular Mg^{2+} activity is primarily determined by an equilibrium with intracellular pools rather than with extracellular Mg^{2+}. A definite resolution of the factors that regulate intracellular Mg^{2+} will require accurate measurements of this parameter. The species variation in response to magnesium is particularly important for interpretation of experiments in myocardial preservation.

Acknowledgments

The helpful comments of G. A. Langer and A. J. Brady are gratefully acknowledged.

This research was supported by Public Health Service grants HL-11351-11-12 and by the American Heart Association, Greater Los Angeles Affiliate and the Castera Foundation.

References

1. ANTONI, H., AND W. DELIUS. Nachweis von zwei Komponenten in der Anstiegsphase des Aktionpotentials von Froschmyokardfascern. *Pfluegers Arch.* 283: 187–202, 1965.
2. ANTONI, H., G. ENGSTFELD, AND A. FLECKENSTEIN. Die Mg^{++}-Lahmung des isoleirten Frosch-myokards. Ein Beitrag zur Frage der Beziehung zwischen Aktionpotential und Kontraktion. *Pfluegers Arch.* 275: 507–525, 1962.
3. ARMBRUST, C. A., AND S. A. LEVINE. Paroxysmal ventricular tachycardia: a study of one hundred and seven cases. *Circulation* 1: 28–40, 1950.
4. BELLER, G. A., W. B. HOOD, JR., T. W. SMITH, W. H. ABELMANN, AND W. E. C. WACKER. Correlation of serum magnesium levels and cardiac digitalis intoxication. *Am. J. Cardiol.* 33: 225–229, 1974.
5. BESCH, H. R., JR., L. R. JONES, J. W. FLEMING, AND A. M. WATANABE. Parallel unmasking of latent adenylate cyclase and (Na^+, K^+)-ATPase activities in cardiac sarcolemmal vesicles. *J. Biol. Chem.* 252: 7905–7908, 1977.
6. BEST, P., S. DONALDSON, AND W. KERRICK. Tension in mechanically disrupted mammalian cardiac cells: effects of magnesium adenosine triphosphate. *J. Physiol. London* 265: 1–17, 1977.
7. BOGURA, K., AND L. WOJTCZAK. Intramitochondrial distribution of magnesium. *Biochem. Biophys. Res. Commun.* 44: 1330–1337, 1971.
8. BOYD, L. J., AND D. SCHERF. Magnesium sulfate in paroxysmal tachycardia. *Am. J. Med. Sci.* 206: 43, 1943.
9. BREMEL, R. D., AND A. WEBER. Calcium binding to rabbit skeletal myosin under physiological conditions. *Biochim. Biophys. Acta* 376: 366–374, 1975.
10. BRIERLEY, G. P., E. MURER, E. BOCHMANN, AND D. E. GREEN. Studies on ion transport. II. The accumulation of inorganic phosphate and Mg ions by heart mitochondria. *J. Biol. Chem.* 238: 3482–3489, 1963.
11. BRINLEY, F. J., JR., A. SCARPA, AND T. TIFFERT. The concentration of ionized magnesium in barnacle muscle fibers. *J. Physiol. London* 266: 545–565, 1977.
12. BRISTOW, M. R., J. R. DANIELS, R. S. KERNOFF, AND D. C. HARRISON. Effect of D 600, practolol, and alterations in magnesium on ionized calcium concentration-response relationships in the intact dog. *Circ. Res.* 41: 574–581, 1977.
13. BRISTOW, M. R., H. D. SCHWARTZ, G. BINETTI, D. C. HARRISON, AND J. R. DANIELS. Ionized calcium and the heart: elucidation of *in vivo* concentration-response relationships in the open-chest dog. *Circ. Res.* 41: 565–574, 1977.
14. CHADDA, K. D., P. K. GUPTA, AND E. LICHSTEIN. Magnesium in cardiac arrhythmia. *N. Engl. J. Med.* 278: 1102, 1972.

15. CHESNAIS, J. M., E. CORABOEUF, M. P. SAUVIAT, AND J. M. VASSAS. Sensitivity to H, Li, Mg ions of the slow inward sodium current in frog fibres. *J. Mol. Cell. Cardiol.* 7: 627–642, 1975.

16. CHIPPERFIELD, B., AND J. R. CHIPPERFIELD. Magnesium and the heart. *Am. Heart J.* 93: 679–682, 1977.

17. COHEN, S. M., AND C. T. BURT. ^{31}P nuclear magnetic relaxation studies of phosphocreatine in intact muscle: determination of intracellular free magnesium. *Proc. Natl. Acad. Sci. USA* 74: 4271–4175, 1977.

18. DONALDSON, S. K. B., P. BEST, AND W. G. L. KERRICK. Characterization of the effects of Mg^{2+} on Ca^{2+}- and Sr^{2+}-activated tension generation of skinned rat cardiac fibers. *J. Gen. Physiol.* 71: 645–655, 1978.

19. DONALDSON, S. K. B., AND W. G. L. KERRICK. Characterization of the effects of Mg^{2+} on Ca^{2+}- and Sr^{2+}-activated tension generation of skinned skeletal muscle fibers. *J. Gen.Physiol.* 66: 427–444, 1975.

20. DÖRING, V., H. G. BAUMGARTEN, N. BLEESE, P. KÁLMAR, H. POKAR, AND G. GERCKEN. Metabolism and structure of the magnesium aspartate-procaine-arrested ischemic heart of rabbit and man. *Basic Res. Cardiol.* 71: 119–132, 1976.

21. DUNNETT, J., AND W. G. NAYLER. Dependence of calcium efflux rate from cardiac sarcoplasmic reticulum vesicles on external calcium concentration. Effect of magnesium ion. *J. Physiol. London* 266: 79P–80P, 1977.

22. ENDO, M. Calcium release from the sarcoplasmic reticulum. *Physiol. Rev.* 57: 71–108, 1977.

23. ENGBAEK, L. The pharmacological actions of magnesium ions with particular reference to the neuromuscular and the cardiovascular system. *Pharmacol. Rev.* 4: 396–413, 1952.

24. ENSELBERG, C. D., H. G. SIMMONS, AND A. A. MINTZ. The effects of magnesium upon cardiac arrhythmias. *Am. Heart J.* 39: 703–712, 1950.

25. FABIATO, A., AND F. FABIATO. Effects of magnesium on contractile activation of skinned cardiac cells. *J. Physiol. London* 249: 497–517, 1975.

26. FABIATO, A., AND F. FABIATO. Calcium and cardiac excitation-concentration coupling. *Ann. Rev. Physiol.* 41: 473–484, 1979.

27. FISHER, J., AND J. ABRAMS. Life-threatening ventricular tachyarrhythmias in delirium tremens. *Arch. Intern. Med.* 237: 1238–1241, 1977.

28. FLINK, E. B., F. L. STUTZMAN, A. R. ANDERSON, T. KONIG, AND R. FRASER. Magnesium deficiency after prolonged parenteral fluid administration and after chronic alcoholism complicated by delirium tremens. *J. Lab. Clin. Med.* 43: 169–183, 1954.

29. FORD, L. E., AND R. J. PODOLSKY. Intracellular calcium movements in skinned muscle fibers. *J. Physiol. London* 223: 21–33, 1972.

30. FREUNDLICH, J. Paroxysmal ventricular tachycardia. *Am. Heart J.* 31: 557, 1946.

31. GARFINKLE, D., M. C. KOHN, AND M. T. ACHS. Computer simulation of metabolism in pyruvate-perfused rat heart. V. Physiological implications. *Am. J. Physiol.* 237(*Regulatory Integrative Comp. Physiol.* 6): R181–R186, 1979.

32. GOLDMAN, R. H., R. E. KLEIGER, E. SCHWEIZER, AND D. C. HARRISON. The effect on myocardial ^{3}H-digoxin magnesium deficiency. *Proc. Soc. Exp. Biol. Med.* 136: 747–749, 1971.

33. GRANTHAM, J. J., W. H. TU, AND P. R. SCHLOERB. Acute magnesium depletion and excess induced by hemodialysis. *Am. J. Physiol.* 198: 1211–1216, 1960.

34. GRIFFITHS, P. J., H. J. KUHN, AND J. P. RUEGG. Activation of the contractile system of insect fibrillar muscle at very low concentrations of Mg^{2+} and Ca^{2+}. *Pfluegers Arch.* 382: 155–163, 1979.

35. GUPTA, R. K., AND R. D. MOORE. ^{31}P NMR studies of intracellular free Mg^{2+} in intact frog skeletal muscle. *J. Biol. Chem.* 255: 3987–3993, 1980.

36. HARRIS, A. S., A. ESTANDIA, H. T. SMITH, R. E. OHLSEN, T. J. FORD, JR., AND R. F. TILLOTSON. Magnesium sulfate and chloride in suppressoin of ecotopic ventricular tachycardia accompanying acute myocardial infarction. *Am. J. Physiol.* 172: 251–258, 1953.

37. HEARSE, D. J., AND S. M. HUMPHREY. Enzyme release during myocardial anoxia: a study of metabolic protection. *J. Mol. Cell. Cardiol.* 7: 463–482, 1975.

38. HEARSE, D. J., D. A. STEWART, AND M. V. BRAIMBRIDGE. Myocardial protection during ischemic cardiac arrest. The importance of magnesium in cardioplegic infusate. *J. Thorac. Cardiovasc. Surg.* 75: 877–885, 1978.

39. HEATON, F. W., L. N. PYRAH, C. C. BERESFORD, R. W. BRYSON, AND D. F. MARTIN. Hypomagnaesemia in chronic alcoholism. *Lancet* 2: 802–805, 1962.

40. HOFFMAN, B. F., AND P. F. CRANEFIELD. *Electrophysiology of the Heart.* New York: McGraw-Hill 1960, p. 67–69.
41. HOFFMAN, B. F., AND E. E. SUCKLING. Effect of several cations on transmembrane potentials of cardiac muscle. *Am. J. Physiol.* 186: 317–324, 1956.
42. ISERI, L. T., L. C. ALEXANDER, R. S. MCCAUGHEY, A. J. BOYLE, AND G. B. MYERS. Water and electrolyte content of cardiac and skeletal muscle in heart failure and myocardial infarction. *Am. Heart J.* 43: 215–227, 1952.
43. ISERI, L. T., J. FREED, AND A. R. BURES. Magnesium deficiency and cardiac disorders. *Am. J. Med.* 58: 837–846, 1975.
44. JONES, J. E., S. R. SHANE, W. H. JACOBS, AND E. B. FLINK. Magnesium balance studies in chronic alcholism. *Ann. NY Acad. Sci.* 162: 934–946, 1969.
45. KERRICK, W. G. L., AND S. K. B. DONALDSON. The effects of Mg^{2+} on submaximum Ca^{2+}-activated tension in skinned fibers of frog skeletal muscle. *Biochim. Biophys. Acta* 275: 117–122, 1972.
46. KERRICK, W. G. L., AND S. K. B. DONALDSON. The comparative effects of $[Ca^{2+}]$ and $[Mg^{2+}]$ on tension generation in the fibers of skinned frog skeletal muscle and mechanically disrupted rat ventricular cardiac muscle. *Pfluegers Arch.* 358: 195–201, 1975.
47. KLEIGER, R. E., K. SETA., J. J. VITALE, AND B. LOWN. Effects of chronic depletion of potassium and magnesium upon the action of acetylstrophanthidin on the heart. *Am. J. Cardiol.* 17: 520–527, 1966.
48. KLEINFIELD, M., E. STEIN, AND D. AGUILLARDO. Divalent cations on action potentials of dog heart. *Am. J. Physiol.* 211: 1438–1442, 1966.
49. KOVÁCS, T., AND J. M. O'DONNELL. An analysis of calcium-magnesium antagonism in contractility and ionic balance in isolated trabecular muscle of rat ventricle. *Pfluegers Arch.* 360: 267–282, 1975.
50. LANGER, G. A., AND S. D. SERENA. Effects of strophanthidin upon contraction and ionic exchange in rabbit ventricular myocardium: relation to control of active state. *J. Mol. Cell. Cardiol.* 1: 65–90, 1970.
51. LANGER, G. A., S. D. SERENA, AND L. M. NUDD. Cation exchange in heart cell culture: correlation with effects on contractile force. *J. Mol. Cell. Cardiol.* 6: 149–161, 1974.
52. LEVINE, B. A., J. M. THORNTON, R. FERNANDES, C. M. KELLY, AND D. MERCOLA. Comparison of the calcium and magnesium-induced structural changes of troponin-C. *Biochim. Biophys. Acta.* 535: 11–24, 1978.
53. LIM, P., AND E. JACOB. Magnesium deficiency in patients on long-term diuretic therapy for heart failure. *Br. Med. J.* 3: 620–622, 1972.
54. LOEB, H. S., R. J. PETRAS, R. M. GUNNAR, AND J. R. TOBIN. Paroxysmal ventricular fibrillation in two patients with hypomagnesemia. *Circulation* 37: 210–215, 1968.
55. MACINTYRE, I., AND D. DAVIDSON. The production of secondary potassium depletion, sodium retention, nephrocalcinosis and hypercalcaemia by magnesium deficiency. *Biochim. J.* 70: 456, 1958.
56. MARTIN, H. E., C. C. MCCUSKEY, JR., AND N. TUPIKOVA. Electrolyte disturbances in acute alcoholism with particular references to magnesium. *Am. J. Clin. Nutr.* 7: 191–196, 1959.
57. MAXWELL, G. M., R. B. ELLIOTT, AND R. H. BURNELL. Effects of hypermagnesemia on general and coronary hemodynamics of the dog. *Am. J. Physiol.* 208: 158–161, 1965.
58. MELTZER, S. J., AND J. AUER. The antagonistic action of calcium upon the inhibitory effect of magnesium. *Am. J. Physiol.* 21: 400–419, 1908.
59. MILLER, J. R., AND T. R. VAN DELLEN. Electrocardiographic changes following the intravenous administration of magnesium sulfate. *J. Lab. Clin. Med.* 26: 1116–1120, 1941.
60. MORCOS, N. C., AND G. I. DRUMMOND. $(Ca^{2+} + Mg^{2+})$-ATPase in enriched sarcolemma from dog heart. *Biochim. Biophys. Acta* 598: 27–39, 1980.
61. MROCZEK, W. J., W. R. LEE, AND M. E. DAVIDOV. Effect of magnesium sulfate on cardiovascular hemodynamics. *Angiology* 28: 720–724, 1977.
62. NEFF, M. S., S. MENDELSSOHN, K. E. KIM, S. BANACH, C. SWARTZ, AND R. H. SELLER. Magnesium sulfate in digitalis toxicity. *Am. J. Cardiol.* 29: 377–382, 1972.
63. ORENTLICHER, M., P. W. BRANDT, AND J. P. REUBEN. Regulation of tension in skinned muscle fibers effect of high concentrations of Mg-ATP. *Am. J. Physiol.* 233 *(Cell Physiol. 2)*: C127–C134, 1977.
64. PADDLE, B. M., AND N. HAUGAARD. Role of magnesium in effects of epinephrine on heart contraction and metabolism. *Am. J. Physiol.* 221: 1178–1184, 1971.

65. PAGE, E., AND P. I. POLIMENI. Magnesium exchange in rat ventricle. *J. Physiol. London.* 224: 121–139, 1972.

66. POLIMENI, P., AND E. PAGE. Magnesium in heart muscle. *Circ. Res.* 33: 367–374, 1973.

67. POOLE-WILSON, P. A., AND G. A. LANGER. Glycoside inotropy in the absence of an increase in potassium efflux in the rabbit heart. *Circ. Res.* 37: 390–395, 1975.

68. PORTZEHL, H., P. ZAORALEK, AND J. GAUDIN. Activation by Ca^{++} of the ATPase of extracted muscle fibrils with variation of ionic strength, pH and concentration of MgATP. *Biochim. Biophys. Acta* 189: 440–448, 1969.

69. POTTER, J. D. Effect of Mg^{2+} on Ca^{2+} binding to myosin. *Federation Proc.* 34: 671, 1975.

70. POTTER, J. D. The Ca^{2+} binding properties of bovine cardiac troponin-C. *Biophys. J.* 17: 118a, 1977.

71. POTTER, J. D., AND J. GERGELY. The calcium and magnesium binding sites on troponin and their role in the regulation of myofibrillar adenosine triphosphatase. *J. Biol. Chem.* 250: 4628–4633, 1975.

72. POTTER, J. D., J. C. SEIDEL, P. LEAVIS, S. S. LEHRER, AND J. GERGELY. Effect of Ca^{2+} binding on troponin-C. Changes in spin label mobility, extrinsic fluorescence, and sulfhydryl reactivity. *J. Biol. Chem.* 251: 7551–7556, 1976.

73. REUBEN, J. P., P. W. BRANDT, M. BERMAN, AND H. GRUNDFEST. Regulation of tension in the skinned crayfish muscle fiber. *J. Gen. Physiol.* 57: 385–407, 1971.

74. SCHAER, H. Antagonistische Wirkungen von Magnesium, Calcium und Natriumionen auf die Impulsbildung im Sinusknoten des Meerschweinschenherzens. *Pfluegers Arch.* 298: 359–371, 1968.

75. SEAMON, K. B., D. J. HARTSHORNE, AND A. A. BOTHNER-GY. Ca^{2+} and Mg^{2+} dependent conformations of troponin-C as determined by ^{1}H and ^{19}F nuclear magnetic resonance. *Biochemistry* 16: 4039–4046, 1977.

76. SEELING, M. S., AND H. A. HAGGTVERT. Magnesium interrelationship in ischemia heart disease. *Am. J. Clin. Nutr.* 27: 59–79, 1974.

77. SELLER, R. H. The role of magnesium in digitalis toxicity. *Am. Heart J.* 82: 551–556, 1971.

78. SELLER, R. H., J. CANGIANO, K. E. KIM, S. MENDELSSOHN, A. N. BREST, AND C. SWARTZ. Digitalis toxicity and hypomagnesemia. *Am. Heart J.* 79: 57–68, 1970.

79. SELLER, R. H., O. RAMIEREZ, A. N. BREST, AND J. H. MOYER. Serum and erythrocytic magnesium levels in congestive heart failure. Effect of hydrochlorothiazide. *Am. J. Cardiol.* 17: 786–791, 1966.

80. SETA, K., R. KLEIGER, E. E. HELLERSTEIN, B. LOWN, AND J. J. VITALE. Effect of potassium and magnesium deficiency on the electrocardiogram and plasma electrolytes of pure-bred beagles. *Am. J. Cardiol.* 17: 516–519, 1966.

81. SHINE, K. I., AND A. M. DOUGLAS. Magnesium effects on ionic exchange and mechanical function in rat ventricle. *Am. J. Physiol.* 227: 317–324, 1974.

82. SHINE, K. I., AND A. M. DOUGLAS. Magnesium effect on rabbit ventricle. *Am. J. Physiol.* 228: 1545–1554, 1975.

83. SHINE, K. I., A. M. DOUGLAS, AND E. E. RAU. Improved mechanical recovery of ischemic myocardium with amino acids. *Circulation* 58: II-98, 1978.

84. SHINE, K. I., A. M. DOUGLAS, AND N. V. RICCHIUTI. Calcium, strontium and barium movements during ischemia and reperfusion in rabbit ventricle. *Circ. Res.* 43: 712–720, 1978.

85. SKOU, J. C. Further investigations on a Mg^{2+} and Na^{+}-activated adenosinetriphosphatase, possibly related to the active, linked transport of Na^{+} and K^{+} across the nerve membrane. *Biochim. Biophys. Acta* 42: 6–23, 1960.

86. SKOU, J. C., K. W. BUTLER, AND O. HANSEN. The effect of magnesium, ATP, P_i, and sodium on the inhibition of the $(Na^{+} + K^{+})$-activated enzyme system by g-strophanthidin. *Biochim. Biophys. Acta* 241: 443–461, 1971.

87. SMITH, P. K., A. W. WINKLER, AND H. E. HOFF. Electrocardiographic changes and concentration of magnesium in serum following injection of magnesium salts. *Am. J. Physiol.* 126: 720–730, 1939.

88. SMITH, W. O., D. J. BAXTER, A. LINDNER, AND H. E. GINN. Effect of magnesium depletion on renal function in the rat. *J. Lab. Clin. Med.* 59: 211–219, 1962.

89. SOLARO, R. J., AND J. SHINER. Modulation of Ca^{2+} control of dog and rabbit cardiac myofibrils by Mg^{2+}. *Circ. Res.* 39: 8–14, 1976.

90. SORDAHL, L. A. Effects of magnesium, ruthenium red and the antibiotic ionophore A-23187 on

initial rats of calcium uptake and release by heart mitochondria. *Arch. Biochem. Biophys.* 167: 104–115, 1975.

91. Spah, F., and A. Fleckenstein. Evidence of a new perferentially Mg-carrying transport system besides the fast Na and the slow Ca channel in the excited myocardial sarcolemma membrane. *J. Mol. Cell. Cardiol.* 11: 1109–1127, 1979.

92. Specter, M. J., E. Schweizer, and R. H. Goldman. Studies on magnesium's mechanism of action in digitalis induced arrhythmias. *Circulation* 52: 1001–1005, 1975.

93. St. Louis, P. J., and P. V. Sulakhe. Adenosine triphosphate-dependent calcium binding and accumulation by guinea pig cardiac sarcolemma. *Can. J. Biochem.* 54: 946–956, 1976.

94. Stanbury, J. B., and A. Farah. Effects of magnesium ion on the heart and on its response to digoxin. *J. Pharmacol. Exp. Ther.* 100: 445–453, 1950.

95. Sullivan, J. F., P. W. Wolpert, R. Williams, and J. D. Egan. Serum magnesium in chronic alcoholism. *Ann. NY Acad. Sci.* 162: 947–962, 1969.

96. Surawicz, B., E. Lepeschkin, and H. C. Herrlich. Low and high magnesium concentration at various calcium levels. Effect on the monophasic action potential, electrocardiogram and contractility of isolated rabbit hearts. *Circ. Res.* 9: 811–818, 1961.

97. Szekely, P. The action of magnesium on the heart. *Br. Heart J.* 8: 115–124, 1946.

98. Szekely, P., and N. A. Wynne. Cardiac arrhythmias caused by digitalis. *Clin. Sci.* 10: 241–253, 1951.

99. Vierling, W., F. Ebner, and M. Reiter. The opposite effects of magnesium and calcium on the contraction of the guinea-pig ventricular myocardium in dependence on the sodium concentration. *Arch. Pharmacol.* 303: 111–119, 1978.

100. Vierling, W., and M. Reiter. Frequency-force relationship in guinea-pig ventricular myocardium as influenced by magnesium. *Arch. Pharmacol.* 289: 111–125, 1975.

101. Vitale, J. J., E. E. Hellerstein, M. Nakamura, and B. Lown. Effects of magnesium-deficient diet upon puppies. *Circ. Res.* 9: 387–394, 1961.

102. Vitale, J. J., H. Velez, C. Guzman, and P. Correa. Magnesium deficiency in cebus monkey. *Circ. Res.* 12: 642–650, 1963.

103. Wacker, W. E. C. The biochemistry of magnesium. *Ann. NY Acad. Sci.* 162: 717, 1969.

104. Watanabe, Y., and L. S. Dreifus. Electrophysiological effects of magnesium and its interactions with potassium. *Cardiovasc. Res.* 6: 79–88, 1972.

105. Weber, A., R. Herz, and I. Reis. Role of magnesium in the relaxation of myofibrils. *Biochemistry* 8: 2266–2271, 1969.

106. Whang, R., and L. G. Welt. Observations in experimental magnesium depletion. *J. Clin. Invest.* 42: 305–313, 1963.

107. Zwillinger, L. Magnesium and the heart. *Klin. Wochenschr.* 14: 1229–1231, 1935.

CHAPTER 6

Reentrant Excitation as a Cause of Cardiac Arrhythmias

ANDREW L. WIT AND PAUL F. CRANEFIELD
Department of Pharmacology, College of Physicians and Surgeons of
Columbia University, New York, New York,
and The Rockefeller University,
New York, New York

The cardiac impulse normally arises in the sinus node and spreads over a well-defined route through the atria, the AV node, the ventricular conducting system, and the ventricle, until the entire heart is activated. The conducting impulse leaves in its wake refractory tissue that cannot be reexcited for 200–500 ms. As a result, when the last remnants of heart muscle are excited the impulse, finding itself completely surrounded by refractory fibers into which it cannot conduct, dies out. Following a period of electrical quiescence during which the heart recovers excitability, a new impulse is initiated in the sinus node. In the normal heart this sequence of events is repeated 50–90 times each minute. Under certain conditions, however, an impulse that has excited the heart once does not die out but finds a pathway of excitable fibers over which it may return to reexcite part or all of the heart. Such "reentry" or circus movement of excitation may occur time and again, for minutes to years. This chapter describes the development of the concepts that have enabled us to understand reentry and its importance as a cause of cardiac arrhythmias.

Mayer, Mines, and Garrey and Circulating Excitation

That a circulating impulse could reexcite the same tissue over and over again was suggested to McWilliam in 1887 (53) by his observations on the nature of ventricular fibrillation. Until that time (and for 20 years thereafter) it was generally believed that all abnormalities in the rhythm of the heart were due to enhanced impulse initiation de novo from hyperexcitable heart muscle. Mc-William writes in his article "Fibrillar Contraction of the Heart": "For apart from the possibility of rapid spontaneous discharges of energy by the muscular fibres, there seems to be another probable cause of continued and rapid movement [observed during fibrillation]. The peristaltic contraction travelling along such a structure as that of the ventricular wall must reach adjacent muscle bundles at different points of time, and since these bundles are connected by anastomosing branches, the contraction would naturally be propagated from one contracting fibre to another over which the contraction wave *had already passed*. Hence, if the fibres are sufficiently excitable and ready to respond to contraction waves reaching them there would evidently be a more or less rapid series of contractions in each muscular bundle" (53).

The suggestion that an impulse could circulate in the heart was revived 20 years later when Alfred Mayer began his study of the causes of rhythmic pulsations in the bell of the medusa (41–43). Mayer described a method by which the disks of *Cassiopea*, deprived of marginal sense organs, may be made to pulsate indefinitely in seawater. Mayer cut the subumbrella tissue into a ring (Fig. 1), stimulated the ring at one point, and noted that if two waves of contraction of equal magnitude start at the same point and progress in opposite directions around the ring, only one contraction of the ring occurs. However, by applying pressure near the site of stimulation a strong contraction could be induced to progress in one direction around the ring, the area of compression preventing a strong wave from progressing in the other direction. The wave conducting in one direction returned to its point of origin and then conducted around the ring again. It was observed that the "single wave going constantly in one direction around the circuit may maintain itself for days at uniform rate" (43). Mayer recognized the importance of the relationship between path length, conduction velocity, and the refractory period (which he referred to as a need for the tissue to rest) for the ability of circulating excitation to occur or persist. He stated that "the wave will maintain itself indefinitely provided the circuit be long enough to permit each and every point in it to remain at rest for a certain period of time before the return of the wave through the circuit" (43). Another important observation made in Mayer's studies was that "the point which was stimulated and from which the contraction wave first arises is of no more importance in maintaining rhythmical movement than any other point in the ring" (43); i.e., stimulation did not set up a rhythmic focus at the site at which it was applied, a possibility that has not been eliminated in many more recent studies. Mayer demonstrated this by cutting away the tissues around the site at which the stimulus was applied— "yet the wave continues unhindered throughout the circuit." As the final proof that the impulse was circulating around the ring causing it to pulse, he showed that it stops instantly when the circuit is interrupted. Mayer also repeated some of the experiments on large rings cut from the hearts of loggerhead turtles. At the conclusion of these studies he provides an interesting comment: "In nature the structure of pulsating organs and their manner of stimulation are designed especially to prevent such a circuit wave from taking possession of the organ. Each pulsation of the heart or of the medusa is a thing separate and distinct from the contraction which preceded or

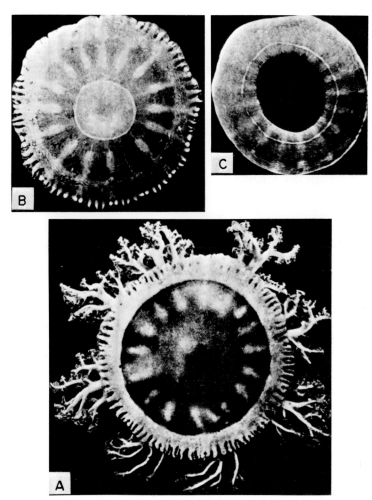

Fig. 1. A–C: preparation of the ring from *Cassiopea xamachana* as described by Mayer. A: aboral view of intact animal. B: oral view after the stomach and mouth-arms were removed. C: ring of subumbrella tissue made by cutting off the marginal sense organs and removing the center of the disk. When this ring was stimulated at one point a contraction wave traveled around it and returned to the point of origin. [From Mayer (41).]

from that which is to follow it. Indeed the heart or pulsating medusa contains within itself the means to prevent any single pulsation wave from coursing constantly in one direction through the tissue. [Yet] it is remarkable that these isolated circuit waves moving constantly in one direction through a circuit are not met with in nature" (43). In all probability they are met with and cause countless cardiac arrhythmias that may be incompatible with life.

Between 1912 and 1914 George Ralph Mines made definitive contributions to the development of the concept that reentry can cause cardiac arrhythmias. Mines (46), in the course of experiments on the two-chambered hearts of cold-blooded animals, observed a phenomenon that he called reciprocating rhythm. The isolated preparations were quiescent unless stimulated; after cessation of a period of stimulation a quick, reciprocating movement of auricle and ventricle appeared: "Let us start with the auricular excitation. This is transmitted over the slowly conducting tissue to the ventricle. But immediately after, the excitation of the ventricle is transmitted back to the auricle" (46). Mines recognized that in the amphibian heart the connection between auricle and ventricle was never a single muscle fiber but always a number of fibers and that "a slight difference in the rate of recovery of two divisions of the A-V connexion might determine that an extrasystole of the ventricle provoked by a stimulus applied to the ventricle shortly after activity of the A-V connexion should spread up to the auricle by that part of the A-V connexion having the quicker recovery process and not by the other part. In such a case, when the auricle became excited by this impulse, the other portion of the A-V connexion would be ready to take up the transmission back to the ventricle. Provided the transmission in each direction was slow, the chamber at either end would be ready to respond (its refractory phase being short) and thus the condition once established would tend to continue" (46). To test the validity of this hypothesis, he cut rings composed of auricle and ventricle from the tortoise heart (Fig. 2). These rings resembled the rings Mayer cut from jellyfish. After stimulation of any part he could sometimes observe the wave of contraction spread continuously around the ring in one direction (46, 47). He therefore concluded that the reciprocating rhythm may "reasonably be regarded as due to circulating excitation." Mines also made another important observation: "While the cycle was being regularly repeated [i.e., during continuous circulation of the impulse], the application of an external stimulus to either of the chambers, if out of phase with the cycle, stopped the contractions, showing that they were not originated by an automatic rhythm in any part of the preparation but were due to a wave of excitation passing slowly round and round the ring of tissues" (46). This assertion rested on the assumption that a single stimulus could not terminate an automatic focus but could terminate a circulating impulse by making the pathway refractory to the circulating wave. In modern cardiology electrical stimuli from pacemakers are often used to stop a tachycardia believed to result from circulating impulses.

Mines' summary of his theory of circulating impulses in the heart (parts of which were derived also from the studies of Mayer and of Garrey) is reproduced in Figure 2 and includes all the salient features considered today to be necessary for reentry to occur. He recognized that the initial circulating impulse must travel in only one direction around the closed circuit from the point of its initiation, so that block in the other direction must exist at the time of impulse initiation. This allows a segment of the loop to remain unexcited and provides a return pathway for the impulse to its point of origin. The block must also be unidirectional or transient for continuous circulation to occur, since the returning impulse must be able to conduct through the region of block to reach its site of origin. Mines also stated the necessary relationship between conduction velocity and refractoriness. He recognized that if the impulse conducted too rapidly around the ring in one direction, it would return to its point of origin before this region recovered excitability and would die out: "But if, on the other hand, the wave is slower and [the refractory period] is shorter, the excited state will have passed off at the region where the excitation started before the wave of excitation reaches this point on the circle at the completion of its revolution.

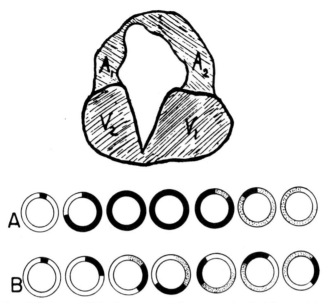

Fig. 2. *Top:* Mines' diagram of his ring preparation comprised of the auricle and ventricle of the tortoise, in which he observed reciprocating rhythm. During this rhythm, contraction proceeded in the order V_1, V_2, A_1, A_2. *Bottom A* and *B:* Mines' explanation of the mechanism for circulating excitation. *A* shows the events that occur if the impulse travels too rapidly in one direction around the loop, unidirectional block being present in the other direction. The conducting impulse, represented by the darkly shaded area, is initiated by a stimulus and travels rapidly around the loop only to return to its point of origin prior to recovery of excitability of this region (3rd and 4th diagrams from left). As a result, the impulse dies out as shown in the 5th and 6th diagrams, and the tissue remains refractory (*stippled area*). In the 7th diagram the tissue begins to regain excitability as shown by disappearance of the *stippled area*. *B* shows that reentry will occur if conduction is slowed and duration of refractory period is decreased. Stimulated impulse (*darkly shaded area*) conducts slowly in one direction, leaving in its wake refractory tissue (*stippled area*). Since conduction is slow and the refractory period is short, by the time the impulse has returned to its site of origin (5th and 6th diagrams from the left) this region has recovered excitability so the impulse can continue to circulate (7th diagram). [From Mines (46).]

Not only so, but there will have been time for the excitability of the muscle to return to something near the value it had at the time of the first excitation. Under these circumstances, the wave of excitation may spread a second time over the same tract of tissue; once started in this way it will continue unless interfered with by some external stimulus (46). It is therefore readily apparent that by 1914 the exact physiological conditions necessary for reentry to occur had been completely described.

Mines also emphasized the probable importance of circulating excitations as a cause of clinical abnormalities in the rhythm of human hearts. He suggested that conditions which produce fibrillation and other arrhythmias slow conduction and shorten refractoriness and thereby favor circulating impulses (46, 47). We now know that many of these changes in electrical properties do occur in diseased tissue.

While Mines was publishing the results of his studies, Walter Garrey (30) was investigating the nature of auricular fibrillation. Many of his experiments were identical to those of Mines, and he seemed annoyed that Mines was given most of the credit for the reentry hypothesis. We have described Garrey's contributions below in the section on atrial fibrillation.

Proving Reentry is a Cause of Cardiac Arrhythmias

The studies of Mayer, Mines, and Garrey clearly elucidated the mechanisms by which an excitatory wave could continue to propagate and reexcite (reenter) tissue it had previously excited. In retrospect, the certainty of their conclusions is derived from the simplicity of their experimental models, which were simple and large rings of excitable tissue. The major difficulties in experimental design and analysis fell on subsequent investigators whose aim it was to demonstrate that circulating excitation was a cause of abnormal rhythms in the intact heart. The experimental models became more complex and it became more difficult to adhere to the criteria for proving the presence of circulating excitation as established by Mines (46, 47). These criteria were: 1) An area of unidirectional block must be demonstrated so that the impulse conducts in one direction. This is not easily accomplished in complex tissues. 2) The movement of the excitatory wave should be observed to progress through the pathway, to return to its point of origin, and then to again follow the same pathway. Mayer, Mines, and Garrey could follow the movement of the slow contractile wave visually, and since the wave was conducting through a circumscribed pathway, it was not difficult to follow it around that pathway. In subsequent attempts to demonstrate circulating excitation in the intact heart or complex isolated cardiac preparations, however, the pathway of conduction was not circumscribed, the spread of a contraction could not be followed visually, and electrical recordings had to be utilized. This has always provided less satisfying results, since electrical activity can be recorded only from a small percentage of sites on the tissues that may be involved in passing on the circulating wave. Mines pointed out that "ordinary graphic records either mechanical or electrical are of no value in attesting occurrence of a true circulating excitation since the records show a rhythmic series of waves and do not discriminate between a spontaneous series of beats and a wave of excitation which continues to circulate because it always finds excitable tissue ahead of it. The chief error to be guarded against is that of mistaking a series of automatic beats originating in one point in the ring and travelling around it in one direction only owing to a complete block close to the point of origin of the rhythm on one side of this point" (47). This problem is obvious in many of the subsequent studies: even though excitation progresses in the sequence expected of circulating excitation, it is often impossible to rule out spontaneous impulse initiation in the ring. This leads to the last and perhaps the most important criterion for demonstrating circulating excitation as described by Mines: 3) "The best test for circulating excitation is to cut through the ring at one point. If impulses continue to arise in the cut ring, circus movement as a cause can be ruled out" (47). This cannot always be done in the intact heart or in complex isolated preparations.

Circulating Excitation in Atria as a Cause of
Flutter, Fibrillation, and Tachycardia

Flutter. Thomas Lewis was one of the first to try to demonstrate that reentrant activation was a cause of the specific clinical arrhythmia atrial flutter. Lewis had earlier believed that "flutter may consist of simple paroxysmal tachycardia arising in a pacemaker" (39) and had attempted to disprove Mines' view of circus movement as being its cause: "We began fortunately, by attempting to follow the path taken by the excitation waves in flutter (in the dog heart); it was not until we had seen pure and long continued flutter that the full significance of experiments on ring preparations was recognized. We were driven slowly to the conclusion that pure flutter consists essentially of simple circus movement . . . in coming to that conclusion we were guided . . . by the dramatic experiments which Mines and others had conducted upon rings of muscle . . . We felt able to conclude that pure flutter is comparable to this ring experiment and that in flutter a wave circulates around a natural opening in the muscles of the auricle" (39).

The experiments that yielded this conclusion were done on the in situ canine heart in which atrial flutter was induced by rapid atrial stimulation (40). Direct bipolar leads were used to record sequentially from 4–11 sites on the right atrium. The times of activation at each site were related to the total atrial activation on a surface electrocardiogram, and the direction of the propagating wave front was determined by the shape of the electrogram. Lewis thus observed electrical activity rather than a wave of contraction. In some of his studies he demonstrated that the wave of excitation during flutter traveled from the inferior vena cava, up the taenia terminalis, around the superior vena cava, and across the interauricular band to the left auricular appendix (Fig. 3). Unfortunately, activation of the body of the left atrium could not be determined in these experiments because the heart was exposed from the right. When a little later a new wave appeared behind the inferior vena cava and repeated the course of the first, the question arose: "Is the new wave a continuation of the old, if so a circus is proved" (40). To prove that the new wave was a continuation of the old, Lewis measured that distance on the left side of the superior vena cava and right pulmonary vein through which a reentering impulse had to travel and multiplied that distance by the supposed conduction velocity of the impulse. He thereby predicted the time at which the circulating impulse should arrive at the inferior vena cava, and this proved identical to the observed time of activation of this site. Lewis therefore felt that a circus movement was proved beyond doubt. He concluded that the rapid stimulation that induced the flutter caused conduction in one direction around the ring of muscle that circumscribed the venae cavae, because fibers in one direction failed to respond to the stimuli while excitation occurred in the other direction. The long path length around the cavae, the slow conduction in partially refractory tissue, and the short refractory period of the atrial muscle all facilitated the continuation of the reentry (40).

The limitations of these experiments are obvious. The data fall far short of meeting Mines' criteria for circulating excitation. The course of the impulse could not be tracked with precision: Lewis could not follow the spread of the

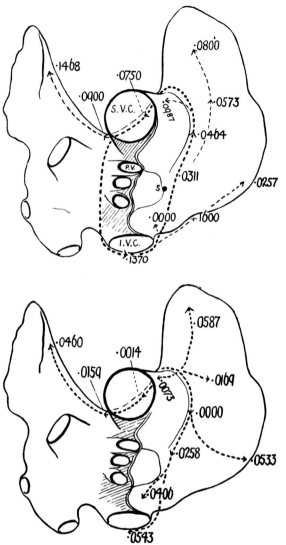

Fig. 3. Diagrams of atrial excitation from Lewis' study on canine atrial flutter. SVC, superior vena cava; IVC, inferior vena cava; PV, pulmonary veins; S, site at which stimulus that induced the flutter was applied. Numbers on each diagram indicate the time of activation of that site, in seconds, as determined with a local electrogram, relative to the beginning of a flutter wave on the surface ECG. For example, in the *top panel*, the site marked .0000 was activated simultaneously with the beginning of each flutter wave, while the site marked .0311 was activated 0.0311 s after the beginning of the flutter wave. *Top panel* shows readings obtained during a period of flutter. *Dashed line* and *arrows* represent the path Lewis believed the circulating excitation wave pursued (around the venae cavae), although the times for left atrial activation were not obtained. *Bottom panel* shows readings and path of the excitation wave during sinus rhythm. It is quite different from activation during flutter. [From Lewis (39).]

wave of excitation around the ring because of the limited number of points from which he could record. Nor was the ring severed to abolish the circulating wave. Lewis' studies were influential, but his results and hypotheses on the mechanism of atrial flutter, never completely accepted, have caused a great deal of controversy during the past 50 years.

Other investigators have extended Lewis' studies and have succeeded in demonstrating more convincingly that circus movement can occur around the venae cavae in experimental animals. In this regard, the experiments of Rosenblueth and his colleagues (57, 78) are classic. Wiener and Rosenblueth believed that the difficulty that Lewis had in initiating sustained flutter by rapid stimulation was due to the difficulty in initiating one-way propagation around the dual obstacle of the two cavae. They state "Even though a one-way impulse may meet one of the cavae end on and thus propagate unidirectionally around it, the presence of a bridge of conducting tissue between the two vessels will usually lead to two-way conduction around the second cava with cancellation of the two wave fronts. This argument led us to believe that a block of the conduction of impulses through intercaval auricular bridges would ensure the ready and regular appearance of experimental flutter" (57). In dogs in which this intervenous bridge was destroyed, continuous flutter was induced by a short period of rapid stimulation. Although the course of the excitation wave was not investigated in the original experiments of Rosenblueth and Garcia Ramos (as it was in later similar studies by others, as described below), the most important criteria of Mines for establishing circus movement was fulfilled; the circulating impulse could be stopped by interrupting the pathway. This was done by extending the lower limit of the obstacle to the auriculoventricular groove so that the obstacle was no longer surrounded by intact auricular tissue. As soon as the obstacle was no longer entirely surrounded by conducting tissue, the flutter disappeared and could no longer be reinitiated (57).

These studies were later extended by Kimura et al. (37) and by Hayden et al. (33), both of whom recorded bipolar electrograms from the left and right atria while inducing flutter in animals with an intercaval crush. The sequence at which electrical activity appeared at their electrodes in many experiments was consistent with an excitation wave moving up the left atrium and down the right; whereas in other studies the wave moved in the opposite direction. Extension of the intercaval crush to the right AV junction or AV groove stopped the flutter in their experiments also.

In all these studies on atrial flutter in animals four major criteria were used to show that a circulating excitation was occurring and that impulse initiation was not due to a rapidly discharging focus. First, flutter was started by rapid stimulation, it persisted after the stimulus was terminated, and the pathway of flutter was not related to the site of stimulation. This observation was most easily explained by Lewis' theory of differences in refractoriness in different regions of the atria causing one-way conduction (39). At the time these studies were performed most investigators did not believe that rapid ectopic activity from a focus could be initiated by such electrical stimulation. Recent studies, however, reviewed elsewhere (14), have shown that rapid impulse initiation can be induced by stimulation in fibers with delayed afterdepolarizations. Therefore, initiation of sustained activity by stimulation is not in itself a sufficient criterion to prove reentry. Second, reentry was suggested when the sequence of arrival of electrical activity at recording electrodes placed at various sites in the atria was consistent with the expected sequence in a reentrant pathway. This too is an inadequate criterion (as pointed out by Mines), since an impulse arising in a single focus might conduct over the atrium in one direction, activating it much as a circulating wave might. In addition, in all the studies the number of sites from which recordings were obtained was very small, and it cannot be certain that the impulse was propagated in a direct line between recording electrodes. Therefore, the real pathway of propagation may not always have been determined. Flutter could always be terminated by reapplying electrical stimulation as predicted by Mines' original observations. However, rhythmic activity sustained by afterdepolarizations can also be terminated by this maneuver (83). Finally, in many of these studies prevention of flutter by cutting the reentrant pathway is strongly suggestive of circulating excitation. When all the characteristics of this experimental arrhythmia are considered, the explanation that it is caused by circulating excitation is tenable. However, whether flutter occurring spontaneously in humans is caused by a similar mechanism is by no means solved.

Recently Boineau et al. (8) have had the opportunity to study *naturally* occurring flutter in a dog, which may be analagous to naturally occurring flutter in humans. In this study the pathway of impulse propagation was determined in great detail by using an electrode array on the right atrium that contained 72 bipolar electrode pairs plus an additional 24 bipolar electrodes from other regions not covered by the array. Twenty-four bipolar potentials were recorded simultaneously, and rapid switching during the flutter permitted successive sampling between each of four sets of 24 points. Patterns of excitation consistent with circus movement were seen on the right atrium; circus movement did not occur around the venae cavae but did occur in regions where local refractory periods were heterogeneous. Boineau and associates postulated that hypoplasia of the atria led to the slowing of conduction, which was necessary for circus movement (8).

Fibrillation. It has often been suggested that atrial fibrillation is caused by circulating excitations, but the spread of electrical activity over the atrium during fibrillation is so complex that it is virtually impossible to satisfy any of the criteria necessary to prove reentry.

Garrey (30, 31) was the first to suggest that atrial fibrillation was caused by circus movement. His arguments in favor of circus movement as the cause of atrial fibrillation rest on several different types of observations. First, simply from watching the exposed fibrillating atrium of an animal heart he could not see how fibrillation could be caused by anything but circus movements. Garrey says, "Riotous and chaotic as this fibrillation appears to be, analysis indicates that it may be aptly spoken of as a contractile maelstrom, for it appears that contractions are not independent of each other but that the contractile impulse travels in a ringlike circuit repeatedly returning to and involving a given region after completion of each circuit" (31). Second, Garrey could start fibrillation with a single stimulus, just as he could start circus movement in a ring. His third line of evidence favoring circus movements came from cutting fibrillating chambers of the heart into pieces of varying size and determining the relationship of tissue mass to presence or absence of fibrillation. These experiments were designed to show that fibrillation could not possibly be caused by single or multiple ectopic foci and must, therefore, be caused by circus movements. His answer to Rothberger's tachysystole theory that atrial fibrillation is caused by a single ectopic focus that fires at extraordinarily rapid rates was to cut the fibrillating tissue into four equal parts and show that "each continues to fibrillate although obviously only one of these pieces can contain the original, hypothetical tachysystolic pacemaker.... At least three of the pieces should have stopped fibrillating instantly." In response to Englemann's theory that fibrillation is caused by foci firing independently throughout the atria, Garrey cut many small pieces of fibrillating muscle from the atrium and observed that each severed piece ceased to fibrillate. He concluded that the individual fibers were not independently rhythmic (31). By these experiments Garrey showed that a minimal amount of tissue was necessary for fibrillation to be maintained. Therefore, he felt that circus movements had to be the mechanism. The circus pathway, however, had to be much more complex to account for the rapid chaotic atrial activation: "The impulse is diverted into different paths weaving and interweaving through the tissue mass, crossing and recrossing old paths again to course over them or to stop short as it impinges on some barrier of refractory tissue" (31). However, it is not possible to follow the course of individual excitation waves through the atrium to demonstrate that they are continuously coursing over circuitous pathways. Garrey introduced the idea that differences in refractoriness and excitability in different parts of the musculature might maintain circus movement in the absence of permanent block: "Natural rings are not essential for maintenance of circus contractions."

When Lewis (38) eventually adhered to the circus-movement theory of atrial fibrillation, he thought fibrillation was caused by a single circulating wave rather than the multiple waves invoked by Garrey. Lewis regarded the existence of a single circulating wave not as a hypothesis but as a fact established by his own studies using the same techniques that he had applied in his investigations on flutter. He relied heavily on the pattern of excitation at a limited number of direct leads on the fibrillating right atrium of canine hearts. Lewis describes each of his experiments in minute detail, and the results are too voluminous to summarize completely. However, he claimed that in most instances, by proper placement of the direct leads, he could demonstrate a "mother" wave circulating around the venae cavae as he had done in the flutter experiments, even when the surface electrocardiogram showed atrial fibrillation (38, 39). Often there were irregularities in direction and incidence of the electrogram deflections that were explained on the basis of occasional shifts in the central reentrant path. Recordings from direct leads outside the central pathway were even more irregular in direction and incidence and were similar in form to recordings obtained when the atrium was stimulated rapidly with electric shocks. Therefore Lewis thought the rapidly circulating central wave shoots off impulses at a rapid frequency, which activate regions lying outside it. The rapidity of activation of these outlying regions causes their irregular activation, because rapid activation causes impluses to arise in partially refractory tissue (38). Garrey was not convinced, however. He reminded his readers of his experiments in which the atria could be cut into two to four pieces of equal size, all of which continued to fibrillate. How could this be explained if there were only one circulating central ring? The same objections to the Lewis experimental approach that we pointed out for the flutter studies apply, and with even greater force, to the analysis of as complex a phenomenon as fibrillation. The limited number of direct leads do not permit detailed analysis of how the atrium is activated. Fibrillation was not terminated by disrupting the central circulating pathway. Even now, 55 years later, there have been no decisive improvements in the direct data derived from experiments on atrial fibrillation, and both the Garrey and Lewis hypotheses remain possible explanations.

Moe (48–50) has used computer simulation to analyze the manner in which circulating excitations could give rise to the sort of behavior seen by Garrey. According to his multiple-wavelet hypothesis, fibrillation may be initiated by a rapid succession of impulses arising by any mechanism (automaticity or reentry). The persistence of fibrillation, however, results from fractionation of these impulses when they are initiated in partially and irregularly excitable tissue. This fractionation occurs because of the marked inhomogeneity of refractoriness in atrial muscle fibers (1). When the atria are rapidly stimulated, the initial impulses conduct away from the site of stimulation at greater speed through fibers that recover early and more slowly through fibers that recover later. Each successive impulse spreads more irregularly than the preceding one, since a short cycle length accentuates the dispersion of refractoriness. The waves of excitation soon degenerate into "independent daughter wavelets," which result when an impulse caused by the initiating stimuli finds both refractory and excitable tissue in its path. The impulse then courses through the excitable tissue, thus bypassing the refractory tissue, and then returns to excite the initially refractory tissue at a time when it has recovered excitability. Numerous reentrant paths are thereby set up around islands of refractory tissue, but these islands continually shift location, causing the paths to shift as well (48, 50). Persistence of fibrillation depends on the average total number of wavelets that the tissue can support. Any factor that increases the number of wavelets perpetuates the arrhythmia, whereas reduction of the number favors the chance for recovery. The mass of the tissue, the refractory period, and the conduction velocity of its fibers all determine the number of wavelets. The hypothesis assumes that a small mass of tissue cannot support a number of wavelets sufficient to maintain fibrillation, although individual circus movements may still be supported. Unfortunately, as Moe concluded, "direct test of this hypothesis is difficult if not impossible in living tissue" (48).

Although experimental demonstration of circulating wavelets in fibrillating atria may not be possible, recent papers by Allessie and co-workers have directly demonstrated circulating excitations in isolated, superfused rabbit left atria (3–5). In the course of these studies many of the phenomena essential to the hypotheses on atrial fibrillation of Garrey and of Moe were shown. Repetitive activity was induced by a single, properly timed, premature stimulus applied during a regular driven rhythm. Extracellular electrical activity was recorded from at least 300 sites during the initiating impulse and during the tachycardia to determine the sequence of atrial activation. Although the basic technique is that introduced by Lewis and subsequently used by others, the extraordinary large number of points on the atrial tissue from which activity could be recorded permitted the course of the excitation wave to be followed with much greater certainty. Whereas the impulse spread radially away from the stimulating electrode during the basic drive, after the premature stimulus the impulse was propagated in a different way. The premature impulse that initiated repetitive activity blocked in one direction in fibers with long refractory periods and followed a course in fibers with shorter refractory periods along a circular pathway, returning to the initial point of block after excitability had recovered (Fig. 4). The impulse could then continue to circulate for many revolutions. No anatomical obstacle was needed for the impulse to circulate. The circular pathway could have a circumference as small as 6–8 mm (3–5). Impulses spread centripetally from the circulating vortex, toward the center, and the rate of rise and amplitude of the action potential gradually decreased as the center was approached. Fibers in the center of the circulating wave showed only local responses. These central fibers were activated twice during each revolution of the impulse, which prevented them from initiating action potentials and prevented the impulse from short-circuiting the circle (Fig. 4).

From their data Allessie and co-workers have developed the leading-circle model for reentry (5). In this model the length of the circuit is completely defined by the conduction velocity and refractory period of the fibers composing it, rather than by an anatomical obstacle as in Mines' original circus-movement studies. The interested reader should consult Allessie's original paper (5) for a detailed comparison of the two different mechanisms for reentry.

These phenomena, elegantly demonstrated by Allessie and his collaborators in isolated tissue, may explain many of the phenomena observed in the intact heart during atrial flutter and fibrillation. The small circumference around which reentry can take place by the leading-circle mechanism also clearly indicates that in mapping experiments, such as Lewis performed in the intact heart, a failure to find a grossly circuitous pathway for impulse conduction does not rule out reentry. If a 6- to 8-mm circle is present with radial propagation away from the circuit to the periphery, the map of the spread of activation would resemble that obtained if excitation spread away from an ectopic focus.

In addition to the heterogeneities in refractory periods causing the circular conduction of premature impulses that lead to reentry as in the experiments of Allessie et al. cited above (5), Spach et al. have shown data that suggest that the orientation of the cardiac fibers in relation to the direction of propagation of premature impulses may be an important determinant of sites of conduction block and slow conduction leading to reentry (68). Propagation of an early premature impulse can become decremental and stop in a direction parallel to the long axis of fiber length while at the same time it continues in the direction perpendicular to the long axis (68). The influence of tissue geometry must, therefore, be considered in all studies on the mechanism of reentrant excitation.

Circulating Excitation in AV Junction

AV Node. We now turn to the possibility that cardiac fibers connecting the atria and ventricles are a site for circulating excitation. That reentry can occur in this region was originally suggested by Mines' studies on reciprocating rhythm in two-chambered amphibian hearts, which we have described above (46). Mines' speculative discussion on its mechanism can stand today completely intact, having been substantiated by more modern experimental approaches. It has been shown that reciprocal excitation may occur in the AV node of the mammalian heart by a similar process.

The impetus to elucidate the mechanism for reentry in the mammalian AV

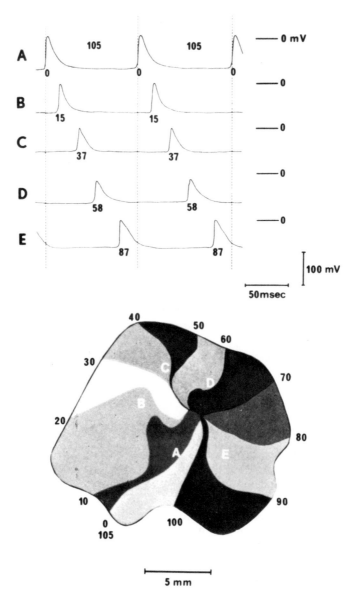

Fig. 4. Map of the spread of excitation in a piece of isolated left atrial muscle during a period of sustained tachycardia from the studies of Allessie, Bonke, and Schopman. Spread of activation was constructed from time measurements of the action potentials of 94 different fibers. The impulse is continuously rotating in a clockwise direction, as shown at the *bottom* (each number and different shading indicates the time, in milliseconds, during which a given region is activated). Activation proceeds from 0 to 105 ms and one complete revolution takes 105 ms. At the *top*, the transmembrane potentials of 5 fibers are shown that lie along the circular pathway (A–E). The numbers next to each action potential indicate the time of activation during the circulating excitation. [From Allessie et al. (5) with permission of the American Heart Association.]

node arose from the interest of clinical electrocardiographers in reciprocal rhythms (19, 77). The first laboratory experiments on reciprocal rhythm in mammalian hearts were those of Scherf and Shookhoff in 1926 (65). Quinine was given intravenously to delay conduction across the AV junction in the canine heart. The ventricles were driven at a regular rate, and it was demonstrated that the last of a series of stimulated ventricular beats that conducted to the atria was followed by a nonstimulated ventricular beat. That this nonstimulated beat arose because the last stimulated ventricular impulse that conducted to the atria return to the ventricle was deduced from analysis of the electrocardiogram. The nonstimulated ventricular depolarization had the same configuration as a ventricular depolarization of sinus origin, but it was not a sinus beat. The nonstimulated ventricular depolarization occurred only when the preceding, stimulated one conducted to the atrium and did not occur if the impulse had blocked in the AV junction; moreover, the nonstimulated ventricular depolarization occurred only when the previously stimulated impulse conducted across the AV junction with long delay (60, 65). To explain this "return extrasystole" or "Umkehr-Extrasystole" (a term that was introduced in their report), Scherf and Shookhoff drew upon Mines' ideas. It was assumed that if several stimulated ventricular extrasystoles followed each other in quick succession, some fibers of the AV conduction system with a shorter refractory phase than others conducted the impulse to the atrium, whereas fibers with a longer refractory period did not. Thus a "longitudinal functional dissociation" in the AV connection was created, which permitted return extrasystoles to occur (Fig. 5). The term "longitudinal functional dissociation" is widely used today to describe events leading to reentry in the AV junction; that it originated in their paper is often forgotten. Scherf and Shookhoff made one additional important observation in these studies: the P-R interval of the return extrasystole was inversely proportional to the length of the preceding R-P interval; i.e., as the preceding ventricular impulse took longer to conduct to the atrium, the return impulse conducted more rapidly from atrium to ventricle. They therefore concluded that at least part of the same conduction path was used by the impulse in both directions. When the impulse conducted more slowly to the atrium, this part of the pathway was allowed more time to recover excitability and therefore a more rapid return conduction occurred (65).

Similar studies were later reported by Rosenblueth (56) who called the nonstimulated ventricular discharges of atrial origin ventricular "echoes." Rosenblueth could not produce the same phenomenon while driving the atrium; in his experiments stimulated atrial impulses that conducted slowly to the ventricles were not followed by nonstimulated atrial activity (atrial echoes). He therefore disagreed with Scherf's functional longitudinal dissociation hypothesis and offered his own (Fig. 5). In this hypothesis the paths for VA and AV conduction were anatomically separate. VA conduction takes place only through the node, whereas AV conduction takes place through an auricular muscle bridge that makes a synaptic junction with fibers of the His bundle. This atrium-His junction he supposed to be "polarized," that is, to allow transmission from atrium to His bundle (A-H) but not from His bundle to atrium (H-A). If the propagation in the node is slow, the atrial impulse will be capable of stimulating His elements at the end of their functional refractory period and an echo will ensue (Fig. 5). In fact, later studies by Moe et al. (51) and by Wallace and Daggett (70) did demonstrate the existence of atrial echoes and showed that they had characteristics similar to ventricular echoes. Nonstimulated retrograde activation of the atria by an impulse seemingly propagated

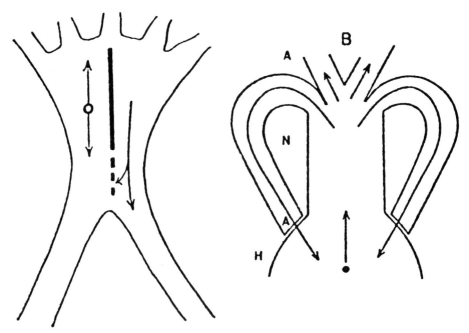

Fig. 5. Comparison of Scherf's and Rosenblueth's proposed mechanisms for return extrasystoles or ventricular echoes. Diagram at *left* is from Scherf's study and depicts the AV node. *Heavy line in center* indicates boundary of functional longitudinal dissociation. An impulse originating in the AV node at *small circle* is conducted up to atrium and down to ventricles in a portion of the node as shown by *arrows attached to circle*. The impulse then returns from the atrium somewhere in upper portion of the node and uses other side of the dissociated node for orthograde conduction (*arrows to right of boundary line*). Diagram at *right* is from Rosenblueth's study. B, auricle; N, AV node; A, auricular muscle bridge that makes synaptic connection with H, the His bundle. A spontaneous impulse that originates at *filled circle* in His bundle (H) conducts upward to atrium (B), through the node, as indicated by *arrow emanating from circle*. Instead of returning to ventricle through the node, the impulse conducts through the auricular muscle bridge (A) to AV ring, where it crosses a synaptic connection to directly reexcite His bundle. There is no functional longitudinal dissociation of the AV node. [*Left* from Scherf and Cohen (62); *right* from Rosenbluth (56).]

from the ventricle occurred after stimulated atrial activity that conducted to the ventricles with long delays. Hence Rosenblueth's concept of a polarized connection was not tenable.

More recent studies by Moe and his collaborators (44, 52) have done much to support the hypothesis that functional longitudinal dissociation in the AV conducting pathway caused echoes or return extrasystoles. Their results strongly suggest that functional longitudinal dissociation occurs within the AV node. Many of these studies were performed on donor-perfused canine hearts in which electrical activity could be recorded from the bundle of His and from the atrium at the atrial margin of the AV node. Premature stimuli applied to the His bundle initiated impulses (H_2) that conducted slowly to the atrium, and when a critical degree of delay occurred nonstimulated. His bundle responses (ventricular echoes) followed that arose in the atrium, as indicated by changes in polarity of the His bundle spike (Fig. 6). The placement of the recording

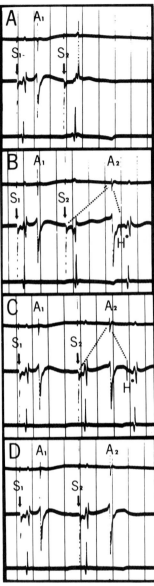

Fig. 6. *A–D*: records of ventricular echoes in isolated canine heart from the study by Mendez, Han, de Jalon, and Moe. *Panels A–D*: top trace is atrial electrogram; *middle trace* is His bundle electrogram; and *bottom trace* is ventricular electrogram. In *panel A*, a premature stimulus (S$_2$) is applied to His bundle 285 ms after regular drive stimulus (S$_1$). Premature impulse does not conduct to atria. Electrogram recorded from His bundle region shows a ventricular spike after S$_1$ followed by an atrial spike, and only a ventricular spike after S$_2$. Although the His bundle is activated by each stimulus, its electrogram is obscured by the stimulus artifact. In *panels B* and *C*, S$_2$ is applied 305 ms and 360 ms after S$_1$. S$_2$ conducts to atrium with a long delay and atrial activation (A$_2$) is followed by His-bundle and ventricular activation (H•, ventricular echo). In *D*, when S$_2$ occurs 364 ms after S$_1$, its conduction to the atrium (A$_2$) is not delayed enough to cause a ventricular echo. [From Mendez et al. (44) with permission of the American Heart Association.]

electrodes enabled localization to the AV node of the conduction delay, which seemingly was related to the echo. The slower the H-A conduction, the more rapid the A-H conduction, as was also noted by Scherf and Shookhoff (65). That there was functional longitudinal dissociation of the node during these echo responses was ingeniously demonstrated as follows: an H_1-H_2 interval was initially selected at which H_2 conducted very slowly to the atrium to elicit an A_2 response which was then followed by a nonstimulated activation of the His bundle (the echo). A stimulus was then applied to the atrium to excite it just prior to the arrival of the retrograde impulse from the His bundle (just prior to A_2). It was demonstrated that this stimulated atrial impulse conducted to the bundle, i.e., the His response occurred earlier than the expected echo response. The conclusion was that the atrial impulse must have taken a pathway not engaged by the impulse initiated in the His bundle, which at that time was conducting toward the atrium. Thus, "two responses can travel in opposite directions on a collision course, yet one of them can complete the journey" (44). For this to occur, impulses must be conducting in separate pathways within the AV node. Other studies by Mendez et al. (44) suggested that the atrium was the link between the functionally dissociated pathways through which the impulse conducted from the retrograde path to the antegrade path so it could return to its site of origin. In addition the site of coalescence of the antegrade and retrograde pathways on the ventricular side, called the final common pathway, was suggested to be in the AV node. The model for reentry in the AV node that was derived from these experiments is shown in Figure 7. This model explains 1) the occurrence of echoes only after premature impulses with short coupling intervals, because only these impulses will block in fibers with the longest refractory periods while conducting through other fibers with shorter refractory periods; 2) the necessity for long VA conduction times to permit recovery of the final common pathway; 3) the reciprocal relationship between retrograde and antegrade conduction during initiation of echoes; impulses that conduct rapidly in a retrograde direction return quickly to the final common pathway before the pathway has completely recovered excitability and therefore conduct through it slowly; impulses that conduct slowly retrograde return to the final common pathway after its complete recovery of excitability permits rapid conduction. The same model equally well explains atrial echoes.

It is apparent that studies on reentry in the mammalian AV node in situ do not lend themselves to the kind of experimental approaches that Mines considered necessary to demonstrate reentry beyond a reasonable doubt. Electrical activity cannot be recorded from the node in situ, so one cannot follow the course of impulse propagation within the node. Since the reentrant pathway within the node cannot be localized by direct techniques, it is not possible to sever it and demonstrate the disappearance of the putative reentrant impulses. Therefore, other interpretations of the results of these in situ studies are possible. Scherf eventually changed his position and suggested the possibility that there is "in the return extrasystole no return of one and the same impulse but a new impulse is created in the AV node by an after-potential which reaches threshold" (62). It has been demonstrated that nondriven activity in atrial fibers can be triggered by a premature impulse that is followed by an afterdepolarization, and such behavior is presumably possible in the AV node (80, 81). The precise relationship between the degree of AV conduction delay and occurrence

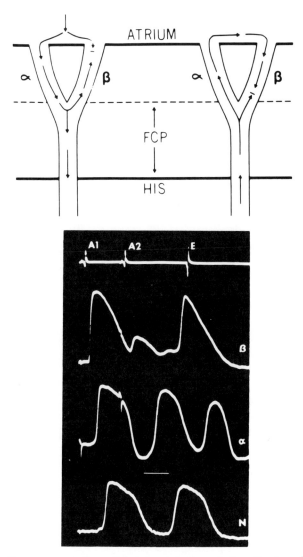

Fig. 7. Schematic representation of intranodal dissociation (*top*) and microelectrode recordings (*below*) from cells in the nodal reentrant pathway from the study by Mendez and Moe on isolated rabbit AV node. In *top panel*, diagram to *left* illustrates sequence of events during an atrial echo and action potentials *below* were recorded during that echo. In diagram a premature atrial response (A₂) fails to penetrate the β-path but propagates through the α-pathway. In microelectrode records note that A₂, the atrial electrogram, does not evoke a response in the β-cell but does evoke a response in the α-cell. The premature impulse propagates to final common pathway (FCP) where it activates the N cell and then returns to atrium via the now recovered β-path. Note late activation of the β-cell in microelectrode record. Atrium is then reexcited. Atrial echo is labeled E on atrial electrogram. In *top panel*, diagram to *right* shows pattern of propagation during a ventricular echo. The premature impulse conducts through the final common path to upper nodal region where it blocks in the β-cells but conducts to node through the α-cells. It then returns to activate the β-cells, final common pathway, and ventricle. Microelectrode recordings are not shown for this sequence. [From Mendez and Moe (45) with permission of the American Heart Association.]

of echoes, however, strongly favors the reentrant mechanism, as Mendez et al. (44) have argued. Even if it is conceded that echoes are caused by reentry, one cannot claim that the existence of functional longitudinal dissociation has been proved beyond doubt. The only plausible alternative to functional longitudinal dissociation requires the antegrade impulse to give rise to another impulse that conducts retrograde over the same pathway, i.e., to show "true reflection" (13). It is not known for certain whether this can occur in the AV node (44).

Similar studies on propagation of premature impulses through the AV node in human hearts have also been accomplished during the past 5 years. Not only has it been shown that ventricular echoes probably result from reentry in the AV node by functional longitudinal dissociation, but that many paroxysmal atrial tachycardias are caused by reciprocating impulses in the AV node (12, 55).

In the studies on the in situ heart the AV node must be treated as a "black box," but microelectrode techniques have permitted detection of some of the events occurring within the node during echoes in isolated, superfused preparations of rabbit atrium. Watanabe and Dreifus (73) were the first to demonstrate inhomogenous activation of nodal cells during reciprocal rhythms. Mendez and Moe (45) studied in detail propagation of premature atrial and ventricular impulses in the AV node during the induction of atrial and ventricular echoes. They found that cells in the upper region of the node show functional longitudinal dissociation during propagation of early premature impulses (Fig. 7). One group of cells, which they called alpha (α), fired in response to any impulse that successfully traversed the node, whether in a normograde or retrograde direction. Another group of cells, called beta (β), had a lower margin of safety; whereas early premature impulses activated the α-cells, only local responses occurred in the β-cells (Fig. 7). AV nodal cells in at least part of the midnode (N region) and in the lower node (NH region) did not show this longitudinal dissociation. All cells in these areas were activated by propagating premature impulses. When an atrial echo occurred after an early atrial premature impulse, the impulse conducted into the AV node to excite first the α-cells in the upper node (but not the β-cells), then the N cells, returning to excite the β-cells and finally the atrium as the echo. The impulse did not have to propagate to the His bundle for an atrial echo to occur, since the final common pathway appeared to be in the N and NH region of the node. Functional longitudinal dissociation of α- and β-cells was also shown during ventricular echoes (45). Therefore, the data from this study supported the validity of the model postulated by Scherf and Shookhoff (65) and later by Moe and his collaborators (44).

Janse et al. (34) have extended these studies by recording potentials sequentially from 54 cells during echoes and supraventricular tachycardia in the isolated rabbit atrium using a 10-microelectrode brush. During the tachycardia the sequence of activation of these cells was in agreement with a circus movement, although most impaled fibers appeared to be part of the returning pathway and potentials were recorded from only two cells in the antegrade pathways.

The major limitation of microelectrode studies on the AV node is still the paucity of cells from which action potentials can be recorded during reentry. The complete pathway through which the circulating impulse propagates has not been traced. Studies using extracellular recording cannot detect activity that propagates very slowly over a pathway that may be as small as a few fibers and have made little contribution to the solution of this problem. Also, in the isolated preparations it remains impossible to sever the reentrant pathway and terminate the echoes. Unless this later intervention can be accomplished, the possibility of a triggered impulse conducting in a circuitous pathway can never be completely eliminated.

Many of these major limitations have been overcome in the recent study of Allessie and Bonke on the sinus node (2), which has clearly shown reentry in this structure by a mechanism identical to the one postulated for the AV node by Mendez and Moe (45). Allessie and Bonke also used an isolated rabbit preparation and recorded electrical activity from 130 sites within the node (although not simultaneously) during the occurrence of repetitive responses induced by a stimulated premature atrial impulse (2). This stimulated impulse caused functional longitudinal dissociation of the node with block of the impulse in one region and slow conduction through another region. The impulse then returned to the region of initial block after the cells recovered excitability and continued on to reexcite the atrium.

Preexcitation. Although there are several different kinds of preexcitation syndromes (26), we will confine our discussion on reentry to the Wolff-Parkinson-White (WPW) syndrome (85). In hearts with WPW the sinus impulse conducts from atrium to ventricle both through the AV node and through an accessory muscular bridge. The experimental evidence, derived mostly from studies on the human heart, shows that both AV connections are used as part of the reentrant pathway (26, 28, 74) that is involved in the genesis of the supraventricular tachycardia associated with WPW. This substantiates early speculations of Mines, for when Kent demonstrated the multiple muscular connections between auricles and ventricles in human hearts (36), which we now know to cause the WPW syndrome, Mines wrote that this finding confirmed his postulate that tachycardias are caused by circulating excitation and said: "Suppose that for some reason an impulse from the auricle reached the main A-V bundle but failed to reach this right lateral connexion. It is possible then that the ventricle would excite the ventricular end of this right lateral connexion, not finding it refractory as normally it would at such a time. The wave spreading then to the auricle might be expected to circulate around the path indicated" (47). Recent clinical studies have shown that this is what actually happens, as described below.

A difference in refractoriness between the fibers in the accessory pathway and the normal AV conducting pathway is critical to the initiation of circus movement. As shown by a number of investigators, the effective refractory period of the AV node and accessory pathway can be determined by recording the electrocardiogram and a His-bundle electrogram, stimulating the atria at a regular rate, and studying the conduction of stimulated atrial impulses from atria to ventricles (26, 28, 54, 74, 75). The effective refractory period of the accessory pathway in the antegrade direction is usually longer than the effective refractory period of the normal AV conducting system. Often premature impulses occurring late in the cycle will conduct into the ventricles either mainly through the accessory path or through both normal and accessory AV connections, as indicated by the WPW form of the QRS complex on the ECG. The His bundle is also activated after the beginning of ventricular activation. Early premature impulses block in the accessory pathway while still conducting through the normal AV conducting system; when this occurs, the QRS of the premature impulse is normal and the His bundle is activated normally, i.e., 40–60 ms prior to the earliest evidence of ventricular activation. The long refractory period of the accessory pathway enables circulating excitation to be initiated by premature atrial or ventricular impulses, as first shown by Durrer et al. (21). This has been investigated chiefly in humans, both by studies using catheter electrodes to stimulate and record from the heart and by detailed epicardial mapping of the surgically exposed heart (10, 28, 29, 54, 72). The premature atrial impulse that initiates circus movement and atrial tachycardia often conducts to the ventricle only through the normal AV connection; it is blocked in the accessory pathway, which has a long refractory period. The ventricles are activated in a normal sequence. During the first impulse of the atrial tachycardia thus induced, retrograde activation results from the impulse returning via the bypass tract to the atria. This circular sequence of activation is then continuously repeated; the impulse travels from atrium to ventricle via the AV node, excites the ventricular end of the bypass connection, conducts through it, and then activates the atrium retrogradely, beginning from the atrial end of the bypass tract (Fig. 8). The mapping data that have demonstrated this fulfill one of Mines' criteria for demonstration of circulating excitation but does not provide definitive proof of reentry, since it is still possible that the impulse originating in the atrium conducts around the circuit back only to the AV groove and stops there. A new impulse might then be initiated in the atrium and follow the same pathway. Evidence that conduction across the AV groove through an accessory path leads to true circus movement is provided by the fact that tachycardia cannot be initiated after surgical sectioning of the accessory pathway in this region (28, 29).

Circulating Excitation in Ventricles

Although it had been suggested much earlier that reentry might cause ventricular premature depolarizations, tachycardias, and fibrillation, the first significant effort devoted to the study of reentry in ventricular tissue was not reported by Schmitt and Erlanger (67) until 1928. They depressed conduction in strips of turtle ventricle by passing them through rubber curtains in a multicompartmented tissue chamber so that different segments of the strip could be exposed to different depressant agents. In some experiments the muscle passed through five chambers; conduction was depressed by elevated KCl in the fourth and fifth chambers but was normal in the others. Conduction of the impulse was monitored by observing the timing of successive contractions in each chamber after one end of the strip was stimulated. When one end of the strip was stimulated under these conditions, a contraction wave was propagated slowly in one direction and a return wave often appeared at the distal end, causing a second nonstimulated contraction that was propagated in the opposite direction. To explain this observation, it was assumed that depression of a segment of muscle dissociated the muscle longitudinally, so that the impulse traveled over one pathway when conducting down the muscle bundle and then returned through another pathway. Unidirectional block in the antegrade direc-

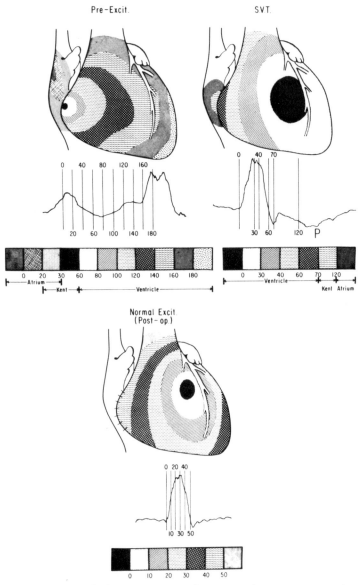

Fig. 8. Sequence of epicardial excitation in a human with type B WPW during preexcitation (*top left*), reentrant supraventricular tachycardia (SVT, *top right*) and after surgical section of the Kent bundle (*below*), from the study of Boineau and Moore. Isochronous maps of activation sequence were determined from local activation times of bipolar electrograms. The earliest recorded time was defined as *time zero*. Sequence key below each map indicates relative activation times, and these intervals are also indicated in the accompanying lead II ECGs. Note that during preexcitation at the *top left*, the earliest site of ventricular activation (30–60 ms) is at the AV groove where the accessory pathway is located. During SVT, ventricular activation is relatively normal, but atrial activation begins at the AV groove (70–120 ms) immediately after ventricles are activated in this region (60–70 ms). This is the retrograde limb of the reentrant path. After the accessory path is cut, ventricular activation during sinus rhythm is normal and tachycardia does not occur. [From Boineau and Moore (10) with permission of the American Heart Association.]

tion was also assumed to be present in the pathway through which the impulse returned (Fig. 9, *lower* diagram). Schmitt and Erlanger's analysis of conditions determining reentry in strips of muscle suggested to them a simple hypothetical explanation for ventricular extrasystoles of the coupled type in the mammalian heart that is based upon the arrangement of the ventricular conduction system. An impulse that enters a loop composed of peripheral Purkinje fiber bundles and ventricular muscle can conduct around it to reemerge as a reentrant excitation if there is slow conduction in the loop and a strategically located area of unidirectional conduction block. This mechanism is shown in detail in Fig. 9. A similar explanation for reentry secondary to unidirectional block in a peripheral twig of the conducting system had been advanced by Wenckebach and Winterberg (76) in 1927.

Although the concept of reentry in the peripheral Purkinje system periodically reappeared in the literature as a possible explanation of ventricular extrasystoles, it also met with strenuous objections. One such objection was that of Scherf and Schott (64), who were convinced that in the original experiments by Schmitt and Erlanger on turtle ventricle, the impulse conducting slowly in one direction somehow initiated a new impulse that conducted back in the opposite direction. Other objections were based on the fact that since the circumference of circuits involving peripheral Purkinje fiber twigs were probably no longer than 10–30 mm, it was hardly conceivable that the impulse could conduct through them slowly enough to outlast the refractory period of the ventricle and thereby permit reentry to occur. These objections were dispensed with by the studies of Cranefield and co-workers (15, 16), who produced localized depression in segments of canine Purkinje fiber bundles by encasing them in high-K^+ agar. This markedly reduced maximum diastolic potential and caused action potentials with very slow upstrokes to appear. Conduction velocity was reduced to 0.01–0.1 m/s, slow enough to permit reentry in small twigs of the conducting system. A continuation of these studies in the same laboratory eventually demonstrated reentry in small loops of the canine and bovine conducting systems (82, 84). Either the entire loops were depressed by high K^+ in combination with epinephrine, or only a segment of the loop was depressed by encasing it in high-K^+ agar with epinephrine. Action potentials were recorded from three to five strategically located sites around the loop. In some experiments a regularly driven impulse entering the loop activated the recording sites in a sequence that was compatible with conduction around the loop in one direction (82). The site at which the impulse entered the loop was then reactivated a second time as the impulse returned. In other experiments a single applied premature impulse conducted around the loop for up to eight revolutions (Fig. 10). In several of the experiments it was also demonstrated that cutting the loop abolished the repetitive responses, further strengthening the reentry hypothesis. Unidirectional conduction block was shown to be located at strategic sites in the loop after it was cut. In other studies on unbranched bundles of Purkinje fibers depressed in a similar manner, a phenomenon similar to that observed by Schmitt and Erlanger occurred (84). When a stimulated impulse conducted along the bundle in a direction in which very slow conduction occurred, a nonstimulated impulse appeared and conducted in the opposite direction. Wit and Cranefield called this phenomenon reflection (84). The

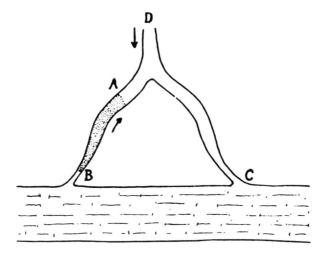

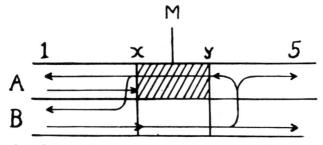

Fig. 9. Proposed mechanisms for reentry in the ventricular conducting system from the study of Schmitt and Erlanger. The *upper* diagram shows a Purkinje fiber bundle D that divides into 2 branches, B and C. These 2 branches are connected distally by ventricular muscle. They describe reentry in this loop as follows: "Under normal conditions the impulse from D reaches B and C at approximately the same time throwing the ventricular musculature at these points into contraction at almost the same instant." However, when the stippled segment A-B is an area of unidirectional conduction with conduction blocked in the normal direction, "an impulse coming from D would be blocked at A and would die out but by way of the other terminal branch it would reach and stimulate the ventricular musculature at C. The excitation from the ventricular fibers would then reenter the Purkinje system at B and traverse the region of injury but at so slow a rate that by the time it arrived at A, the uninjured fibers there would have recovered refractoriness and would again be excited. This excitation would immediately spread through the conducting system and reexcite the entire ventricular musculature." In the *lower* diagram A and B indicate 2 parallel muscle fibers with lateral connections. The shaded region (x-y) is an area of unidirectional conduction with conduction in the antegrade direction blocked in this fiber. The impulse propagating from left to right along A and B blocks in A at the *shaded area*, continues to propagate in B beyond this area, then crosses over to A via lateral connections and propagates slowly back through this region to reexcite the *left* part of the bundle (A and B). [From Schmitt and Erlanger (67).]

interpretations of these results and those of Schmitt and Erlanger were similar; the nonstimulated impulse was assumed to be caused by reentry due to longitudinal dissociation of the bundle, although there was no positive proof that this was the mechanism. Cranefield also suggested an alternative explanation

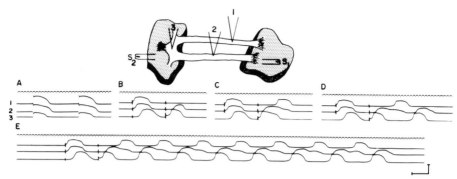

Fig. 10. Sustained circus movement of excitation in a loop comprised of Purkinje fiber bundles and ventricular muscle isolated from a calf heart, from the study of Wit, Cranefield, and Hoffman. Diagram at *top* shows preparation and location of the 3 microelectrodes and stimulating electrodes. Records in B–E show responses obtained after preparation was exposed to high K^+ and epinephrine. Initial set of responses in B–E arose when drive was applied at S_1; activation appeared more or less simultaneously at sites 1 and 2 but later at 3, suggesting that impulse spread simultaneously down both bundles. Application of a premature stimulus at S_2 in B caused excitation to travel from S_2 to site 3 to 2: in C, after the premature stimulus, excitation traveled from S_2 to site 3 to 2 to 1; in D, excitation evoked prematurely at S_2 traveled S_2 to site 3 to 2 to 1 to 3 to 2; in E, excitation after the premature stimulus traveled as in D and made several more full circuits in the counterclockwise direction. Calibration: vertical = 100 mV; time marks at 100-ms intervals. [From Wit, Cranefield, and Hoffman (82) with permission of the American Heart Association.]

for the phenomenon of reflection (13). He noted that "in many of our experiments we find that the rapid upstroke within the depressed segment arises after the rapid upstroke of the normal fiber excited by propagation into it from the depressed segment. This observation raises the possibility that the reflected impulse that travels slowly backward through the depressed segment is evoked by retrograde depolarization of the cells within the depressed segment by the rapid upstroke of the cells beyond (13)." Reflection of this type could occur in a single fiber and does not require longitudinal dissociation. Such a mechanism for reflection has been studied recently by Antzelevitch et al. (6), who showed that the electrotonic transmission of an impulse through an inexcitable segment of a Purkinje fiber bundle in a sucrose gap resulted in reflected responses when the delay in excitation of the region beyond the gap exceeded the refractory period of the fibers proximal to the gap. The region distal to the gap apparently excited the proximal region by an electrotonic interaction without circus movement occurring in the inexcitable depressed segment.

In relation to the previously mentioned studies of Cranefield et al. (15, 16), they also described the properties of the action potential that permitted reentry to occur. Reentry caused by slow conduction and unidirectional block was observed when resting membrane potential was reduced enough to block the fast inward current responsible for the normal upstroke. Conduction then resulted from a slow upstroke caused by current flowing through the slow channel; such action potentials are called slow responses. The studies leading to their characterization are described in detail in the monograph by Cranefield,

(13) and slow responses are the subject of a great deal of subsequent research (86).

Other studies on isolated preparations have suggested that reentry can occur via pathways comprised of specialized ventricular conducting fibers. Sasyniuk and Mendez (59) showed that nonstimulated impulses, probably caused by reentry, can occur because of local alterations in the refractory periods of Purkinje fibers in regions in which conduction is blocked. Friedman et al. (27) showed reentry to occur in the subendocardial Purkinje fibers that survive on the endocardial surface of anteroseptal infarcts. These fibers have long but unequal refractory periods. Early premature impulses are blocked in regions with the longest refractory periods but proceed slowly through regions with shorter refractory periods, eventually returning to their point of origin through the areas initially blocked.

Although the studies on isolated preparations proved that reentry can occur in Purkinje fibers, it is much more difficult to demonstrate reentry in the Purkinje system in the in situ heart. As yet no studies have shown reentry in peripheral Purkinje twigs mainly because electrical activity from such peripheral twigs cannot be readily recorded in situ from the number of sites that would be necessary to map the sequence of activation. There are so many of these twigs in the heart that finding the one from which extrasystoles are arising also presents a formidable problem. Experiments by Moe et al. (51) have provided the best evidence that reentry can occur in the ventricular specialized conducting tissue of the intact heart, but in those experiments reentry probably utilized both bundle branches rather than peripheral twigs as major conduction pathways. In this study of functional bundle branch block by Moe et al. (51) early premature impulses were induced in the His bundle. Such impulses blocked in the right bundle branch but often conducted through the left bundle. Since the right ventricle was not activated via the right bundle branch, the impulse from the left ventricle spread into the right ventricle and caused retrograde activation of the right bundle. Occasionally, the His bundle was also reexcited in a retrograde direction after right ventricular activation. In these experiments the spread of the impulse was not determined in detail, so the exact reentrant pathway is still open to question. However, nonstimulated responses of the His bundle after early premature stimuli did not occur if either of the bundle branches was sectioned, thereby strongly suggesting that they were part of a reentrant pathway.

Reentry may occur not only in the specialized conducting system but also in ventricular muscle. Such reentry requires either alteration in the refractory period, depression of conduction, or both. Wallace and Mignone (71) produced alterations in ventricular refractoriness in localized regions of the in situ canine heart by circulating cold water through a coil applied to the epicardium of the left ventricle. This prolonged the refractory period of the ventricular muscle immediately under the coil by several hundred milliseconds, prolonged the refractory period rather less in the midmyocardial well, and not at all in the endocardium. Under these conditions early premature impulses evoked by stimulating an uncooled portion of the ventricle activated the endocardium, normally blocked in the cooled epicardial region, and conducted slowly into the midportion of the wall. This was followed by reexcitation of the endocardium coincident with the onset of a nonstimulated ventricular premature depolarization. Nonstimulated responses did not occur if the stimulated premature impulse occurred so late in the cycle that it did not block in the epicardium. The nonstimulated extrasystoles were interpreted as resulting from reentry caused by a combination of conduction block in the epicardium and slowed conduction through midmyocardial cells. The induction of "automatic" activity was deemed unlikely by Wallace and Mignone because hypothermia suppresses spontaneous phase 4 depolarization. Although this is true, cooling might also cause early afterdepolarizations that can lead to extrasystoles. In fact, Scherf et al. (61) had earlier demonstrated that focal cooling caused extrasystoles and interpreted them as resulting from afterpotentials rather than from reentry. Therefore, although the occurrence of extrasystoles after local changes in temperature is suggestive of reentry, other possible mechanisms exist, and many of the criteria needed to prove reentry cannot be satisfied.

Alterations in conduction in ventricular myocardium that may lead to reentry do occur in regions rendered ischemic by coronary artery occlusion. Several different groups of investigators have demonstrated that after occlusion of a coronary artery, electrograms recorded from muscle in the ischemic region rapidly decrease in amplitude and increase in duration (9, 22, 66, 69). Eventually they become markedly fragmented, and at this time ventricular extrasystoles occur. A logical interpretation of these events is that the low-amplitude, fragmented activity is caused by slowing of conduction in the ischemic region, probably because muscle cells lose resting potential and generate action potentials with very slow upstrokes (18). Areas of unidirectional block and islands of inexcitable tissue may also occur. The impulse may therefore conduct slowly through the ischemic region while normal myocardium recovers excitability, reemerging at this time to reexcite the ventricles. This hypothesis was originally suggested by Ashman and Hull (7). The recording of fragmented, low-amplitude activity after ischemia, although consistent with the hypothesis, is but scanty evidence that reentry is occurring. Harris and Rojas (32) in 1943 argued that if reentry is occurring at this time after a coronary occlusion, continuous electrical activity should be recorded in the ischemic region from the time the impulse enters it until it reemerges to cause reexcitation of the ventricles. Although they could not demonstrate such continuous activity, Waldo and Kaiser (69) later showed that it was present on the epicardial surface of the canine heart within 10 min of coronary artery ligation.

The recent study of Janse et al., supports this earlier proposal that reentry is an important cause of the arrhythmias that occur soon after a coronary artery occlusion (35). They recorded simultaneously from 60 unipolar electrodes that were either on the epicardial surface or were intramurally in regions of the isolated, perfused porcine heart made ischemic by acute coronary occlusion. The pattern of activation of spontaneously occurring premature impulses was mapped and it was found that the initial beats of tachycardia may arise in Purkinje tissue in normal regions, close to the border zone with the ischemic area. Janse et al. suggested that the initial impulses may have been excited by the current of injury that flows across this border zone. They also found patterns of activation during subsequent beats of the tachycardia suggesting that reentry was occurring. Figure 11, which is from the study of Janse et al.,

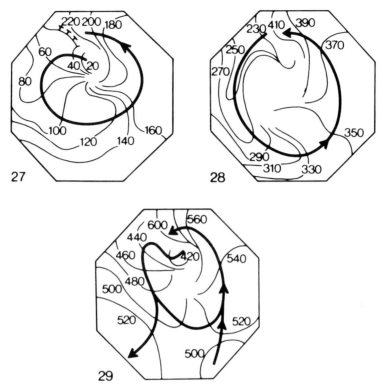

Fig. 11. Activation pattern of epicardial surface of left ventricle in porcine heart during ventricular tachycardia, 4 min after left anterior descending coronary artery occlusion from the study of Janse et al. Recordings from 60 electrodes within area outlined provided data used to construct isochrones. Activation sequence beginning at 20 ms and progressing to 220 ms in *top left* diagram suggests a circus movement as indicated by *arrow*. *Top right* diagram shows that activation continues in circular pattern around ischemic region and returns to its initial point (isochrones 230 to 410 ms). Diagram *below* shows next time period, from 420 to 600 ms; circular conduction continues. [From Janse et al. (35) with permission of the American Heart Association.]

shows circus movement on the epicardial surface of the ischemic area where activity circulated around a central area of block in the order of 1–2 cm. Reentry was not always demonstrated during arrhythmias. Sometimes it was not possible to tell whether there was continuous propagation of a wavefront during supposed circus movements, for example, when adjacent isochrones were separated by 50–70 ms.

El-Sherif et al. (23, 24) have shown in 3- to 7-day-old infarcts in the canine heart that continuous electrical activity on the epicardium is associated with ventricular extrasystoles that therefore may be caused by reentry. In these studies a long electrode with multiple recording poles along its shaft was applied over a large area on the epicardial surface of the infarct (Fig. 12). The electrode thereby sensed electrical activity over a wide area, and a composite electrogram of the whole region was recorded. When such an electrode records from nonischemic myocardium a single spike is registered because of the rapidity at which the normally conducted impulse activates the area under it. In the

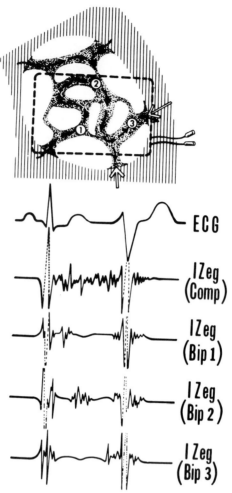

Fig. 12. Electrograms recorded from ischemic canine myocardium several days after coronary artery occlusion, from the study of El Sherif, Hope, Scherlag, and Lazzara. *Top* diagram shows schematic postulate of pathway for impulse propagation in ischemic region. This region is covered by their composite electrodes in the area enclosed by *dashed lines. Tracings* beneath diagram show ECG, a composite electrode record (IZeg-Comp), and 3 close bipolar recordings from localized regions in the ischemic zone (IZeg-Bip 1-3). Since the composite electrode covers entire reentrant pathway, it shows continuous electrical activity during diastolic interval between normal ventricular beat (*left*) and supposed reentrant ventricular beat (*right*). Only part of this electrical activity is recorded in each bipolar electrogram. [From El-Sherif et al. (24) with permission of the American Heart Association.]

ischemic zone, however, multiple asynchronous spikes were often recorded, presumably because of delayed conduction from one region under the electrode to another. The number of multiple asynchronous spikes could often be increased by increasing the heart rate, and when this was done activity was seen during the entire diastolic interval. The induction of this sort of continuous electrical activity was associated with the appearance of ventricular extrasystoles and tachycardia. It was presumed that "the composite electrode was able to depict the electrical activity of the entire reentrant pathway in the form of a continuous series of multiple asynchronous spikes," and the authors state that "the demonstration of continuous electrical activity that regularly and predictably bridged the entire diastolic interval between initiating and reentrant beats as well as between consecutive reentrant beats constitutes the necessary missing link long sought to document reentry." Although the presence of continuous electrical activity is highly suggestive of reentry, few of the criteria needed to prove its occurrence, as we have discussed above, were fulfilled in these studies. Indeed, continuous electrical activity during diastole might be caused by mechanisms other than just slow and depressed conduction associated with reentry. For example, if the basic impulse that enters the ischemic region excites ventricular muscle cells that then initiate an action potential characterized by a long series of early afterdepolarizations, continuous electrical activity might be seen in the extracellular records. Similarly, basic impulses conducting slowly into the center of an ischemic region might trigger fibers with delayed afterdepolarizations, and the triggered activity might conduct slowly out of the ischemic region to activate surrounding myocardium. Continuous activity might be seen in extracellular records. Janse's studies also showed continuous activity caused by slow conduction of the impulse in a random manner through the ischemic region without any evidence of reentrant excitation (35). Therefore the recording of continuous activity is not sufficient evidence to prove the occurrence of reentry.

El-Sherif et al. (25) and Wit et al. (79) have mapped excitation patterns in regions of infarction and have shown circus movement of excitation during those arrhythmias that can be induced in the canine ventricles by stimulated premature impulses 3–7 days after coronary artery occlusion. These arrhythmias may arise from the epicardial surface of the infarct. The unique structure of this region largely eliminates the transmural conduction patterns that so often confound the interpretation of results of studies utilizing epicardial ventricular mapping; on the epicardial surface of such infarcts a rim of surviving cells as little as 1–10 cells thick survives, with connections to deeper muscle only around the border (79). The muscle directly beneath this surviving rim is necrotic, therefore conduction over the infarct occurs mainly in a horizontal direction, since deep connections in the middle of the infarcted region are lacking. Figure 13 shows an example of circus movement in this region from the study of Wit and associates, in which electrograms were recorded simultaneously from 192 bipolar electrodes, located on a region of ischemic epicardium measuring 3.5 by 4.5 cm. It is obvious that during a stimulated premature impulse the excitation wave that blocks along one margin of the infarct does conduct slowly around the other margins and then returns to reexcite the region where conduction block occurred, causing a ventricular premature depolarization.

In none of these recent studies by Janse et al. (35), El-Sherif et al. (24) or Wit et al. (79) utilizing epicardial mapping was the proposed reentrant circuit cut with resultant abolition of the arrhythmia. Therefore all Mines' criteria for proving the occurrence of reentry were not fulfilled.

Conclusion

Reentry secondary to circus movement of excitation has been invoked to explain nearly every known arrhythmia of the heart including atrial and

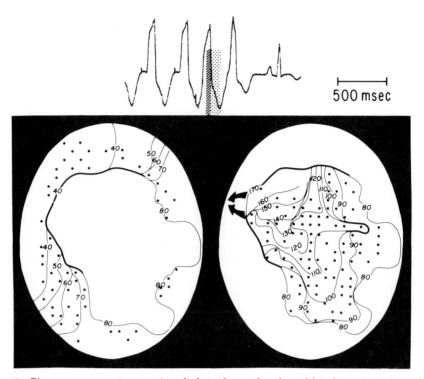

Fig. 13. Circus movement on epicardial surface of 3-day-old infarct in canine heart caused by stimulated premature ventricular impulse. At *top* the electrocardiogram is shown during basic drive of ventricles (first 2 QRS complexes from *left*) and stimulated premature ventricular impulse (*shaded area*), which is followed by nonstimulated ventricular response. Ovals represent area of epicardium on anterior surface of left ventricle covered by electrode array that was 3.5 × 4.5 cm and that contained 192 bipolar electrodes. Each oval shows activation of region during a different time period. At *left margin of oval* electrode array was left anterior descending coronary artery (LAD), *above* was base of heart, at *right* the margin on the lateral left ventricle, and *below* was the margin toward the apex. Each *dot* represents site at which an electrogram was recorded. *Left oval* shows activation during initial 80 ms after stimulated premature impulse and corresponds to *striped area* on the ECG above. Impulse entered recording field along margin toward the LAD at 30–40 ms and then blocked along 40- and 50-ms isochrones indicated by *heavy black line*. Activation proceeded around margins of the infarct, and central region was not activated during this initial 80 ms. *Oval at right* shows pattern of activation during next 90 ms, indicated by *stippled region* on ECG above. Activation occurred from apical margin and margin on the lateral left ventricle back toward the LAD. Margin of the infarcted region along the LAD was reexcited after 170 ms, indicated by *arrows*. [Adapted from Wit et al. (79).]

ventricular extrasystoles, tachycardia, flutter, and fibrillation. That such reentry can occur in simple rings of excised tissue has been established beyond any reasonable doubt, and its presence in the heart has been strongly suggested by a number of studies. Yet nearly every arrhythmia attributed to circus movement could as well result from other causes, among them triggered activity. Difficult as it is to doubt that circus movement causes arrhythmias, it is just as difficult to prove that it does.

References

1. ALESSI, R., M. NUSYNOWITZ, J. A. ABILDSKOV, AND G. K. MOE. Nonuniform distribution of vagal effects on the atrial refractory period. *Am. J. Physiol.* 194: 406–410, 1958.
2. ALLESSIE, M. A., AND F. I. M. BONKE. Direct demonstration of sinus node reentry in the rabbit heart. *Circ. Res.* 44: 557–569, 1979.
3. ALLESSIE, M. A., F. I. M. BONKE, AND F. SCHOPMAN. Circus movement in rabbit atrial muscle as a mechanism of tachycardia. *Circ. Res.* 33: 54–62, 1973.
4. ALLESSIE, M. A., F. I. M. BONKE, AND F. J. G. SCHOPMAN. Circus movement in rabbit atrial muscle as a mechanism of tachycardia. II. Role of nonuniform recovery of excitability in the occurrence of unidirectional block, as studied with multiple microelectrodes. *Circ. Res.* 39: 168–177, 1976.
5. ALLESSIE, M. A., F. I. M. BONKE, AND F. J. G. SCHOPMAN. Circus movement in rabbit atrial muscle as a mechanism of tachycardia. III. The "leading circle" concept: a new model of circus movement in cardiac tissue without the involvement of an anatomical obstacle. *Circ. Res.* 41: 9–18, 1977.
6. ANTZELEVICH, C., J. JALIFE, AND G. K. MOE. Characteristics of reflection as a mechanism of reentrant arrhythmias and its relationship to parasystole. *Circulation* 61: 182–191, 1980.
7. ASHMAN, R., AND H. HULL. *Essentials of Electrocardiography.* New York: Macmillan, 1947, p. 203.
8. BOINEAU, J. P., R. B. SCHUESSLER, C. R. MOONEY, C. B. MILLER, A. C. WYLDS, R. D. HUDSON, J. M. BORREMANS, AND C. W. BROCKUS. Naturally and evoked atrial flutter due to circus movement in dogs: role of abnormal atrial pathways, slow conduction, non-uniform refractory period distribution and premature beats. *Am. J. Cardiol.* 45: 1167–1181, 1980.
9. BOINEAU, J. P., AND J. L. COX. Slow ventricular activation in acute myocardial infarction. A source of reentrant premature ventricular contractions. *Circulation* 48: 703–713, 1973.
10. BOINEAU, J. P., AND E. N. MOORE. Evidence for propagation of activation across an accessory atrioventricular connection in types A and B preexcitation. *Circulation* 41: 375–397, 1970.
11. BURCHELL, H. B., R. L. FRYE, M. W. ANDERSON, AND D. C. McGOON. Atrioventricular and ventriculoatrial excitation in Wolff-Parkinson-White syndrome (type B): temporary ablation at surgery. *Circulation* 36: 663–672, 1967.
12. COUMEL, P. Mechanism of supraventricular tachycardia. In: *His Bundle Electrocardiography and Clinical Electrophysiology,* edited by O. S. Narula. Philadelphia: Davis, 1975.
13. CRANEFIELD, P. F. *Conduction of the Cardiac Impulse.* Mount Kisco, N. Y.: Futura, 1975.
14. CRANEFIELD, P. F. Action potentials, afterpotentials and arrhythmias. *Circ. Res.* 41: 415–423, 1977.
15. CRANEFIELD, P. F., AND B. F. HOFFMAN. Conduction of the cardiac impulse. II. Summation and inhibition. *Circ. Res.* 28: 220–233, 1971.
16. CRANEFIELD, P. F., H. O. KLEIN, AND B. F. HOFFMAN. Conduction of the cardiac impulse. I. Delay, block and one-way block in depressed Purkinje fibers. *Circ. Res.* 28: 199–219, 1971.
17. CRANEFIELD, P. F., A. L. WIT, AND B. F. HOFFMAN. Genesis of cardiac arrhythmias. *Circulation* 47: 190–204, 1973.
18. DOWNAR, E., M. J. JANSE, AND D. DURRER. The effect of acute coronary artery occlusion on subepicardial transmembrane potentials in the intact porcine heart. *Circulation* 56: 217–224, 1977.
19. DRURY, A. N. Paroxysmal tachycardia of A-V nodal origin exhibiting retrograde heart block and reciprocal rhythms. *Heart* 11: 405–415, 1924.
20. DURRER, D., AND J. P. ROOS. Epicardial excitation of the ventricles in a patient with Wolff-Parkinson-White syndrome. *Circulation* 35: 15–21, 1967.
21. DURRER, D., L. SCHOO, R. M. SCHUILENBURG, AND H. J. WELLENS. The role of premature beats in the initiation and termination of supraventricular tachycardia in the Wolff-Parkinson-White syndrome. *Circulation* 36: 644–662, 1967.
22. DURRER, D., R. T. VAN DAM, G. E. FREUD, AND M. J. JANSE. Reentry and ventricular arrhythmias in local ischemia and infarction of the intact dog heart. *Koninkl. Med. Akad. Wetenschap., Proc., Ser. Cx* 74: 321–330, 1971.
23. EL-SHERIF, N., R. R. HOPE, B. J. SCHERLAG, AND R. LAZZARA. Reentrant ventricular arrhythmias in the late myocardial infarction period: 2. Patterns of initiation and termination of reentry. *Circulation* 55: 702–718, 1977.
24. EL-SHERIF, N., R. R. HOPE, B. J. SCHERLAG, AND R. LAZZARA. Reentrant ventricular arrhythmias

146   A. L. Wit and P. F. Cranefield

in the late myocardial infarction period: 1. Conduction characteristics in the infarction zone. *Circulation* 55: 686–702, 1977.

25. EL-SHERIF, N., A. SMITH, AND K. EVANS. Canine ventricular arrhythmias in the late myocardial infarction period: 8. Epicardial mapping of reentrant circuits. *Circ. Res.* 49: 255–265, 1981.

26. FERRER, M. I. *Pre-excitation Including the Wolff-Parkinson-White and other Related Syndromes.* Mount Kisco, N. Y.: Futura, 1976.

27. FRIEDMAN, P. L., J. R. STEWART, AND A. L. WIT. Spontaneous and induced cardiac arrhythmias in subendocardial Purkinje fibers surviving extensive myocardial infarction in dogs. *Circ. Res.* 22: 612–626, 1973.

28. GALLAGHER, J. J., M. GILBERT, R. H. SEVENSON, W. C. SEALY, J. KASELL, AND A. G. WALLACE. Wolff-Parkinson-White syndrome: the problem, evaluation and surgical correction. *Circulation* 51: 767–785, 1975.

29. GALLAGHER, J. J., W. C. SEALY, A. G. WALLACE, AND J. KASSELL. Correlation between catheter electrophysiologic studies and findings on mapping of ventricular excitation in the WPW syndrome. In: *The Conduction System of the Heart,* edited by H. J. J. Wellens, K. I. Lie, and M. J. Janse. Leiden: Stenfert Kroese, 1976, p. 588–612.

30. GARREY, W. E. The nature of fibrillary contraction of the heart. Its relation to tissue mass and form. *Am. J. Physiol.* 33: 397–408, 1914.

31. GARREY, W. E. Auricular fibrillation. *Physiol. Rev.* 4: 215–250, 1924.

32. HARRIS, A. S., AND A. G. ROJAS. The initiation of ventricular fibrillation due to coronary occlusion. *Exp. Med. Surg.* 1: 105–111, 1943.

33. HAYDEN, W. G., E. J. HURLEY, AND D. A. RYTAND. The mechanism of canine atrial flutter. *Circ. Res.* 20: 496–505, 1967.

34. JANSE, M. J., F. J. L. VAN CAPELLE, G. E. FREUD, AND D. DURRER. Circus movement within the AV node. *Circ. Res.* 28: 403–414, 1971.

35. JANSE, M. J., F. J. L. VAN CAPELLE, H. MORSINK, A. G. KLÉBER, F. WILHS-SCHOPMAN, R. CARDINAL, C. NAUMANN D'ALNONCOURT, AND D. DURRER. Flow of "inury" current and patterns of excitation during early ventricular arrhythmias in acute regional myocardial ischemia in isolated porcine and canine hearts; evidence for two different arrhythmogenic mechanisms. *Circ. Res.* 47: 151–165, 1980.

36. KENT, A. F. S. A conducting path between the right auricle and the external wall of the right ventricle in the heart of the mammal. *J. Physiol. London* 48: 22, 1914.

37. KIMURA, E., K. KATO, S. MURAO, H. AJIOAKA, S. KOYAMA, AND Z. OMIYA. Experimental studies on the mechanism of the circular flutter. *Tôhoku J. Exp. Med.* 60: 197–207, 1954.

38. LEWIS, T. Observations upon flutter and fibrillation. Part IV. Impure flutter; theory of circus movement. *Heart* 7: 293–331, 1920.

39. LEWIS, T. *The Mechanism and Graphic Registration of the Heart Beat.* London: Shaw, 1925.

40. LEWIS, T., H. S. FEIL, AND W. D. STROUD. Observations upon flutter and fibrillation. Part II. The nature of auricular flutter. *Heart* 7: 191–233, 1920.

41. MAYER, A. G. Nerve conduction in *Cassiopea xamachana. Carnegie Inst. Wash., Papers, Dept. Marine Biol., 1917.* XI, p. 1–35.

42. MAYER, A. G. Rhythmical pulsation in Scyphomedusae. *Carnegie Inst. Wash. Publ. No. 47, 1906.*

43. MAYER, A. G. Rhythmical pulsation in Scyphomedusae: II. *Carnegie Inst. Wash., Papers, Tortugar Lab.* 1: 113–131. *Carnegie Inst. Wash. Publ. No. 102, part VII, 1908.*

44. MENDEZ, C., J. HAN, G. DE JALON, AND G. K. MOE. Some characteristics of ventricular echoes. *Circ. Res.* 16: 562–581, 1965.

45. MENDEZ, C., AND G. K. MOE. Demonstration of a dual A-V nodal conduction system in the isolated rabbit heart. *Circ. Res.* 29: 378–392, 1966.

46. MINES, G. R. On dynamic equilibrium in the heart. *J. Physiol. London* 46: 350–383, 1913.

47. MINES, G. R. On circulating excitations in heart muscles and their possible relation to tachycardia and fibrillation. *Trans. Roy Soc. Can. Ser. 3, Sect IV* 8: 43–52, 1914.

48. MOE, G. K. On the multiple wavelet hypothesis of atrial fibrillation. *Arch. Int. Pharmacodyn.* 140: 183–188, 1962.

49. MOE, G. K. Evidence for reentry as a mechanism for cardiac arrhythmias. *Rev. Physiol. Biochem. Pharmacol.* 72: 56–66, 1975.

50. MOE, G. K., AND J. A. ABILDSKOV. Atrial fibrillation as a self-sustaining arrhythmia independent of focal discharge. *Am. Heart. J.* 58: 59–70, 1959.

51. MOE, G. K., C. MENDEZ, AND J. HAN. Aberrant AV impulse propagation in the dog heart: a study of functional bundle branch block. *Circ. Res.* 16: 261–286, 1965.

52. Moe, G. K., J. B. Preston, and H. Burlington. Physiologic evidence for a dual AV transmission system. Circ. Res. 14: 357–375, 1957.
53. McWilliam, J. A. Fibrillar contraction of the heart. J. Physiol. London 8: 296–310, 1897.
54. Narula, O. S. Wolff-Parkinson-White syndrome: a review. Circulation 47: 872–887, 1973.
55. Rosen, K. M., P. Denes, D. Wu, and R. C. Dhingra. Electrophysiological diagnoses and manifestation of dual AV nodal pathways. In: The Conduction System of the Heart, edited by H. J. J. Wellens, K. I. Lie, and M. J. Janse. Leiden: Stenfert Kroese, 1976.
56. Rosenblueth, A. Ventricular "echoes." Am. J. Physiol. 195: 53–60, 1958.
57. Rosenblueth, A., and J. Garcia Ramos. Studies on flutter and fibrillation. II. The influence of artificial obstacles on experimental auricular flutter. Am. Heart. J. 33: 677–684, 1947.
58. Rytand, D. A. The circus movement (entrapped circuit wave) hypothesis and atrial flutter. Ann. Int. Med. 65: 125–159, 1966.
59. Sasyniuk, B. I., and C. Mendez. A mechanism for reentry in canine ventricular tissue. Circ. Res. 28: 3–15, 1973.
60. Scherf, D. An experimental study of reciprocating rhythm. Arch. Int. Med. 67: 372–382, 1941.
61. Scherf, D., S. Blumenfeld, M. Golbey, C. Ladopoulos, and F. Roth. Experimental studies on arrhythmias caused by focal cooling of the heart. Am. Heart J. 47: 218, 1955.
62. Scherf, D., and J. Cohen. The Atrioventricular Node and Selected Cardiac Arrhythmias. New York: Grune, 1964.
63. Scherf, D., A. I. Schaffer, and S. Blumenfeld. Mechanism of flutter and fibrillation. Arch. Int. Med. 91: 333–352, 1953.
64. Scherf, D., and A. Schott. Extrasystoles and Allied Arrhythmias. New York: Grune, 1953.
65. Scherf, D., and C. Shookhoff. Experimentelle Unterusuchungen ueber die "Umkehr-Extrasystole" (reciprocating beat). Wien Arch. Inn. Med. 12: 501–514, 1926.
66. Scherlag, B. J., R. H. Helfant, J. I. Haft, and A. N. Damato. Electrophysiology underlying ventricular arrhythmias due to coronary ligation. Am. J. Physiol. 219: 1665–1671, 1970.
67. Schmitt, F. O., and J. Erlanger. Directional differences in the conduction of the impulse through heart muscle and their possible relation to extrasystolic and fibrillary contractions. Am. J. Physiol. 87: 326–347, 1928–1929.
68. Spach, M. S., W. T. Muller, D. B. Geselowitz, R. C. Barr, J. M. Kootsey, and E. A. Johnson. The discontinuous nature of propagation in normal canine cardiac muscle: evidence for recurrent discontinuities of intracellular resistance that affect the membrane currents. Circ. Res. 48: 39–54, 1981.
69. Waldo, A. L., and G. A. Kaiser. Study of ventricular arrhythmias associated with acute myocardial infarction in the canine heart. Circulation 47: 1222–1228, 1973.
70. Wallace, A. G., and W. M. Daggett. Reexcitation of the atrium. "The echo phenomenon." Am. Heart J. 67: 661–666, 1964.
71. Wallace, A. G., and R. J. Mignone. Physiologic evidence concerning the reentry hypothesis for ectopic beats. Am. Heart J. 72: 60–70, 1966.
72. Wallace, A. G., W. C. Sealy, J. J. Gallagher, and J. Kassell. Ventricular excitation in the Wolff-Parkinson-White Syndrome. In: The Conduction System of the Heart, edited by H. J. J. Wellens, K. I. Lie, and M. J. Janse. Leiden: Stenfert Kroese, 1976, p. 613–632.
73. Watanabe, J., and L. S. Dreifus. Inhomogeneous conduction in the AV node. Am. Heart J. 70: 505–514, 1965.
74. Wellens, H. J. J. The electrophysiologic properties of the accessory pathway in the Wolff-Parkinson-White syndrome. In: The Conduction System of the Heart, edited by H. J. J. Wellens, K. I. Lie, and M. J. Janse. Leiden: Stenfert Kroese, 1976, p. 567–587.
75. Wellens, H. J. J. Contribution of cardiac pacing to our understanding of the Wolff-Parkinson-White syndrome. Br. Heart J. 37: 231, 1975.
76. Wenckebach, K. F., and H. Winterberg. Die Unregelmässige Herztätigkeit. Leipzig: Engelmann, 1927.
77. White, P. D. The bigeminal pulse in atrioventricular rhythm. Arch. Int. Med. 28: 313–318, 1921.
78. Wiener, N., and A. Rosenblueth. Mathematical formulation of the problem of conduction of impulses in a network of connected excitable elements, specifically in cardiac muscle. Arch. Inst. Cardiol. Mex. 16: 205–209, 1946.
79. Wit, A. L., M. A. Allessie, F. I. M. Bonke, W. Lammers, J. Smeets, and J. J. Fenoglio, Jr. Electrophysiological mapping to determine the mechanism of experimental ventricular tachycardia initiated by premature impulses: experimental approach and initial results demonstrating reentrant excitation. Am. J. Cardiol. In press.

80. WIT, A. L., AND P. F. CRANEFIELD. Triggered activity in cardiac muscle fibers of the simian mitral valve. *Circ. Res.* 38: 85–98, 1976.

81. WIT, A. L., AND P. F. CRANEFIELD. Triggered and automatic activity in the canine coronary sinus. *Circ. Res.* 41: 435–445, 1977.

82. WIT, A. L., P. F. CRANEFIELD, AND B. F. HOFFMAN. Slow conduction and reentry in the ventricular conducting system. II. Single and sustained circus movement in networks of canine and bovine Purkinje fibers. *Circ. Res.* 30: 11–22, 1972.

83. WIT, A. L., D. C. GADSBY, AND P. F. CRANEFIELD. Electrogenic sodium extrusion can stop triggered activity in the canine coronary sinus. *Circ. Res.* In press.

84. WIT, A. L., B. F. HOFFMAN, AND P. F. CRANEFIELD. Slow conduction and reentry in the ventricular conducting system. I. Return extrasystole in canine Purkinje fibers. *Circ. Res.* 30: 1–10, 1972.

85. WOLFF, L., J. PARKINSON, AND P. D. WHITE. Bundle branch block with short P-R interval in healthy young people prone to paroxysmal tachycardia. *Am. Heart. J.* 5: 685–704, 1930.

86. ZIPES, D. P., J. C. BAILEY, AND V. ELHARRAR (editors). *The Slow Inward Current and Cardiac Arrhythmias.* The Hague: Nijhoff, 1980.

CHAPTER 7

Cardiac Electrophysiological Alterations During Myocardial Ischemia

VICTOR ELHARRAR AND DOUGLAS P. ZIPES
Krannert Institute of Cardiology, Department of Medicine, and the
Department of Pharmacology, Indiana University School of Medicine,
and Veterans Administration Hospital, Indianapolis, Indiana

For many years electrophysiologists have produced regional myocardial ischemia in different animal models and have investigated the resultant cardiac electrophysiological changes and arrhythmias. Recently interest in this area has become more intense, and in the past several years the contributions of many have advanced our knowledge of cardiac electrophysiological alterations during ischemia. The purpose of this communication is 1) to review the effects of ischemia on the specialized ventricular conduction system, 2) to review the effects of ischemia on the ventricular myocardium, 3) to review the cellular ionic changes resulting from myocardial ischemia, 4) to consider concepts involved in determination of the ventricular fibrillation threshold, and 5) to apply these data to an understanding of the pathophysiology of the ventricular arrhythmias caused by ischemia. In the context of this communication, ischemia is defined as a regional deprivation of blood flow to the cardiac muscle whether or not such deprivation is severe enough to cause infarction.

Effects of Ischemia on Ventricular Conduction System

Subendocardial Purkinje Fibers. Bipolar extracellular endocardial recordings have been used to document in vivo after coronary artery ligation the survival of subendocardial Purkinje fibers within the ischemic region. Prior to coronary arterial occlusion, such endocardial recordings often show two components: a rapid initial component attributed to activation of subendocardial Purkinje fibers and a second high-amplitude component considered to be due to activation of myocardial cells in the vicinity of the electrodes. These two components are differentially affected by chronic and acute ischemia. In chronic infarcts the Purkinje component, having recovered from its early changes, remains relatively intact, whereas the deflection attributed to local muscular activation is lost (26, 36, 37, 69, 70) and is replaced by a small, slow deflection that lasts throughout the entire duration of the QRS complex. This latter activity is attributed to activation of distant, still viable myocardium and has also been recorded from sponges replacing a portion of the left ventricular wall (25). In acute myocardial ischemia the deflection associated with Purkinje activity usually, though not always (21), becomes diminished in amplitude and occasionally slightly delayed in onset (70, 112).

More recently important information on these changes has been derived from intracellular studies of infarcted, isolated, and superfused myocardium obtained from dogs after coronary artery occlusion. Observations were made on tissue excised at various times after coronary artery occlusion: 20 min or less (114), 30 min (70), 24 h (36, 69), and 10 days (70). Transmembrane recordings indicated that significant changes in intrinsic cellular activity did not occur until after approximately 20 min of arterial occlusion had elapsed. This finding is interesting in view of the fact that, during the first 20 min after coronary artery occlusion, the in situ canine heart develops severe rhythm disturbances (114). The lack of significant changes in intrinsic cellular activity may be attributed to the rapid reversal in vitro of the early electrophysiological alterations induced by an acute ischemic episode. After 30 min the maximum diastolic potential and the amplitude and upstroke velocity of Purkinje fiber action potentials recorded in the ischemic zone were reduced compared with those recorded in the normal zone. The duration of action potentials of Purkinje and muscle cells were shorter in the ischemic zone (70), and diastolic depolarization was minimal (Fig. 1A).

Friedman et al. (36, 37) have studied the isolated and superfused infarcted myocardium obtained from dogs in which a two-stage ligation of the left anterior descending coronary artery (LAD) was performed 24 h prior to the study. These dogs developed extensive infarction of the anterior left ventricular wall, left anterior papillary muscle, and interventricular septum, and they developed ventricular arrhythmias consisting of single and multiple ventricular extrasystoles and episodes of ventricular tachycardia. The time of study thus corresponds to the time at which late arrhythmias appear after coronary artery occlusion. They investigated the characteristics of the subendocardial fibers and determined the number of layers from which electrical activity could be obtained by advancing the microelectrode vertically through the subendocardium. These investigators found that action potentials could be recorded as deep as 15–20 fibers beneath the endocardial surface in noninfarcted myocardium, and activity characteristic of ventricular muscle was usually found after the microelectrode penetrated the first 4–5 layers. In infarcted myocardium, however, action potentials could be recorded from only the first one or two cell

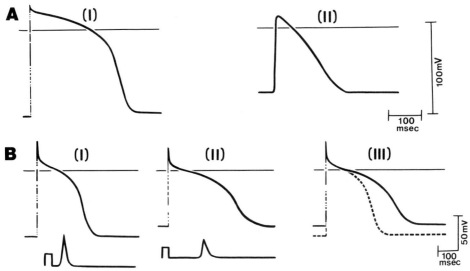

Fig. 1. *A*: intracellular action potentials from subendocardial Purkinje fibers located in normal zone (I) and ischemic zone (II) of a preparation excised 20 min after occlusion of left anterior descending coronary artery (LAD) in dog. Recording from ischemic zone was made 13 min after excision. *B*: action potentials recorded from Purkinje fibers in normal preparation (I) and from ischemic zone in infarcted preparation obtained 24 h after occlusion of LAD in dog (II); (III): superimposition of *traces* I and II. From *top* to *bottom* each *trace* represents zero potential, transmembrane potential, and upstroke velocity of phase 0 (cal. 200 V/s). [*A* from Lazzara et al. (70) with permission of the American Heart Association, Inc.; *B* from Friedman et al. (36) with permission of the American Heart Association, Inc.]

layers, and no ventricular muscle action potentials were found. Characteristics of the action potentials recorded from Purkinje fibers in the ischemic zone were profoundly altered. The maximum diastolic potential, action potential amplitude, and maximum rate of rise of phase 0 were reduced; phase 4 was increased; and the duration of the action potential was greatly prolonged (Fig. 1*B*). There was a wide variability in these action potential characteristics, but when the data were displayed in histogram form it was apparent that the cell population from the infarcted zone was different from that of the normal zone, although the histograms overlapped somewhat. A similar study by Lazzara et al. (69) also performed 24 h after coronary artery occlusion has confirmed these observations. In this latter study, however, more marked depression of the action potential characteristics and a more marked increase in the slope of phase 4 depolarization were noted in recordings from cells within the infarct zone, particularly early after the preparation was isolated and superfused. These alterations were different from those observed after 30 min of coronary artery occlusion, at which time diastolic depolarization was minimal and action potential duration was shorter. During both periods the amplitude of resting and action potentials was reduced.

Ten days after coronary artery occlusion, intracellular Purkinje fiber action potentials did not differ from those recorded in the normal zone except for occasional cells that showed slightly decreased resting potential and somewhat shorter action potential duration (70).

From these data it thus appears that a narrow inner shell of Purkinje tissue, one to three fibers thick, survives after myocardial infarction in the dog. Several factors may contribute to the preservation of subendocardial Purkinje fibers. They may derive their oxygen and nutrient needs by diffusion from the blood within the ventricular cavity and even by retrograde perfusion through ventricular sinusoidal channels. Conduction fibers, as opposed to contractile fibers, have a high glycogen content, suggesting a lesser dependence on oxidative metabolism. The oxygen requirement of Purkinje fibers is only one-fifth that of working contractile fibers (104), and this may further enhance the ability of Purkinje fibers to survive interruption of their arterial blood supply. The possibility that epicardium and endocardium exhibit intrinsically different electrophysiological responses to ischemia was examined in canine Purkinje fibers and in pieces of ventricular endocardium and epicardium (41). Superfusion of these tissues in vitro with altered Tyrode's solution (pH 6.8, $[K]_0$ 8 mM, and $Po_2 < 50$ mmHg) led to inexcitability of the epicardial pieces, whereas activity of endocardial muscle and Purkinje fibers was preserved, although depressed. Therefore the greater resistance of endocardial muscle and Purkinje fibers to combined hypoxia, acidosis, and hyperkalemia may contribute to the relative preservation of the electrical activity in the endocardium as compared to the epicardium in the early period after coronary artery occlusion. Despite these factors, substantial alterations in the electrical activity of Purkinje fibers were observed both in vivo and in vitro following 30 min of coronary occlusion (70). Alterations in cellular ionic environment may contribute to these changes but were not solely responsible for them, since recovery in vitro sometimes required several hours. Rapid recovery following superfusion would be expected if these changes were due only to alteration of the cellular ionic environment.

Although in vitro studies of the myocardium infarcted in situ have provided invaluable observations, this approach has some inherent limitations, since it eliminates the special conditions that produced the infarct zone. In vitro studies do not take into account possible influences of the autonomic nervous system and blood-borne factors such as metabolic byproducts produced by infarcted tissue. Also there may have been some features in cells in the ischemic zone that disappeared after the tissue was isolated and placed in a tissue bath. For example, Lazzara et al. (70) noted in 1-day-old myocardial infarction that superfusion, even with Tyrode's solution poor in oxygen and free of dextrose, led to recovery toward normal of Purkinje fibers' resting and action potential amplitude, upstroke velocity, and automaticity. This progressive recovery of the isolated preparations was also observed to a lesser degree by Friedman et al. (36).

Proximal Portions of Ventricular Conduction System. The alterations produced by myocardial infarction in the proximal portions of the ventricular conduction system have received little experimental attention, since the LAD is the most commonly occluded coronary artery. LAD ligation is usually performed distal to the first septal branch so that only the peripheral branches of the conduction system are affected. Recently, however, El-Sherif et al. (32) induced in dogs ischemic conduction disorders in the proximal His-Purkinje system by ligation of the anterior septal artery. These conduction disturbances, manifested after 20 min to 2 h of occlusion of the anterior septal artery, ranged from a simple prolongation of the conduction time to complete block of conduction and were localized in the His bundle, right and left bundle branches, singly or in combination. Using microelectrode techniques, Lazzara et al. (71) have further studied in vitro the cellular basis of these conduction disturbances. They found in the His bundle and the bundle branches partially depolarized cells that generated action potentials of diminished amplitude and upstroke velocity. An important finding was that the duration of these action potentials was not prolonged, but that refractoriness often was prolonged beyond the time for completion of repolarization. Postrepolarization refractoriness has been documented to be present in cells of the AV node (85) and may also be a feature of slow channel-dependent action potentials generated in Purkinje fibers (24) (see section **Cellular Environment and Metabolism in Myocardial Ischemia**). However, since the refractory period of the right bundle branch in this study was determined by extracellular stimulation of the left bundle branch, it is possible that the impulse blocked at a site proximal to the impaled cell and that the refractory period measured was that of that pathway traveled and not that of the cell impaled in the right bundle branch. These derangements in cellular electrophysiology resulted in slow, intermittent, and continuous conduction failure. Moreover it was found that these conduction defects were strongly rate dependent, probably because of the anomalous relationship between duration of refractoriness and heart rate; i.e., the refractory period prolonged rather than shortened at faster rates. It is of interest that these conduction disturbances are usually transient (32, 61, 111).

There is little similarity between electrophysiological changes found in the proximal portion of the His-Purkinje system and those reported to occur in the peripheral Purkinje system. The reasons remain unclear but may relate in part to the fact that the nonbranching portion of the His bundle is not a superficial structure in the dog. The relative sensitivity of the proximal portion of the His-Purkinje system as compared to the peripheral portion was initially suggested from experiments in which ischemia was produced by cross-clamping the aorta while recording electrical activity of the His bundle, the left bundle branch, and the peripheral Purkinje system (4). The peripheral Purkinje system was most resistant, since it continued to discharge throughout 40 min of ischemia at a time when all other electrical activity had ceased.

Effects of Ischemia on Ventricular Myocardium

TQ-ST Segment Deviations, Q Waves. Pardee (92) first suggested that ST segment changes in the surface ECG indicated the presence of myocardial ischemia. This observation was later confirmed by Wilson et al. (130), who recorded electrical activity of the epicardium over the ischemic zone after coronary artery occlusion. Unipolar epicardial recordings over the ischemia area consistently showed elevation of the ST segment and increase in the R wave, sometimes described as giant R waves (28, 96). It has been postulated that changes in the ST segment during myocardial ischemia resulted from a current of injury flowing across the boundary between ischemic and normal zones (94, 99) because of the potential gradient between these two zones and the fact that cells remain electrically connected by low resistance pathways or nexuses. An electrical potential difference between ischemic and nonischemic zones exists both during diastole and systole. It exists in diastole because cells within the ischemic area are depolarized relative to cells in the normal zone, and this leads to a shift in the TQ segment. During systole the gradient of potential is reversed and is due to the faster repolarization of cells within the ischemic zone. This results in shift of the ST segment opposite to that of the TQ segment. Conventional electrocardiography does not allow differentiation of TQ and ST segment changes. Studies by magnetocardiography (18) and direct-current electrocardiography (99) have indicated that deviation of both segments indeed occur during myocardial ischemia. The biophysical basis for TQ-ST segment changes are obviously more complex. Both size and shape of the ischemic tissue as well as the location of the electrode relative to that tissue are important factors influencing the direction and magnitude of ST changes recorded (27, 56, 57). Ekmecki et al. (28) found that the magnitude of the ST elevation decreases as the recording electrode is moved from the center to the periphery of the ischemic zone. During subendocardial ischemia, ST depression is recorded with an epicardial electrode over the ischemic area. This ST depression results from the reciprocal effect of subendocardial current of injury, which produces ST elevation at the endocardium. ST depression would be recorded by an endocardial electrode during subepicardial infarct, which results in ST elevation at the epicardium. Primary ST depression (i.e., not resulting from reciprocal changes due to ST elevation) may also be recorded during mild ischemia (28, 93). However, the electrophysiological basis for the primary ST depression remains unclear.

These changes in ST segment develop rapidly following coronary occlusion (Fig. 2), reach a maximum after approximately 10 min, and remain constant for as long as 2 h (80). Maroko et al. (80) developed a method to evaluate the degree of myocardial injury as well as to predict the later development of cellular damage using unipolar epicardial electrocardiograms recorded at various sites surrounding the occluded coronary artery. They postulated that the magnitude as well as the number of sites demonstrating ST elevation reflect the degree of myocardial ischemia. Depression of myocardial creatine phosphokinase activity measured at 24 h correlated well with the magnitude of ST segment elevation observed after 15 min of myocardial ischemia (53, 80). ST segment changes were also found to correlate reasonably well with alterations in regional blood flow and myocardial gas tensions. Simultaneous determination of ST segment changes and regional blood flow in the left ventricle using radioactive labeled microspheres showed that most but not all sites overlying low-flow zones demonstrated significant ST segment changes (9). ST segment changes also correlated well with intramyocardial oxygen tension (65, 100) and carbon dioxide tension (65). However, recent theoretical and experimental analysis of TQ-ST segment changes in the pig model (56) have presented evidence that ST segment elevation recorded in epicardial ECG may in fact be inversely related to the infarct size and that any alteration of the TQ-ST segment cannot be interpreted, with complete assurance, to result from an alteration in the extent of the ischemic injury. Thus the assumptions upon which the findings of ST segment mapping are related to the infarct size have been questioned and must undergo reevaluation (12, 35).

In chronic myocardial infarction, unipolar epicardial mapping consistently

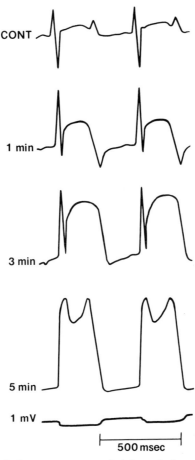

Fig. 2. Unipolar epicardial electrogram recorded from ischemic zone before (cont: control) and at various times after occlusion of left anterior descending coronary artery in dog.

demonstrates Q waves over areas above and slightly larger than the infarcted area. QS complexes are found only at sites where the infarction is transmural, whereas QR complexes are found at sites where the infarct is nontransmural. The appearance of pathologic Q waves is related to a significant replacement of myocardium by electrically silent fibrotic tissue (127).

Delayed Activation of Ischemic Myocardium. After approximately 2 or 3 min of coronary artery occlusion, various changes are noted in bipolar electrograms recorded epicardially from the ischemic zone and at its boundaries. Initially a transient increase in conduction velocity, manifested by earlier onset of the electrogram in the ischemic zone, is observed (38, 57). Following this, bipolar electrograms recorded from the ischemic epicardium usually exhibit loss of amplitude, increase in duration, and delayed activation in relation to the onset of the QRS complex or electrogram recorded from the nonischemic zone (Fig. 3). Other changes include the appearance of one or more spikes following the initial major complex. This event suggests that some of the fibers that were

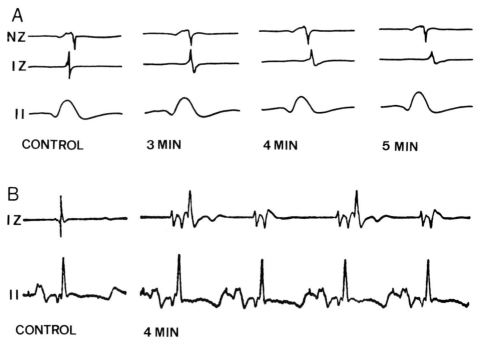

Fig. 3. *A:* progressively delayed activation of ischemic zone epicardium produced by occlusion of left anterior descending coronary artery (LAD) in dog while pacing atrium at cycle length of 500 ms. NZ and IZ are, respectively, epicardial bipolar electrograms recorded from a normal zone and from ischemic zone. *Bottom trace:* ECG (II). *B:* bipolar epicardial electrogram recorded from ischemic zone (*upper trace*) before and 4 min after occlusion of LAD in dog while pacing atrium at cycle length of 500 ms. Note alteration of electrogram configuration during ischemia indicating 2:1 conduction block into ischemic zone recording site. *Bottom trace:* ECG (II).

normally activated during the primary complex did not discharge initially but fired synchronously later, during the following spikes. A single delayed spike would then reflect synchronous activation of that group of fibers by a slowly propagated wave front. Fractionation of bipolar complex into multiple asynchronous spikes has been reported (11, 26, 122, 128) and interpreted to suggest marked desynchronization of activation within this region. The extensive degree of fractionation as well as the presence of irregular activity continuing well past electrical systole may suggest that localized fibrillation is actually occurring in the ischemic zone. If so, then a question arises concerning the mechanisms that keep such electrical activity confined, at least temporarily. We have reported that during myocardial ischemia, the prolongation of the conduction time between two electrodes may be greater in one direction than in another (29).

The importance of these observations lies in the fact that electrical activity from the ischemic region can become delayed to such a degree that it may outlast repolarization of neighboring sites and reactivate them. Examples representing an association between locally delayed excitation and the occurrence of ventricular extrasystoles in experimental myocardial infarction have been presented (11, 102).

Only recently have attempts been made to quantitate the extent of conduction delay recorded in the ischemic zone electrogram and to correlate it with the incidence of ventricular tachyarrhythmias that occurred at the time of coronary occlusion. Williams et al. (128) found that this conduction delay followed an exponential curve from the time of ligation of the LAD to the time of occurrence of ventricular arrhythmias and that the occurrence of maximum conduction delay invariably preceded the onset of ventricular tachycardia. All the determinants of the extent and time course of delayed activation of the ischemic zone have not yet been elucidated, but heart rate and the degree of ischemia appear important. Decreasing the heart rate or reducing the degree of the ischemia decreases the extent of conduction delay and retards or even prevents the onset of ventricular tachycardia (30, 102).

Studies of the effects of various antiarrhythmic drugs on the extent of conduction delay and incidence of ventricular arrhythmias have indicated that neither lidocaine, procainamide, nor propranolol, at doses within the therapeutic range, affected the time to onset of ventricular arrhythmias during acute ischemia. Hope et al. (58) noted that the extent and time course of conduction delay in the epicardial ischemic zone were not altered by procainamide or propranolol. The extent of conduction delay was, however, greater in the presence of lidocaine. On the other hand Kupersmith et al. (68) found that propranolol reduced the mean number of ventricular beats but further slowed conduction in the ischemic zone. Lidocaine, at therapeutic serum concentration, was also reported to prolong the ischemia-induced conduction delay (67). The discrepancy between these two studies regarding the effects of propranolol and lidocaine may stem from the different modes of drug administration. Hope et al. (58) administered the drugs prior to coronary occlusion, whereas Kupersmith et al. (67, 68) administered the drugs after coronary occlusion. The first mode of drug administration may be analogous to that of a patient already receiving an antiarrhythmic agent at the time of myocardial infarction, whereas the second mode of drug administration may be more relevant to clinical situations when a drug is administered to a patient after myocardial infarction. In either instance the concentration of the drug within the ischemic myocardium, as well as the effects of different concentrations on the ischemic versus the nonischemic myocardium, may have an important bearing on the observed electrophysiological and antiarrhythmic results and may differ depending on the mode of drug administration.

We recently studied the effects of aprindine, quinidine, verapamil, and isoproterenol on the extent of ischemia-induced conduction delay and the incidence of ventricular arrhythmias after coronary occlusion (37). Therapeutic doses of the drugs were given prior to coronary occlusion while maintaining a constant heart rate by atrial or ventricular pacing. Aprindine significantly increased the extent of conduction delay as well as the incidence of ventricular arrhythmias, whereas verapamil had the opposite effect (Fig. 4). Neither quinidine nor isoproterenol had a significant effect on either the extent of conduction delay or the incidence of ventricular arrhythmias. The lack of correlation between the drug effects on conduction delay and incidence of arrhythmias in previous studies may have been due to the fact that the drugs, at concentrations used, only minimally altered the extent of conduction delay and that other possible antiarrhythmic effects of the drugs could counterbalance the slight

CONTROL   ISCHEMIA

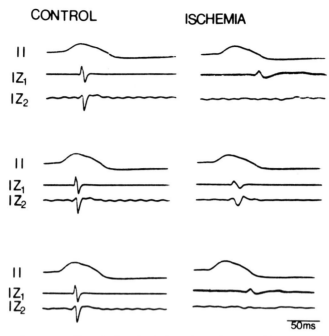

Fig. 4. Effects of verapamil on ischemia-induced conduction delay. IZ_1 and IZ_2 are two epicardial bipolar electrograms obtained from ischemic zone while pacing base of right ventricle at 2 Ht. *Left* and *right panels* were obtained prior to and 5 min after coronary occlusion; before (*top panels*), 11 min after (*middle panels*), and 47 min after (*bottom panels*) verapamil administration (0.2 mg/kg). Note that 11 min after verapamil administration extent of conduction delay at IZ_1 and IZ_2 was less compared to before and 47 min after verapamil administration. [From Elharrar et al. (31).]

increase in conduction delay that they produced. Although the extent of conduction delay is probably not the only determinant of ventricular arrhythmias and may not bear a primary relationship to the generation of ventricular premature beats, it is probably one of the significant factors during acute ischemia. Drugs that significantly depress conduction in the ischemic myocardium may predispose to the development of ventricular arrhythmia, whereas those that significantly improve conduction may be protective.

Excitability and Refractoriness of Ischemic Myocardium. Regardless of the method of stimulation used to measure changes in diastolic excitability, various studies have shown that tissue made ischemic gradually becomes inexcitable. In those areas that retained some degree of excitability, Brooks et al. (13) reported an eightfold increase in the diastolic threshold and also noted that in some instances an initial decrease in excitability threshold persisted for the first few minutes after coronary artery occlusion. Similar observations were also reported by Tsuchida (119). In each case, however, the time required for these changes to occur was considerably longer than the time to onset of the early arrhythmias following coronary artery occulsion. Recently we reinvestigated the changes in diastolic excitability following acute coronary ligation using a threshold tracking pacemaker that allowed precise measurement of the diastolic threshold approximately every 4 s (29). In both the ischemic epicardium and ischemic endocardium we found that an initial decrease in excitability threshold occurred between 1 and 3 min after LAD occlusion. This decrease was then followed by a rapid increase in the excitability threshold. It was not uncommon to observe a 10-fold increase in the diastolic excitability threshold after 5 min of coronary occlusion. We also found that a gradient of increasing excitability threshold existed from normal to ischemic tissue, passing through a heterogenous border zone that manifested some areas that had decreased excitability threshold and other areas that had increased excitability threshold (Fig. 5).

Changes in excitability probably influence other electrophysiological properties. For example, the increase in the conduction velocity of the ischemic myocardium in the first few minutes after coronary occlusion (38, 57) can be related to the concomitant decrease in excitability threshold that occurs at that time. Also the changes in ventricular fibrillation threshold determined from the epicardial ischemic zone appear influenced by the change in excitability threshold (see below). Finally changes in excitability may influence refractory period duration. It is well established that the durations of the absolute and relative refractory periods are shortened in the ischemic tissue (13, 71, 78, 119). Shortening by as much as 40–50 ms has been reported (13). In these studies the refractory period was determined using a stimulus with a voltage or current that was 1 to 1½ times the diastolic threshold measured after ischemia. However, using a stimulus intensity equal to twice the diastolic threshold measured before ischemia, we found that the refractory period duration lengthened after coronary artery occlusion (29). The physiological relevance of these observations on refractory period duration is not clear, since any method used to determine refractory period duration does not duplicate the natural event during which a wave front traveling in a three-dimensional syncytium activates an area.

How these alterations in refractory period duration and in excitability relate to the genesis of arrhythmias is also not clearly established. It is possible that the inhomogeneity in the magnitude of these changes among closely adjacent sites rather than the actual direction of these changes is instrumental in the genesis of arrhythmias, as hypothesized by Han (42, 43).

Resting and Action Potential of Ischemic Cells. Action potentials recorded from the ventricular epicardial surface using suction microelectrodes (13, 67, 68) or from within a single cell using floating microelectrodes (60, 62, 93) have indicated that the action potential duration shortened following coronary artery occlusion. Kardesh et al. (62) found the resting potential to decrease by 15 mV after 6 min of complete ischemia and also noted that electrical activity ceased when the resting potential had decreased to −54 mV. Prinzmetal et al. (93) observed that cells within a severely ischemic myocardium became depolarized, and this loss of resting membrane potential correlated with depression of the T-Q segment of the unipolar electrogram recorded over these cells. In contrast, mild ischemia caused an increase in the transmembrane resting potential that was manifested in the overlying surface electrogram by elevation of the T-Q segment. These differences were attributed to differing metabolic changes produced by mild and severe ischemia (93).

Using a pig model, Janse and Downar (60) studied the action potential changes that occurred after coronary artery occlusion. They used pig hearts because

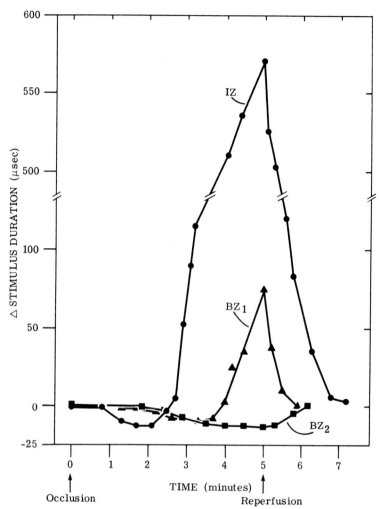

Fig. 5. Excitability changes, expressed as changes in threshold stimulus duration (Y axis) during successive 5-min occlusions of left anterior descending coronary artery (LAD) in dog while stimulating ischemic zone (IZ) and 2 sites of border zone (BZ$_1$, BZ$_2$). These curves were obtained by unipolar cathodal pacing of ventricle at 2 V and at a cycle length of 450 ms using an automatic threshold-following pacemaker (ATFP 1476H, Medtronic, Inc.) to determine diastolic excitability. When measured from IZ, an initial decrease in threshold stimulus duration is followed by an increase. Similar changes but with a slower time course are observed in BZ$_1$, whereas in BZ$_2$ only initial decrease in threshold stimulus duration is observed. BZ$_2$ was located 3 mm from BZ$_1$. No changes in threshold stimulus duration are observed when stimulating from normal zone (not shown). Note that threshold stimulus duration did not decrease below control values following reperfusion. [From Elharrar et al. (29) with permission of the American Heart Association, Inc.]

their coronary circulation resembles more closely that of man (101). These investigators demonstrated in vivo dramatic changes in action potentials recorded from the subepicardium within a time span corresponding to the genesis of early arrhythmias. Loss of resting potential and increase in activation time followed the initial diminution in action potential duration, amplitude, and upstroke velocity. Postrepolarization refractoriness developed and led to alternation of high- and low-amplitude action potentials and 2:1 responses until the cell became totally unresponsive. These alterations gave rise to a higher degree of inhomogeneity than expected from the so-called "dispersion of refractoriness." At a time when postrepolarization refractoriness developed and the cell could still give rise to 1:1 response, the refractory period duration actually prolonged to values exceeding control. This observation is interesting in view of the fact that previous in vivo studies have consistently demonstrated a decrease in refractory period duration early after coronary artery occlusion. The discrepancy between intracellular and extracellular findings in this respect may relate to the fact that a varying number of cells within the field of current generated by the stimulus are activated and the refractory period measured by detecting the extracellular response corresponds to those cells with the shortest refractory period duration within that field. During ischemia, cells with widely different refractory period durations coexist and those with short refractory periods determine the refractory period of the stimulated tissue. Also, since alteration in action potential duration and refractoriness may exist, the cycle during which stimulation occurs may influence the refractory period determination. Additional factors such as the decrease of tissue impedance and the use of increased stimulus strength to compensate for the loss of excitability intervene to increase the field of current flow.

Cellular Environment and Metabolism in Myocardial Ischemia

The ventricular myocardium deprived of its oxygen quickly undergoes metabolic and electrophysiological changes, the nature of which depends on whether hypoxia occurs alone or in combination with blood flow reduction (i.e., ischemia). Reduction in oxygen availability results in cessation of ATP production through β-oxidation of fatty acids and through the Krebs cycle and enhances glycogenolysis and glycolysis. The main sources of ATP in these conditions are from anaerobic glycolysis and from conversion of creatine phosphate to ATP. In the early stage of hypoxia the ATP content falls only slightly due to the large capacity of the creatine phosphate pool. Because pyruvic acid, the end product of glycolysis, cannot enter the Krebs cycle it is converted into lactic acid and accumulates both within the cell and in the extracellular space to a varying degree, depending on blood flow. Accumulation of lactic acid results in intracellular and extracellular acidosis and also in formation of CO_2 by displacement of the bicarbonate equilibrium (17, 89, 91, 133).

Ionic shifts, including loss of K^+ and Mg^{2+}, inorganic phosphates, and increase in intracellular Na^+ and Ca^{2+}, occur soon after coronary artery occlusion (16). Net loss of K^+ and accumulation of Na^+ by the ischemic cell has been ascribed to the lack of energy for ion pumping. The mechanisms relating oxygen deprivation to alterations in intracellular and extracellular Na^+ and K^+ are not fully understood, and information is lacking as to precisely when the K^+ loss begins in relation to the hypoxic or ischemic episode, how it relates to the reduced ATP production, whether the ionic shift is significant by 5–10 min following coronary occlusion, and whether these changes can explain the early arrhythmias that follow coronary occlusion.

The electrophysiological alterations observed during ischemia can be considered to result from the effects of multiple factors among which are hypoxia, intracellular and extracellular acidosis, increased extracellular K^+, lactate, and many other metabolic byproducts that may accumulate in the extracellular space because of the reduced blood flow. Because ischemia involves a reduction in blood flow and hypoxia does not, the resultant metabolic and electrophysiological effects of the two conditions are quite different.

Hypoxia, whether generalized or regional (by perfusion of a coronary artery with venous blood), fails to produce arrhythmias in vivo (3, 24, 47, 59). In vitro hypoxia decreases the duration and amplitude of the action potential plateau of ventricular muscle, effects that are greater at more rapid rates (19, 117). Moreover, hypoxia has little effect on resting membrane potential until after several hours of oxygen deficit. McDonald and MacLeod (82) suggested that the resting potential of anoxic muscle is controlled by at least two components, one dependent on K^+ distribution and the other on the activity of an electrogenic pump. It appears that the rate of ATP production by anoxic muscle is sufficient to maintain the activity of the electrogenic pump and that its activity might in fact be stimulated by the increased intracellular sodium concentration, thus preventing reduction in resting membrane potential despite continuous loss of K^+ and gain of Na^+. There is evidence to suggest that these changes in electrical activity occur under conditions in which the glycolytic activity and therefore energy production are suboptimal. Alterations produced by prolonged anoxic incubation in a medium containing 5 mM glucose can be reversed by raising the glucose concentration to 50 mM, a maneuver that increases the rate of ATP production and, thus, the contribution of the electrogenic pump to the resting potential (82).

Decrease in action potential duration and plateau amplitude has been related to the loss of slow inward Ca^{2+} current (19). Decrease in the slow inward current, by itself or in combination with an increased repolarizing outward K^+ current, could hasten cellular repolarization. Hypoxia alone, with its accompanying lactate overproduction, intracellular acidosis, and ion shift, fails to reproduce effects similar to those produced by ischemia.

Besides reduction in oxygen availability, reduced blood flow results in a reduction of nutrient availability and also accumulation of metabolic byproducts into the extracellular space. Such byproducts may react on the hypoxic cell to further alter its electrophysiological and metabolic characteristics. Among several factors, elevation of extracellular K^+ concentration, acidosis, and lactate accumulation have received particular attention.

The ischemic muscle loses a large amount of K^+ rapidly, and various investigators (33, 48, 75) have suggested that release and accumulation of K^+ in the extracellular space may be arrhythmogenic. The pattern and time course of changes in K^+ concentration in blood samples obtained from a vein draining the ischemic area suggest a causal relationship (48). Also, regional infusion of KCl into a coronary artery produces biphasic changes in excitability (29) as well as regional conduction disturbances, ectopic beats, and ventricular fibrillation similar to those changes resulting from ischemia (33, 75, 76). Hill and Gettes (52) used flexible K^+-sensitive electrodes to monitor the extracellular K^+ activity within the ischemic myocardium of the domestic pig. Approximately 15 s after coronary artery occlusion the K^+ activity in the center of the ischemic area rose from 3.5 mM to 8 mM (corresponding to a concentration of 11 mM). They also reported a greater rate of increase of the K^+ activity in the subendocardium than in the subepicardium, which leads to transmural inhomogeneity. Downar et al. (27) recently reported values as high as 8–9 meq/liter in venous blood draining an ischemic area, in agreement with those levels reported by Harris (48). Yet, superfusing isolated myocardial strands with normal venous blood containing 8 meq/liter of K^+ fails to produce changes similar to those observed by superfusing with venous blood draining the ischemic area. A K^+ concentration of 12–16 meq/liter would be necessary to induce these changes (27).

The effects of intracellular acidosis per se are difficult to evaluate because it is not feasible to control experimentally the intracellular pH. Various investigators (10, 14, 19, 50) have examined the effect of acidic media on the electrophysiological properties of ventricular and Purkinje fibers. The extrapolation of the findings of these studies to the ischemic cell has been questioned (129), since ischemic cells have an increased amount of acidic metabolites, so that intracellular pH may fall more than extracellular pH. In their study on the effect of extracellular acidosis (pH 6.6) on canine Purkinje fibers, Coraboeuf et al. (19) observed repolarization abnormalities in some fibers. These abnormalities consisted of partial depolarization that appeared at different levels in the descending phase of the action potential. More commonly extracellular acidosis prolonged the action potential duration, an effect opposite to that produced by acute ischemia or hypoxia. The relative contribution of hypoxia, acidosis, hyperkalemia, and elevated lactate to the genesis of arrhythmias is not established. The combined effects of these factors have recently been studied in isolated, perfused pig hearts (88). Perfusion of the LAD with hypoxic, glucose-free, high-K^+ (10 mM), and acidic (pH 6.8) Tyrode's solution led to action potentials similar in configuration to those seen in ischemia. The combined effects of these factors have also been studied in vitro using venous blood draining the ischemic area and coronary sinus blood altered to simulate ischemic blood by increasing its K^+ concentration and lowering its pH and Po_2. Superfusing myocardial strands with such coronary sinus blood failed to reproduce in vitro electrophysiological alterations similar to those produced by superfusing with blood draining the ischemic area (28). The conclusions from these two studies are contradictory. In the first study (88), however, Po_2 was lower, and the observations pertain to the epicardial aspect of the heart. Therefore additional factor(s) beside hyperkalemia, hypoxia, and acidosis may be present in blood draining the ischemic area and may be responsible for electrophysiological abnormalities.

Recently fatty acid and prostaglandin metabolites have been implicated in the genesis of ischemic arrhythmias. Phospholipids, in particular phosphatidyl choline and phosphatidyl ethanolamine, are major constituents of the cellular membrane. By deacylation they give rise to lysophosphatidyl choline and to lysophosphatidyl ethanolamine. It appears that their reacylation is inhibited in the ischemic myocardium. Because of this inhibition and the reduced blood flow, these compounds accumulate in the ischemic myocardium. Corr et al. (21)

studied the effects of these lysophospholipids on canine Purkinje fibers, using concentrations similar to those found in extracts of ischemic myocardium. They noted profound alterations of the cellular electrophysiology closely resembling the derangements seen in the ischemic tissue. These authors suggest that these phospholipids may be of major importance as biochemical mediators of ischemic arrhythmias.

Relatively little is known about the control of ionic permeabilities in ischemic conditions. The genesis of the action potential has been ascribed to gating reactions of various membrane channels. These channels differ with respect to their ionic specificity, their time constant of activation and inactivation, and their voltage dependency (81, 116, 123). Na^+ inward current through the fast channel provides the rapid depolarization, which in turn is largely responsible for the spread of excitation throughout the myocardial syncytium. Inward current of Ca^{2+} (and Na^+) ions through the slow channel is responsible for maintaining the long plateau characteristic of cardiac action potentials and plays an important role in excitation-contraction coupling. Repolarization is due to inactivation of the slow channel as well as to the development of an outward current, mostly carried by K^+ ions through channels that show delayed rectification (116, 123).

During myocardial ischemia, alterations in the properties of these channels in addition to changes in ionic concentration across the membranes may result initially in shortening of the action potential duration and decreased plateau amplitude. A decreased inward Ca^{2+} current and/or an increased outward current is assumed to explain these findings (19). Whatever the mechanisms, a decrease of the plateau duration limits the duration of Ca^{2+} influx, which may be related to a decrease in the mechanical tension developed. The fast inward Na^+ current may also be altered for various reasons, including cellular depolarization, accumulation in the cell of Na^+ ions that alter the Na equilibrium potential, and also from alteration in properties of fast channels.

The slow response may be revealed when fast channels are blocked or inactivated and slow channels are still operative. For example, increase in K^+ to 22 meq/liter and simultaneous addition of isoproterenol is a common way to induce in vitro slow channel-dependent action potentials (Fig. 6). Other methods include superfusion of fibers with Na-free, elevated-Ca^{2+} solutions (1). Because of its milieu of elevated K^+ that has been released from ischemic cells and the presence of catecholamines released endogenously, ischemia seems conducive to the genesis of slow responses in ventricular and Purkinje fibers. This type of cellular response has been extensively studied recently (24, 132) and is characterized by a low resting potential, small overshoot, and an upstroke velocity of less than 20 V/s. Other characteristics of the slow response include a low safety factor for conduction, a relatively high threshold of excitability, and postrepolarization refractoriness (24, 135). Automaticity is not a feature of the slow response induced in K^+-depolarized, isoproterenol-treated fibers. When induced in Na-free, Ca^{2+}-rich solutions, pacemaker properties are usually present and may be suppressed by elevating the K^+ concentration (127). Thus, even if a slow response could develop in the ischemic tissue, it may not show automatic properties in the presence of high K^+ concentration. Moreover, other factors such as hypoxia, acidosis, and poor energy reserves tend to inhibit the slow

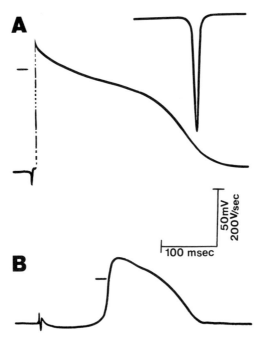

Fig. 6. *A*: action potential recorded from canine Purkinje fiber superfused with Tyrode's solution containing 4 mM KCl. *B*: "slow response" recorded from the same fiber superfused with Tyrode's solution containing 22 mM KCl and 10^{-6} M isoproterenol. *Upper trace*: dv/dt of upstroke. This parameter was less than 15 V/s in case of slow response and could not be recorded. *Lower trace*: transmembrane potential. *Horizontal bar* indicates zero potential.

response. It was recently suggested that the slow channel may shut off soon after ischemias, thereby reducing energy expenditure by the cell (110). Thus, although some conditions during ischemia appear favorable to the genesis and maintenance of the slow response, other factors may tend to inhibit its development.

Ventricular Fibrillation Threshold in Myocardial Ischemia

Quantitative studies on the susceptibility of the ventricle to fibrillate were initiated by Wiggers and Wegria (126), who demonstrated the capability of a single electrical impulse to induce ventricular fibrillation when delivered during early diastole. Since these early studies, two methods have been widely used to evaluate the propensity of the ventricle to fibrillate. In the original method introduced by Wiggers and Wegria (126), a single electrical pulse is delivered to the ventricle during the ascending limb of the T wave to induce ventricular fibrillation. Both the position of the pulse in the cardiac cycle and its intensity had to be controlled and were found to be of critical importance. The need for repeated scans with a single pulse of increasing intensities was eliminated in the method introduced by Han et al. (43), during which gated pulses at a given frequency (usually 100 Hz) were delivered during the T wave (Fig. 7). In both

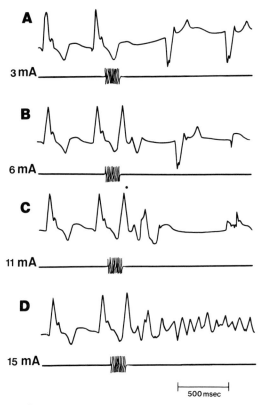

Fig. 7. Determination of ventricular fibrillation threshold with gated pulses of 4 ms duration at frequency of 100 Hz. Train of impulses was 150 ms duration and was timed to span the T wave. As plus amperage is increased from *A* to *D*: no response, one premature ventricular extrasystole, 2 premature ventricular extrasystoles, and finally ventricular fibrillation result. *Upper trace:* ECG (II). *Lower trace:* stimuli.

methods a gradual increase in stimulus intensity resulted in progressive decrease in the coupling interval of the first premature ventricular response, a progressive increase in the number of subsequent spontaneous ventricular responses until ventricular fibrillation ensued. The minimum amount of current required to induce fibrillation is taken as the ventricular fibrillation threshold (VFT) and is considered a quantitative estimate of the propensity of the ventricle to fibrillate.

Various mechanisms have been proposed to explain the repetitive activity that follows the initial premature ventricular beat. Spear et al. (109) have demonstrated that the degree of dispersion of refractoriness following the initial premature beat was determined not only by the coupling interval of that premature beat but also by the amount of current used to initiate it, independent of its coupling interval. Thus the stimulus not only initiated the first premature beat but, but its intensity, also increased the degree of inhomogeneity following that premature beat. When enough inhomogeneity in recovery of excitability was induced, the conditions were established for reentrant excitation to occur, which then may lead to repetitive ventricular activity. The problem becomes much more complex when one considers the initiation of ventricular fibrillation by a train of pulses because of interactions between the effects of sequential stimuli (113). A strong pulse delivered during the vulnerable period may also initiate automatic discharge in some foci that will repetitively induce closely coupled ventricular extrasystoles. Moe et al. (86) have suggested this possibility to explain the fact that the QRS complex during repetitive responses was usually similar to the QRS complex of the first premature beat. The above observations relate to the mechanisms initiating ventricular fibrillation under specific experimental conditions and may or may not be similar to those initiating spontaneous ventricular fibrillation. Moreover, fibrillation may be initiated by one mechanism but perpetuated by others. For example, repetitive ventricular extrasystoles initiated by automatic discharge may precipitate ventricular fibrillation, which is then sustained by reentry.

It is well established that the VFT is decreased during myocardial infarction (15, 42, 107, 127). Some discrepancy, however, exists as to the magnitude and time course of VFT changes following coronary occlusiuon. The site of VFT determination (ischemic zone, border zone, normal zone), as well as the use of bipolar or unipolar mode of stimulation, may partly explain these differences. The changes in VFT are characterized by a rapid decrease within the first 2 min after one-stage coronary artery occlusion. Following the initial decrease, the VFT rises continuously until 10–15 min, when preocclusion values are again recorded. Thereafter, the VFT remains at this level for up to 10 h when measured from the nonischemic zone (7, 83). When measured from the ischemic zone, the VFT increases above the control value following the initial decrease (39, 98), probably because of the loss of excitability of the infarcted myocardium.

In a number of studies (7, 72, 90) there appears to be a good temporal relationship between the VFT and the propensity of the ventricle to fibrillate spontaneously, at least following coronary occlusion. The highest incidence of ventricular fibrillation is observed between 3 and 6 min after coronary occlusion (7). It is also the time at which the VFT is minimum. During the second phase of arrhythmias, which begin after 4–8 h and may last for up to 5 days, spontaneous ventricular fibrillation is a rather rare phenomenon despite marked and frequent arrhythmias. Morover, when the coronary artery is slowly occluded, the arrhythmias and fibrillation characteristic of the first phase do not develop, and no decrease in the VFT is observed (83).

Han and co-workers (42, 43, 46) have related the propensity to fibrillate to the disparity between closely adjacent sites to recover excitability. Dispersion in recovery of excitability has been repeatedly demonstrated in ventricular myocardium (2, 46) and results from a natural variability between individual cells of similar and different types. This is to be expected in any population of biologic units, even in the heart where electrotonic interactions may tend to distribute or equalize the variability in polarization among closely adjacent cells. When an impulse reaches such a matrix of cells at a time when dispersion in recovery of excitability is present, the cells cannot be activated in the same sequence as if the tissue had had a homogeneous recovery of excitability. That impulse may travel rapidly through tissue more fully excitable, propagate slowly through poorly excitable fibers, and fail to excite those fibers still refractory. Such a premature impulse leaves in its wake tissue that remains more inhomogeneous for a longer period of time, i.e., inhomogeneity of refractoriness as well as the duration of the vulnerable period are increased following a premature beat (45,

108). Such asynchrony of excitation and recovery become progressively increased by repetitive stimulation to the point of fractionation of the wave front into many independent wavelets and the development of fibrillation.

Various agents and interventions have been demonstrated to alter the dispersion of excitability and refractoriness of the ventricular myocardium and to alter the VFT in the expected direction (for review see refs. 87, 112, 134). Whether all agents that alter the VFT do so by altering the degree of dispersion among closely adjacent myocardial sites remains uncertain, however.

We (39) have recently obtained experimental evidence that indicates that alterations in myocardial excitability per se will also lead to alterations in the VFT with no demonstrable change in the degree of dispersion of refractoriness or inhomogeneity. For example, when the serum K^+ concentration is increased in 5–6 meq/liter by slow intravenous infusion of KCl in dogs, the VFT decreases as does the diastolic excitability threshold. When the serum K^+ concentration is increased above 7 meq/liter, both the VFT and the diastolic excitability threshold increase in a parallel fashion. We were not able to demonstrate changes in the dispersion of refractoriness at any of these serum K^+ concentrations. Aprindine, a new antiarrhythmic agent that we have extensively investigated (30, 136), increases the diastolic threshold of excitability as well as the VFT, but does not alter the dispersion of recovery of excitability. These examples suggest that the VFT is somehow influenced by the diastolic threshold of excitability, independently of myocardial inhomogeneity. Vagal stimulation has also been reported to increase the VFT without altering myocardial inhomogeneity (63). Acetylcholine, however, does not affect the diastolic threshold of excitability (55).

Myocardial infarction, and more specifically the ischemic zone, represents a good example in which changes in the degree of inhomogeneity in refractoriness and conduction and in excitability interplay to influence the VFT. As mentioned previously, a decrease in the diastolic threshold of excitability is observed in the ischemic zone during the initial 1–3 min after coronary occlusion and later increases quite rapidly. The dispersion of refractoriness is increased between 3 and 7 min, after which it remains at a slightly higher level than control (2). Early after coronary occlusion, both changes in excitability and in dispersion of refractoriness contribute to lower the VFT. After approximately 3–4 min, the VFT, as measured from the ischemic zone, increases and may actually exceed control values (39). The increasing excitability threshold probably contributes significantly to the return of the VFT toward and above control values, despite the increase in dispersion of refractoriness.

It is common to infer that a drug that increases the VFT of the normal myocardium will decrease the incidence of ventricular fibrillation during myocardial ischemia. In our experience, aprindine increases the VFT of the nonischemic myocardium and yet increases the incidence of ventricular fibrillation following acute coronary occlusion (31). According to Hope et al. (58), lidocaine and procainamide, two agents that have been demonstrated to increase the VFT (108), do not prevent or even retard the occurrence of ventricular arrhythmias during acute myocardial ischemia. Kent et al. (64) reported that nitroglycerine infusion during myocardial ischemia increases the VFT to control, nonischemic levels, yet the incidence of spontaneous fibrillation decreased only from 92% to

50% in dogs following acute coronary occlusiuon. Thus the physiological relevance of the VFT is open to question and the efficacy of antiarrhythmic agents as determined by this technique may be misleading. Moreover, the measure of the fibrillation threshold has limited usefulness in understanding the nature and origin of ventricular arrhythmias produced by ischemia.

Pathophysiology of Ventricular Arrhythmias in Myocardial Ischemia

In the canine model, arrhythmias following one-stage occlusion of a major coronary artery occur during two distinct phases. The first phase begins within 1–2 min following coronary artery occlusion and lasts up to 30 min. Arrhythmias occurring during this first phase consist of premature ventricular extrasystoles occurring singly or in groups. The ventricular extrasystoles may progress to ventricular tachycardia and eventually degenerate into ventricular fibrillation. If the dog survives this initial phase, a second phase of arrhythmias develop usually after 8–10 h, reaches its maximum after 12–16 h, and may last for 2–5 days. It is possible that the initial phase may correspond to the prehospital phase of myocardial infarction in man, whereas the second phase or late arrhythmias may correspond to the time when in-hospital arrhythmias develop. One might speculate that the electrophysiological basis of these two phases of arrhythmias differs because the electrophysiological, biochemical, and ultrastructural status of the ischemic ventricular myocardium and Purkinje fibers also differs during these two phases after coronary artery occlusion. If this speculation is true, then the determinants of these arrhythmias as well as their therapeutic management may also differ. Although this assumption is generally accepted, no rationale for the management of these two phases of arrhythmias has emerged from this concept as yet.

Sudden reinstitution of blood flow through a previously occluded coronary artery may also lead to ventricular fibrillation (106, 115). The incidence of this reperfusion arrhythmia increases with the duration of the occlusion, whereas the incidence of those arrhythmias that occur during occlusion decrease with the duration of the occlusion (7). The mechanisms of reperfusion arrhythmias are still obscure but are very likely different from those responsible for the occlusion arrhythmias and may relate to a nonhomogeneous washout of the ischemic zone.

Various mechanisms have been postulated to explain the genesis of arrhythmias in the ischemic myocardium (43, 129, 130). The hypothesis of focal reexcitation is based upon the disparate rate of repolarization of neighboring myocardial fibers, particularly during ischemic conditions when repolarization of ischemic cells becomes greatly accelerated (33, 43). The potential difference between normal and ischemic fibers present during the major part of repolarization results in a systolic current of injury (ST segment shift in the surface ECG) that would depolarize the ischemic cell to threshold, thus reexciting it to produce a closely coupled ventricular premature impulse. This mechanism is considered reentrant, partly because it does not require the presence of automatic properties in the tissues involved. However, it has not been established whether the gradient of potential is steep enough to result in a sufficient current density to reexcite the ischemic cell. A similar gradient of potential exists normally between Purkinje and muscle fibers, but yet does not result in focal reexcitation of muscle fibers (84) because electrotonic mechanisms tend to distribute the differences in action potential duration. Reduction of the systolic current of injury and thus of the ST segment deviation in the surface electro-cardiogram may be an important goal for the management of arrhythmias. Agents and maneuvers that reduce the ST segment deviation may not necessarily do so by reducing the infarct size but may do so by increasing intercellular resistance to current flow (34, 35).

The concept of reentry was appreciated long ago (105). For reentry to occur it is necessary that an area of tissue show unidirectional block but is able to conduct an impulse in the reverse direction. Slow conduction is also a prerequisite for reentry. Loop arrangements are abundant in peripheral regions of the conducting system where Purkinje fibers arborize into many branches before interconnecting with muscular fibers (Fig. 8). Under conditions when an impulse blocks in one of these branches but conducts over adjacent branches and ventricular muscle, the branch that exhibited anterograde block may be reexcited from its muscular end and conduct retrogradely past the point at which block occurred. If the total transit time around the loop exceeds the refractory period of those tissues proximal to the site of block, the ventricle may be reexcited and give rise to a premature ventricular depolarization. If the conduction time around the loop is too rapid, reentry will not result, since the impulse may return to regions it has previously excited before they have recovered excitability. Both unidirectional block and slow propagation velocity are thus necessary for reentrant arrhythmias. Intermittent conduction over such a reentrant circuit would result in occasional extrasystoles, whereas an episode of ventricular tachycardia would result from continuous impulse propagation around the established circuit. Reentry may also occur in unbranched bundles of Purkinje fibers (6). These reentrant mechanisms are not specific to the ischemic heart and may occur under different pathological states not related to ischemia. However, ischemia may bring about the conditions necessary for these mechanisms to occur. Some strands of Purkinje fibers, particularly those beneath the ischemic myocardium, may demonstrate very slow propagation velocity and unidirectional block early after coronary occlusion, not necessarily because of changes in intrinsic properties of the cellular membrane but more likely because of alterations in the cellular ionic environment immediately subjacent. These fibers may generate action potentials that exhibit severely depressed upstroke velocity resulting from an inward current through the slow channel alone (slow response), with or without a variable, but reduced contribution of inward current through the fast channel (depressed fast response).

Recent investigations have suggested another reentrant mechanism whereby slow and inhomogeneous conduction over an ischemic area may permit electrical activity to persist until after normal surrounding areas have recovered excitability (see subsection *Delayed Activation of Ischemic Myocardium*).

If reentrant mechanisms are responsible for the early phase of arrhythmias following coronary artery occlusion, one might speculate that a drug which alters conduction in the ischemic myocardium or depressed Purkinje fibers may either increase or reduce the incidence of early arrhythmias, depending on whether this agent is capable of producing unidirectional block or is capable of

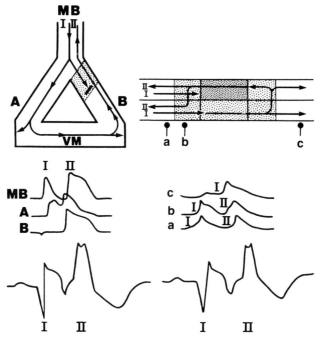

Fig. 8. *Left panel*: reentry in Y-shaped loop composed of main bundle of Purkinje fibers (MB) which divides into 2 branches (A, B) before terminating in ventricular muslce (VM). A severely depressed area demonstrating unidirectional conduction is located in B. Action potentials recorded from MB and from branches A and B are shown *below*. An impulse conducting down main bundle enters both branches A and B, blocks in B, but continues through A into ventricular muscle, from which it invades B in retrograde direction. Impulse then conducts through area of unidirectional block and reemerges in main bundle. *Bottom panel* shows how such events might appear in surface ECG. *Right panel*: reentry over 2 parallel fibers in an unbranched bundle of Purkinje fibers. Upper fiber demonstrates an area of depressed and unidirectional conduction (*shaded area*). Action potentials from sites a, b, and c in lower fiber are shown *below*. An impulse entering both fibers blocks in *shaded area*, but conducts through lower fiber. Impulse enters upper fiber in which it conducts retrogradely and emerges as a reentrant impulse. *Bottom panel* shows how such events might appear in surface ECG. [From Wellens et al. (125) with permission of the American Heart Association, Inc.]

converting a unidirectional block to a bidirectional block. This capability may be related to the drug dosage and concentration in the various tissues involved. An initial report from our laboratoy (31) supports the concept that drugs which increase significantly the ischemia-induced conduction delay increase the incidence of early arrhythmias, whereas drugs which improve conduction have an opposite effect.

Although it seems likely that conduction delay and reentry provide the probable electrophysiological substrate for the development of the early ventricular arrhythmias, abnormal forms of automaticity, particularly those occurring at reduced membrane potentials, may contribute to the development of arrhythmias. There exists no experimental evidencc at the present time to indicate increased degrees of ventricular automaticity in dogs immediately following coronary artery occlusion (72, 102, 103). Purkinje fibers obtained from 20-min-old infarcts do not demonstrate increased automaticity in vitro (67, 114). Therefore, such abnormal forms of automaticity are not likely to be a factor in the genesis of early arrhythmias but appear to play a role in the latter phase of ventricular arrhythmias. Twenty-four hours after coronary artery occlusion, Purkinje fibers exhibit increased phase 4 diastolic depolarization (36, 69), and in vivo canine studies reveal an accelerated intrinsic ventricular rate that may be automatic in origin (102, 103).

Two different mechanisms are thought to be responsible for the property of diastolic depolarization. At high resting membrane potential such as that found in normal Purkinje fibers and atrial plateau fibers, spontaneous diastolic depolarization results from a time- and voltage-dependent reduction in outward current (90). This mechanism is operative between -90 and -60 mV and thus cannot account for the diastolic depolarization observed in fibers with a resting membrane potential of -60 mV or less. At a resting membrane potential below -60 mV, a situation that may be normal (sinus nodal cells, fibers in the mitral and tricuspid valves) or abnormal (Purkinje fibers, atrial and ventricular fibers depolarized below -60 mV), other mechanisms not fully elucidated at the present time seem responsible for the diastolic depolarization. Both forms of automaticity may lead to arrhythmias, and certain antiarrhythmic drugs may be effective against arrhythmias resulting from one type of automaticity and not the other (131, 132).

The volume of myocardium involved in the ischemic process is an important determinant of the incidence and severity of ventricular arrhythmias, partly because of the association between the ventricular ischemic mass and the opportunity for reentrant loops or automatic foci to develop (97). More severe cellular electrophysiological distrubances may also be associated with more extensive infarcts.

Alterations of the autonomic nervous system outflow to the heart appear to exert an important role in the genesis of arrhythmias during myocardial ischemia. Stellate ganglionectomy prior to coronary artery occlusion decreases the incidence of ectopic activity during the initial phase but not during the late phase (23, 49) of arrhythmia development following coronary artery occlusion. The sympathetic efferent outflow to the heart is increased within 30 s of coronary occlusion due to reflexes associated with fear and pain as well as due to cardiac reflexes (78, 120, 121). Increase in extracellular K^+ concentration also has an excitatory effect on these fibers (120). Direct or reflex stimulation of cardiac sympathetic nerves evokes a marked fall in ventricular fibrillation threshold, lowering of ventricular diastolic excitability threshold, slight shortening of the refractory period, and increases in the temporal dispersion of the refractory periods of the ventricular myocardium (44, 54). The increase in temporal dispersion and vulnerability by stellate stimulation appears to result from nonuniform distribution of the adrenergic nerve terminals within the ventricular myocardium. Although controversial for many years, the parasympathetic innervation of the ventricles and the role of such innervation in modulating ventricular contractility and coronary circulation is presently well established (51, 74). Whether or not vagal stimulation and acetylcholine have direct electrophysiological effects on mammalian ventricles has also been controversial. Earlier studies failed to show any appreciable effects on ventricular and Purkinje fibers (55), but recently various authors (5, 118) demonstrated a definite slowing effect of acetylcholine on spontaneously beating canine Purkinje fibers. A recent study has indicated that cervical vagal stimulation under nonischemic conditions and at constant heart rate substantially increased the ventricular fibrillation threshold (63). This protective effect was also demonstrated under ischemic conditions (40, 63). The mechanisms by which this increase in VFT is brought about are not known, however, but were felt to be a direct action of the cholinergic neurotransmitter on ventricular fibers (63). Decreased dispersion of refractoriness in contiguous areas of epicardium by vagal stimulation could not be demonstrated. Evidence that this action of the vagus is significantly related to the prevailing level of adrenergic tone has been obtained (66, 77, 95) and is in agreement with the concept that parasympathetic neuromediator may modulate the activity and effects of the sympathetic limb of the autonomic nervous system (73). It is important for future investigations to define the mechanisms by which the autonomic nervous system influences ventricular arrhythmias that follow coronary artery occlusion.

Application of these concepts to the genesis of arrhythmias in man must be done very cautiously. There is no question that an early phase of arrhythmia development exists in man soon after the onset of symptoms, indicating the presence of an acute myocardial infarction. These early ventricular arrhythmias account for the fact that more than 50% of those who die from an acute myocardial infarction do so in the first few minutes after the attack, prior to being hospitalized. Ventricular fibrillation appears to be the most likely cause of these sudden deaths.

It is of interest that 60% of patients resuscitated from ventricular fibrillation occurring out of the hospital do not evolve an acute myocardial infarction, and these patients appear at a greater risk to experience a second episode of ventricular fibrillation than those who develop an acute myocardial infarction with the first episode of ventricular fibrillation (8). The reason for this is unknown but may relate to the phenomenon of "reperfusion ventricular fibrillation." It is well established in the dog that some animals which do not develop ventricular fibrillation at the time of coronary artery occlusion may do so when reperfusion occurs at the time the occlusion is released. One may speculate that some patients who do not evolve an acute myocardial infarction may have coronary spasm which, upon relaxation, initiates "reperfusion ventricular fibrillation."

Few electrophysiological studies in patients having ventricular tachycardia due to acute myocardial infarction have been reported. In one such study, Wellens et al. (124, 125) noted that ventricular tachycardia could not be initiated or terminated by premature ventricular stimulation. These patients were studied 8–24 h after the onset of symptoms of acute myocardial infarction. This time span might be consistent with the delayed phase of arrhythmias seen in canine studies when enhanced automaticity is thought to play a role in the genesis of ventricular arrhythmias. In patients who had chronic recurrent ventricular tachycardia due to various etiologies but not to myocardial infarction, these authors (124, 125) were able to induce and terminate ventricular tachycardia with premature ventricular stimulation. The significance of these findings is not entirely clear at the present time. The ability to initiate and terminate a tachycardia with premature stimulation is thought to be consistent with reentrant mechanisms, whereas the failure to do so, as seen in patients between 8 and 24 h after acute myocardial infarction, is more consistent with the behavior of an automatic focus.

Conclusion

Knowledge of the cardiac electrophysiological alterations that occur after occlusion of a coronary artery is a prerequisite for understanding the pathophysiology of arrhythmias associated with myocardial infarction. Observations made in various animal models and at various times following occlusion of a coronary artery are invaluable in developing the proper therapeutic management of patients with myocardial infarction. Ideally, such management involves being able to identify the electrophysiological basis of ventricular arrhythmias at a specific time after myocardial infarction and to intervene with appropriate agents and maneuvers designed for that specific event. However, at the present stage of our knowledge and understanding of these alterations and arrhythmias, there is little on which we can capitalize and offer in the way of specific therapeutic interventions, since drugs are not available that act more specifically on ventricular myocardium rather than Purkinje fibers or on automaticity rather than reentry. Moreover the applicability to man of data derived from animal studies is limited by the fact that these data are obtained from animal models in which ischemia is produced acutely, or relatively acutely, in the setting of otherwise normal coronary arteries and myocardium. The process in man occurs spontaneously following a long and gradually developing pathological state that generally affects multiple coronary arteries and often the myocardium as well. It is hoped that new findings and more sophisticated techniques applicable to both man and animal studies may, in the future, yield new insight into the pathophysiology of myocardial infarction.

Acknowledgments

The authors thank Mrs. Shirley Proffitt and Mrs. Laura E. Najar for their secretarial assistance.

This investigation was supported in part by the Herman C. Krannert Fund; by Grants HL-06308, HL-05363, HL-07182, and HL-18795 from the National Heart, Lung, and Blood Institute of the National Institutes of Health; and by the American Heart Association, Indiana Affiliate, Inc.

References

1. ARONSON, R. S., AND P. F. CRANEFIELD. The electrical activity of canine cardiac Purkinje fibers in sodium-free, calcium-rich solutions. *J. Gen. Physiol.* 61: 786–808, 1973.
2. AVITALL, B., S. NAIMI, A. H. BRILLA, AND H. J. LEVINE. A computerized system for measuring dispersion of repolarization in the intact heart. *J. Appl. Physiol.* 37: 456–458, 1974.
3. BADEER, H., AND S. M. HORVATH. Role of acute myocardial hypoxia and ischemic-nonischemic boundaries in ventricular fibrillation. *Am. Heart J.* 58: 706–714, 1959.
4. BAGDONAS, A. A., J. H. STUCKEY, J. PIERA, N. S. AMER, AND B. F. HOFFMAN. Effects of ischemia and hypoxia on the specialized conducting system of the canine heart. *Am. Heart J.* 61: 206–218, 1961.
5. BAILEY, J. C., K. GREENSPAN, M. V. ELIZARI, G. J. ANDERSON, AND C. FISCH. Effects of acetylcholine on automaticity and conduction in the proximal portion of the His-Purkinje specialized conduction system of the dog. *Circ. Res.* 30: 210–216, 1972.
6. BAILEY, J. C., J. F. SPEAR, AND E. N. MOORE. Functional significance of transverse conducting pathways within the canine bundle of His. *Am. J. Cardiol.* 34: 790–795, 1974.

7. BATTLE, W. E., S. NAIMI, B. AVITALL, A. H. BRILLA, J. S. BANAS, JR., J. M. BETE, AND H. J. LEVINE. Distinctive time course of ventricular vulnerability to fibrillation during and after release of coronary ligation. *Am. J. Cardiol.* 34: 42–47, 1974.
8. BAUM, R. S., H. ALVAREZ, AND L. A. COBB. Survival after resuscitation from out-of-hospital ventricular fibrillation. *Circulation* 50: 1231–1235, 1974.
9. BECKER, L. C., R. FERREIRA, AND M. THOMAS. Mapping of left ventricular blood flow with radioactive microspheres in experimental coronary artery occlusion. *Cardiovasc. Res.* 7: 391–400, 1973.
10. BENZING, H., G. GEBERT, AND M. STROHM. Extracellular acid-base changes in dog myocardium during hypoxia and local ischemia, measured by means of glass micro-electrodes. *Cardiology* 56; 85–88, 1971/72.
11. BOINEAU, J. P., AND J. L. COX. Slow ventricular activation in acute myocardial infarction. A source of re-entrant premature ventricular contractions. *Circulation* 48: 702–713, 1973.
12. BRAUNWALD, E., AND P. R. MAROKO. ST-segment mapping. Realistic and unrealistic expectations. *Circulation* 54: 529–532, 1976.
13. BROOKS, C. McC., J. L. GILBERT, M. E. GREENSPAN, G. LANGE, AND H. M. MAZZELLA. Excitability and electrical response of ischemic heart muscle. *Am. J. Physiol.* 198: 1143–1147, 1960.
14. BROWN, R. H., JR., AND D. NOBLE. Effect of pH on ionic currents underlying pace-maker activity in cardiac Purkinje fibres. *J. Physiol. London* 224: 38P–39P, 1972.
15. BURGESS, M. J., J. A. ABILDSKOV, K. MILLAR, J. S. GEDDES, AND L. S. GREEN. Time course of vulnerability to fibrillation after experimental coronary occlusion. *Am. J. Cardiol.* 27: 617–621, 1971.
16. CASE, R. B. Ion alterations during myocardial ischemia. *Cardiology* 56: 245–262, 1971/72.
17. CASE, R. B., M. G. NASSER, AND R. S. CRAMPTON. Biochemical aspects of early myocardial ischemia. *Am. J. Cardiol.* 24: 766–775, 1969.
18. COHEN, D., AND L. A. KAUFMAN. Magnetic determination of the relationship between the S-T segment shift and the injury current produced by coronary artery occlusion. *Circ. Res.* 36: 414–424, 1975.
19. CORABOEUF, E., E. DEROUBAIX, AND J. HOERTER. Control of ionic permeabilities in normal and ischemic heart. *Circ. Res.* 38, Suppl. I: 92–98, 1976.
20. CORABOEUF, E., Y. M. GARGOUIL, J. LAPLAUD, AND A. DESPLACES. Action de l'anoxie sur les potentiels electriques des cellules cardiaques de mammiferes actives et inertes (tissu ventriculaire isole de cobaye). *C. R. Acad. Sci. Paris* 246: 3100–3103, 1958.
21. CORR, P. B., M. E. CAIN, F. X. WITKOWSKI, D. A. PRICE, AND B. E. SOBEL. Potential arrhythmogenic electrophysiological derangements in canine Purkinje fibers induced by lysophosphoglycerides. *Circ. Res.* 44: 822–832, 1979.
22. COX, J. L., T. M. DANIEL, AND J. P. BOINEAU. The electrophysiologic time-course of acute myocardial ischemia and the effects of early coronary artery reperfusion. *Circulation* 48: 971–983, 1973.
23. COX, W. V., AND H. F. ROBERTSON. Effect of stellate ganglionectomy on the cardiac function of intact dogs and its effect on extent of myocardial infarction and on cardiac function following coronary artery occlusion. *Am. Heart J.* 12: 285–300, 1936.
24. CRANEFIELD, P. F. *The Conduction of the Cardiac Impulse.* Mount Kisco, NY: Futura, 1975.
25. DANESE, C. Pathogenesis of ventricular fibrillation in coronary occlusion. Perfusion of coronary arteries with serum. *J. Am. Med. Assoc.* 179: 52–53, 1962.
26. DURRER, D., A. A. W. VAN LIER, AND J. BÜLLER. Epicardial and intramural excitation in chronic myocardial infarction. *Am. Heart J.* 68: 765–776, 1964.
27. DOWNAR, E., M. J. JANSE, AND D. DURRER. The effect of "ischemic" blood on transmembrane potentials of normal porcine ventricular myocardium. *Circulation* 55; 455–462, 1977.
28. EKMEKCI, A., H. TOYOSHIMA, J. K. KWOCZYNSKI, T. NAGAYA, AND M. PRINZMETAL. Angina pectoris. IV. Clinical and experimental difference between ischemia with S-T elevation and ischemia with S-T depression. *Am. J. Cardiol.* 7: 412–426, 1961.
29. ELHARRAR, V., P. R. FOSTER, T. L. JIRAK, W. E. GAUM, AND D. P. ZIPES. Alterations in canine myocardial excitability during ischemia. *Circ. Res.* 40: 98–105, 1977.
30. ELHARRAR, V., P. R. FOSTER, AND D. P. ZIPES. Effects of aprindine HCl on cardiac tissues. *J. Pharmacol. Exp. Ther.* 195: 201–205, 1975.
31. ELHARRAR, V., W. E. GAUM, AND D. P. ZIPES. Effect of drugs on conduction delay and the incidence of ventricular arrhythmias induced by acute coronary occlusion in dogs. *Am. J. Cardiol.* 39: 544–549, 1977.

32. EL-SHERIF, N., B. J. SCHERLAG, AND R. LAZZARA. Conduction disorders in the canine proximal His-Purkinje system following acute myocardial ischemia. I. The pathophysiology of intra-His bundle block. *Circulation* 49: 837–847, 1974.
33. ETTINGER, P. O., T. J. REGAN, H. A. OLDEWURTEL, AND M. I. KHAN. Ventricular conduction delay and arrhythmias during regional hyperkalemia in the dog. Electrical and myocardial ion alterations. *Circ. Res.* 33: 521–531, 1973.
34. FOZZARD, H. A. Validity of myocardial infarction models. *Circulation Suppl.* III, 51: 131–138, 1975.
35. FOZZARD, H. A., AND D. S. DAS GUPTA. ST-segment potentials and mapping. Theory and experiments. *Circulation* 54: 533– 537, 1976.
36. FRIEDMAN, P. L., J. R. STEWART, J. J. FENOGLIO, JR., AND A. L. WIT. Survival of subendocardial Purkinje fibers after extensive myocardial infarction in dogs. *Circ. Res.* 33: 597–611, 1973.
37. FRIEDMAN, P. L., J. R. STEWART, AND A. L. WIT. Spontaneous and induced cardiac arrhythmias in subendocardial Purkinje fibers surviving extensive myocardial infarction in dogs. *Circ. Res.* 33: 612–626, 1973.
38. GAMBETTA, M., AND R. W. CHILDERS. The initial electrophysiologic disturbance in experimental myocardial infarction (Abstract). *Ann. Int. Med.* 70: 1076, 1969.
39. GAUM, W. E., V. ELHARRAR, T. L. JIRAK, AND D. P. ZIPES. Influence of excitability on the ventricular fibrillation threshold. *Am. J. Cardiol.* 40: 929–935, 1977.
40. GILLIS, R. A. Role of the nervous system in the arrhythmias produced by coronary occlusion in the cat. *Am. Heart J.* 81: 677–684, 1971.
41. GILMOUR, R. F., AND D. P. ZIPES. Different electrophysiological responses of canine endocardium and epicardium to combined hyperkalemia, hypoxia and acidosis. *Circ. Res.* 44: 814–825, 1980.
42. HAN, J. Mechanisms of ventricular arrhythmias associated with myocardial infarction. *Am. J. Cardiol.* 24: 800–813, 1969.
43. HAN, J. Ventricular vulnerability during acute coronary occlusioin. *Am. J. Cardiol.* 24: 857–864, 1969.
44. HAN, J., P. G. DE JALON, AND G. K. MOE. Adrenergic effects of ventricular vulnerability. *Circ. Res.* 14: 516–524, 1964.
45. HAN, J., P. D. G. DE JALON, AND G. K. MOE. Fibrillation threshold of premature ventricular responses. *Circ. Res.* 18: 18–25, 1966.
46. HAN, J., AND G. K. MOE. Nonuniform recovery of excitability in ventricular muscle. *Circ. Res.* 14: 44–60, 1964.
47. HARRIS, A. S. Terminal electrocardiographic patterns in experimental anoxia, coronary occlusion and hemorrhagic shock. *Am. Heart J.* 35: 895–909, 1948.
48. HARRIS, A. S. Potassium and experimental coronary occlusion. *Am. Heart J.* 71: 797–802, 1966.
49. HARRIS, A. S., A. ESTANDIA, AND R. F. TILLOTSON. Ventricular ectopic rhythms and ventricular fibrillation following cardiac sympathectomy and coronary occlusion. *Am. J. Physiol.* 165: 505–512, 1951.
50. HECHT, H., AND O. F. HUTTER. Action of pH on cardiac Purkinje fibers. In: *Electrophysiology of the Heart*, edited by B. Taccardi and G. Marchetti. Oxford: Pergamon, 1965, p. 105–123.
51. HIGGINS, C. B., S. F. VATNER, AND E. BRAUNWALD. Parasympathetic control of the heart. *Pharmacol. Rev.* 25: 119–155, 1973.
52. HILL, J. L., AND L. S. GETTES. Effects of acute coronary artery occlusion on local myocardial extracellular K^+ activity in swine. *Circulation* 61: 768–778, 1980.
53. HIRSHFELD, J. W., JR., J. S. BORER, R. E. GOLDSTEIN, M. J. BARRETT, AND S. E. EPSTEIN. Reduction in severity and extent of myocardial infarction when nitroglycerin and methoxamine are administered during coronary occlusion. *Circulation* 49: 291–297, 1974.
54. HOFFMAN, B. F., A. A. SIEBENS, P. F. CRANEFIELD, AND C. McC. BROOKS. The effect of epinephrine and norepinephrine on ventricular vulnerability. *Circ. Res.* 3: 140–146, 1955.
55. HOFFMAN, B. F., AND E. SUCKLING. Cardiac cellular potentials: effect of vagal stimulation and acetylcholine. *Am. J. Physiol.* 173: 312–320, 1953.
56. HOLLAND, R. P., AND H. BROOKS. Precordial and epicardial surface potentials during myocardial ischemia in the pig. A theoretical and experimental analyses of the TQ and ST segments. *Circ. Res.* 37: 471–480, 1975.
57. HOLLAND, R. P., AND H. BROOKS. The QRS complex during myocardial ischemia. An experimental analysis in the porcine heart. *J. Clin. Invest.* 57: 541–550, 1976.
58. HOPE, R. R., D. O. WILLIAMS, N. EL-SHERIF, R. LAZZARA, AND B. J. SCHERLAG. The efficacy of

antiarrhythmic agents during acute myocardial ischemia and the role of heart rate. *Circulation* 50: 507–514, 1974.

59. JACOBSON, E. E., W. SCHIESS, AND G. K. MOE. The effect of hypoxia on experimental ventricular tachycardia. *Am. Heart J.* 64: 368–375, 1962.

60. JANSE, M. J., AND E. DOWNAR. The effect of acute ischaemia on transmembrane potentials in the intact heart. The relation to re-entrant mechanisms. In: *Re-entrant Arrhythmias*, edited by H. E. Kubertus. Baltimore: University Park, 1977, p. 195–209.

61. JULIAN, D. G., P. A. VALENTINE, AND G. G. MILLER. Disturbances of rate, rhythm and conduction in acute myocardial infarction. A prospective study of 100 consecutive unselected patients with the aid of electrocardiographic monitoring. *Am. J. Med.* 37: 915–927, 1964.

62. KARDESCH, M., C. E. HOGANCAMP, AND R. J. BING. The effect of complete ischemia on the intracellular electrical activity of the whole mammalian heart. *Circ. Res.* 6; 715–720, 1958.

63. KENT, K. M., E. R. SMITH, D. R. REDWOOD, AND S. E. EPSTEIN. Electrical stability of acutely ischemic myocardium. Influences of heart rate and vagal stimulation. *Circulation* 47: 291–298, 1973.

64. KENT, K. M., E. R. SMITH, D. R. REDWOOD, AND S. E. EPSTEIN. Beneficial electrophysiologic effects of nitroglycerin during acute myocardial infarction. *Am. J. Cardiol.* 33; 513–516, 1974.

65. KHURI, S. F., J. T. FLAHERTY, J. B. O'RIORDAN, B. PITT, R. K. BRAWLEY, J. S. DONAHOO, AND V. L. GOTT. Changes in intramyocardial ST segment voltage and gas tensions with regional myocardial ischemia in the dog. *Circ. Res.* 37: 455–463, 1975.

66. KOLMAN, B. S., R. L. VERRIER, AND B. LOWN. Effect of vagus nerve stimulation upon excitability of the canine ventricle. Role of sympathetic-parasympathetic interactions. *Am. J. Cardiol.* 37: 1041–1045, 1976.

67. KUPERSMITH, J., E. M. ANTMAN, AND B. F. HOFFMAN. In vivo electrophysiological effects of lidocaine in canine acute myocardial infarction. *Circ. Res.* 36: 84–91, 1975.

68. KUPERSMITH, J., H. SHIANG, R. S. LITWAK, AND M. V. HERMAN. Electrophysiological and antiarrhythmic effects of propranolol in canine acute myocardial ischemia. *Circ. Res.* 38: 302–307, 1976.

69. LAZZARA, R., N. EL-SHERIF, AND B. J. SCHERLAG. Electrophysiological properties of canine Purkinje cells in one-day-old myocardial infarction. *Circ. Res.* 33: 722–734, 1973.

70. LAZZARA, R., N. EL-SHERIF, AND B. J. SCHERLAG. Early and late effects of coronary artery occlusion on canine Purkinje fibers. *Circ. Res.* 35: 391–399, 1974.

71. LAZZARA, R., N. EL-SHERIF, AND B. J. SCHERLAG. Disorders of cellular electrophysiology produced by ischemia of the canine His bundle. *Circ. Res.* 36: 444–454, 1975.

72. LEVITES, R., V. S. BANKA, AND R. H. HELFANT. Electrophysiologic effects of coronary occlusion and reperfusion: observations of dispersion of refractoriness and ventricular automaticity. *Circulation* 52: 760–765, 1975.

73. LEVY, M. N., AND B. BLATTBERG. Effect of vagal stimulation on the overflow of norepinephrine into the coronary sinus during cardiac sympathetic nerve stimulation in the dog. *Circ. Res.* 38: 81–85, 1976.

74. LEVY, M. N., M. NG, R. I. LIPMAN, AND H. ZIESKE. Vagus nerves and baroreceptor control of ventricular performance. *Circ. Res.* 18: 1–106, 1966.

75. LOGIC, J. R. Electrophysiologic effects of regional hyperkalemia in the canine heart. *Proc. Soc. Exp. Biol. Med.* 141: 725–730, 1972.

76. LOGIC, J. R. Enhancement of the vulnerability of the ventricle to fibrillation (VF) by regional hyperkalaemia. *Cardiovasc. Res.* 7: 501–507, 1973.

77. LOWN, B., AND R. L. VERRIER. Neural activity and ventricular fibrillation (VF). *N. Engl. J. Med.* 294: 1165–1175, 1976.

78. MALLIANI, A., P. J. SCHWARTZ, AND A. ZANCHETTI. A sympathetic reflex elicited by experimental coronary occlusion. *Am. J. Physiol.* 217: 703–709, 1969.

79. MANDEL, W. J., M. J. BURGESS, J. NEVILLE, JR., AND J. A. ABILDSKOV. Analysis of T-wave abnormalities associated with myocardial infarction using a theoretic model. *Circulation* 38: 178–188, 1968.

80. MAROKO, P. R., J. K. KJEKSHUS, B. E. SOBEL, T. WATANABE, J. W. COVELL, J. ROSS, JR., AND E. BRAUNWALD. Factors influencing infarct size following experimental coronary artery occlusion. *Circulation* 43: 67–82, 1971.

81. MCALLISTER, R. E., D. NOBEL, AND R. W. TSIEN. Reconstruction of the electrical activity of cardiac Purkinje fibres. *J. Physiol. London* 251: 1–59, 1975.

82. McDonald, T. F., and D. P. MacLeod. Metabolism and the electrical activity of anoxic ventricular muscle. *J. Physiol. London* 229: 559–582, 1973.

83. Meesman, W., H. Gülker, B. Kràmer, and K. Stephan. Time course of changes in ventricular fibrillation threshold in myocardial infarction: characteristics of acute and slow occlusion with respect to the collateral vessels of the heart. *Cardiovasc. Res.* 10: 466–473, 1976.

84. Mendez, C., W. J. Mueller, J. Meridith, and G. K. Moe. Interactions of transmembrane potentials in canine Purkinje fibers and/or Purkinje fiber muscle junctions. *Circ. Res.* 24: 361–372, 1969.

85. Merideth, J., C. Mendez, W. J. Mueller, and G. K. Moe. Electrical excitability of atrioventricular nodal cells. *Circ. Res.* 23: 69–85, 1968.

86. Moe, G. K., A. S. Harris, and C. J. Wiggers. Analysis of the initiation of fibrillation by electrographic studies. *Am. J. Physiol.* 134; 473–492, 1941.

87. Moore, E. N., and J. F. Spear. Ventricular fibrillation threshold. Its physiological and pharmacological importance. *Arch. Int. Med.* 135: 446–453, 1975.

88. Morena, H., M. J. Janse, J. W. T. Fiolet, W. J. G. Krieger, H. Crijns, and D. Durrer. Comparison of the effects of regional ischemia, hyperkalemia and acidosis on intracellular and extracellular potentials and metabolism in the isolated porcine heart. *Circ. Res.* 46: 634–645, 1980.

89. Neely, J. R., and H. E. Morgan. Relationship between carbohydrate and lipid metabolism and the energy balance of heart muscle. *Ann. Rev. Physiol.* 36: 413–459, 1974.

90. Noble, D., and R. W. Tsien. The kinetics and rectifier properties of the slow potassium current in cardiac Purkinje fibres. *J. Physiol. London* 195: 185–214, 1968.

91. Opie, L. H. Effects of regional ischemia on metabolism of glucose and fatty acids. Relative rates of aerobic and anaerobic energy production during myocardial infarction and comparison with effects of anoxia. *Circ. Res.* 38: *Suppl.* I: 52–68, 1976.

92. Pardee, H. E. B. An electrocardiograph sign of coronary artery obstruction. *Arch. Int. Med.* 26: 244–257, 1920.

93. Prinzmetal, M., H. Toyoshima, A. Ekmekci, Y. Mizuno, and T. Nagaya. Myocardial ischemia. Nature of ischemic electrocardiographic patterns in the mammalian ventricles as determined by intracellular electrographic and metabolic changes. *Am. J. Cardiol.* 8: 493–503, 1961.

94. Prinzmetal, M., H. toyoshima, A. Ekmekci, and T. Nagaya. Angina pectoris VI: The nature of ST segment elevation and other ECG changes in acute severe myocardial ischemia. *Clin. Sci.* 23: 489–514, 1962.

95. Rabinowitz, S. H., R. L. Verrier, and B. Lown. Muscarinic effects of vago-sympathetic trunk stimulation on the repetitive extrasystole (RE) threshold. *Circulation* 53: 622–627, 1976.

96. Rakita, L., J. L. Borduas, S. Rothman, and M. Prinzmetal. Studies on the mechanism of ventricular activity. XII. Early changes in the RS-T segment and QRS complex following acute coronary artery occlusion: Experimental study and clinical applications. *Am. Heart J.* 48: 351–372, 1954.

97. Roberts, R., A. Husain, H. D. Ambos, G. C. Oliver, J. Cox, Jr., and B. E. Sobel. Relation between infarct size and ventricular arrhythmia. *Br. Heart J.* 37: 1169–1175, 1975.

98. Roland, J. M., N. Dashkoff, P. J. Varghese, and B. Pitt. Time course of ventricular fibrillation threshold in infarcted and non-infarcted myocardium after coronary ligation (Abstract). *Federation Proc.* 34: 390, 1975.

99. Samson, W. E., and A. M. Scher. Mechanism of S-T segment alteration during acute myocardial injury. *Circ. Res.* 8: 780–787, 1960.

100. Sayen, J. J., G. Pierce, A. H. Katcher, and W. F. Sheldon. Correlation of intramyocardial electrocardiograms with polarographic oxygen and contractility in the non-ischemic and regionally ischemic left ventricle. *Circ. Res.* 9: 1268–1279, 1961.

101. Schaper, W. *The Collateral Circulation of the Heart.* New York: Elsevier, 1971.

102. Scherlag, B. J., N. El-Sherif, R. Hope, and R. Lazzara. Characterization and localization of ventricular arrhythmias resulting from myocardial ischemia and infarction. *Circ. Res.* 35; 372–383, 1974.

103. Scherlag, B. J., R. H. Helfant, J. I. Haft, and A. N. Damato. Electrophysiology underlying ventricular arrhythmias due to coronary ligation. *Am. J. Physiol.* 219: 1665–1671, 1970.

104. Schiebler, T. H., M. Stark, and R. Caesar. Die Stoffwechselsituation des Reizleitungssystems. *Klin. Wochenschr.* 34: 181–183, 1956.

105. Schmitt, F. O., and J. Erlanger. Directional differences in the conduction of the impulse

through heart muscle and their possible relation to extrasystolic and fibrillary contractions. *Am. J. Physiol.* 87: 326–347, 1928/29.

106. SEWELL, W. H., D. R. KOTH, AND C. E. HUGGINS. Ventricular fibrillation in dogs after sudden return of flow to the coronary artery. *Surgery* 38: 1050–1053, 1955.

107. SHUMWAY, N. E., J. A. JOHNSON, AND R. J. STISH. The study of ventricular fibrillation by threshold determinations. *J. Thoracic Surg.* 34: 643–653, 1957.

108. SPEAR, J. F., E. N. MOORE, AND G. GERSTENBLITH. Effect of lidocaine on the ventricular fibrillation threshold in the dog during acute ischemia and premature ventricular contractions. *Circulation* 46: 65–73, 1972.

109. SPEAR, J. F., E. N. MOORE, AND L. N. HOROWITZ. Effect of current pulses delivered during the ventricular vulnerable period upon the ventricular fibrillation threshold. *Am. J. Cardiol.* 32: 814–822, 1973.

110. SPERELAKIS, N., AND J. A. SCHNEIDER. A metabolic control mechanism for calcium ion influx that may protect the ventricular myocardial cell. *Am. J. Cardiol.* 37: 1079–1085, 1976.

111. SPIELMAN, S. R., E. L. MICHELSON, J. F. SPEAR, AND E. N. MOORE. His bundle, bundle branch and Purkinje system conduction in dogs 24 hours after experimental septal infarction (Abstract). *Clin. Res.* 24: 241A, 1976.

112. SURAWICZ, B. Ventricular fibrillation. *Am. J. Cardiol.* 28: 268–287, 1971.

113. TAMARGO, J., B. MOE, AND G. K. MOE. Interaction of sequential stimuli applied during the relative refractory perioid in relation to determination of fibrillation threshold in the canine ventricle. *Circ. Res.* 37: 534–541, 1975.

114. TEN EICK, R. E., D. H. SINGER, AND L. E. SOLBERG. Coronary occlusion: effect on cellular electrical activity of the heart. *Med. Clin. North Am.* 60: 49–67, 1976.

115. TENNANT, R., AND C. J. WIGGERS. The effect of coronary occlusion on myocardial contraction. *Am. J. Physiol.* 112: 351–361, 1935.

116. TRAUTWEIN, W. Membrane currents in cardiac muscle fibers. *Physiol. Rev.* 53: 793–835, 1973.

117. TRAUTWEIN, W., U. GOTTSTEIN, AND J. DUDEL. Der aktionsstrom der myokardfaser im sauer-stoffmangel. *Pfluegers Arch.* 260: 40–60, 1954.

118. TSE, W. W., J. HAN, AND M. S. YOON. Effect of acetylcholine on automaticity of canine Purkinje fibers. *Am. J. Physiol.* 230: 116–119, 1976.

119. TSUCHIDA, T. Experimental studies on the excitability of ventricular musculature in infarcted region. *Jpn. Heart J.* 6: 152–164, 1965.

120. UCHIDA, Y., AND S. MURAO. Potassium-induced excitation of afferent cardiac sympathetic nerve fibers. *Am. J. Physiol.* 226: 603–607, 1974.

121. UCHIDA, Y., AND S. MURAO. Excitation of afferent cardiac sympathetic nerve fibers during coronary occlusion. *Am. J. Physiol.* 226: 1094–1099, 1974.

122. WALDO, A. L., AND G. A. KAISER. Study of ventricular arrhythmias associated with acute myocardial infarction in the canine heart. *Circulation* 47: 1222–1228, 1973.

123. WEIDMANN, S. Heart; electrophysiology. *Ann. Rev. Physiol.* 36: 155–169, 1974.

124. WELLENS, H. J., K. I. LIE, AND D. DURRER. Further observations on ventricular tachycardia as studied by electrical stimulation of the heart. Chronic recurrent ventricular tachycardia and ventricular tachycardia during acute myocardial infarction. *Circulation* 49: 647–653, 1974.

125. WELLENS, H. J., R. M. SCHULLENBURG, AND D. DURRER. Electrical stimulation of the heart in patients with ventricular tachycardia. *Circulation* 46: 216–226, 1972.

126. WIGGERS, C. J., AND R. WEGRIA. Ventricular fibrillation due to single, localized induction and condenser shocks applied during the vulnerable phase of ventricular systole. *Am. J. Physiol.* 128: 500–505, 1939/40.

127. WIGGINS, J. R., AND P. F. CRANEFIELD. Two levels of resting potential in canine cardiac Purkinje fibers exposed to sodium free solutions. *Circ. Res.* 39: 466–474, 1976.

128. WILLIAMS, D. O., B. J. SCHERLAG, R. R. HOPE, N. EL-SHERIF, AND R. LAZZARA. The pathophysiology of malignant ventricular arrhythmias during acute myocardial ischemia. *Circulation* 50: 1163–1172, 1974.

129. WILSON, F. N., F. D. JOHNSTON, AND I. G. W. HILL. The form of the electrocardiogram in experimental myocardial infarction. IV. Additional observations on the later effects produced by ligation of the anterior descending branch of the left coronary artery. *Am. Heart J.* 10: 1025–1041, 1935.

130. WILSON, F. N., A. G. MacLEOD, F. D. JOHNSTON, AND I. G. W. HILL. Monophasic electrical response produced by the contraction of injured heart muscle. *Proc. Soc. Exp. Biol. Med.* 30: 797–798, 1933.

131. WIT, A. L., AND J. T. BIGGER. Possible electrophysiological mechanisms for lethal arrhythmias accompanying myocardial ischemia and infarction. *Circulation* 51–52, Suppl. III: 96–115, 1975.
132. WIT, A. L., AND P. L. FRIEDMAN. Basis for ventricular arrhythmias accompanying myocardial infarction. *Arch. Int. Med.* 135: 459–472, 1975.
133. WOLLENBERGER, A., AND E.-G. KRAUSE. Metabolic control characteristics of the acutely ischemic myocardium. *Am. J. Cardiol.* 22: 349–359, 1968.
134. ZIPES, D. P. Electrophysiological mechanisms involved in ventricular fibrillation. *Circulation* 51–52, Suppl. III: 120–130, 1975.
135. ZIPES, D. P., H. R. BESCH, JR., AND A. M. WATANABE. Role of the slow current in cardiac electrophysiology. *Circulation* 51: 761–766, 1975.
136. ZIPES, D. P., V. ELHARRAR, W. E. GAUM, R. J. NOBLE, P. R. FOSTER, AND A. F. FASOLA. Cardiac electrophysiologic effects of aprindine in dogs. In: *New Aspect of Antiarrhythmic Therapy. Experiences with Aprindin.* Aulendorf, Germany: Editio Cantor, 1976, p. 48–62.

CHAPTER 8

Ventricular Repolarization and Electrocardiographic T Wave Form and Arrhythmia Vulnerability

MARY JO BURGESS
Nora Eccles Harrison Cardiovascular Research and Training Institute,
and Division of Cardiology, Department of Internal Medicine,
University of Utah College of Medicine,
Salt Lake City, Utah

The electrophysiological events of ventricular repolarization determine the form of the ST-T deflection in the body surface electrocardiogram. On the basis of the concordance in polarity of QRS and T deflections in normal electrocardiograms, it has been suggested that ventricular repolarization properties are inhomogeneous (10, 56, 93, 94). Inhomogeneity of ventricular repolarization properties in localized areas of the ventricles has also been related to vulnerability to ventricular arrhythmias (31–33, 35, 36). The study of ventricular repolarization in intact preparations is complicated by the relatively long duration of ventricular repolarization and the fact that a wide variety of factors including heart rate, autonomic tone, and electrolyte balance influence the process. For these reasons ventricular repolarization in intact hearts has been less completely defined than ventricular activation sequence. Some progress has been made, however, in acquiring data about ventricular repolarization and its relation to ECG wave form and ventricular vulnerability to arrhythmias. The data have been acquired by both indirect analysis of electrocardiographic wave form and by various measurement and recording techniques. In the first section of this review, information about normal ventricular repolarization properties is presented. Models relating ventricular transmembrane action potential form to the ST-T deflection of the body surface electrocardiogram are presented in the second section. The third section reviews studies relating disparity of ventricular functional refractory periods and vulnerability to arrhythmias, and the final section discusses methods of analysis of body surface electrocardiograms designed to detect patients at risk of developing ventricular arrhythmias.

Normal Ventricular Repolarization

Two basic approaches have been used to analyze the electrocardiogram. The forward approach uses available information concerning cardiac electrophysiology and the body as a volume conductor to predict body surface P, QRS, and T wave forms. The inverse approach starts with P, QRS, and T wave forms and infers characteristics of the cardiac generator from these wave forms. Although in clinical interpretation of the electrocardiogram the approach is qualitative, the approach is in fact an inverse solution. In this section normal ventricular repolarization is discussed first in terms of inferences made about the process from body surface electrocardiograms, second in terms of inferences made from cardiac surface electrograms, and finally in terms of extrastimulus techniques for measuring the time of recovery of excitability.

In this review the term ventricular repolarization properties refers to the duration of repolarization. In the sense used, repolarization properties are independent of activation sequence and are analogous to action potential duration. The term recovery time refers to the time of completion of repolarization and is influenced by both repolarization properties and activation sequence. For example, a portion of the ventricle with short repolarization properties, that is, short action potential durations, could have a recovery time later than another portion of the ventricle with long repolarization properties provided the area with short repolarization properties was activated sufficiently later than the area with long repolarization properties.

Body Surface Electrocardiograms. Waller (87) in 1887 was the first to note that the QRS and T deflections of normal mammalian electrocardiograms had the same polarity. Five years later Bayliss and Starling (10) suggested that the concordance of the QRS and T deflections of dogs' electrocardiograms indicated that repolarization properties were longer at the base of the ventricle than at the ventricular apex. Later Wilson et al. (93) suggested that the concordance of the QRS and T deflections indicated that the normal sequence of ventricular recovery differed from the sequence of activation. He further noted that the ventricular gradient pointed in a base-to-apex direction. Wilson et al. (94) defined the ventricular gradient as the area of the QRST deflection. QRST areas were measured in any two standard leads, projected on the triaxial reference system, and expressed as a vector. He suggested the orientation of the QRST axis, or gradient vector, could be due to longer repolarization properties on the endocardium than on the epicardium of the apex of the ventricles, longer repolarization properties at the base than at the apex of the ventricular septum, longer repolarization properties on the epicardium than on the endocardium of the basal portion of the ventricles, or longer repolarization properties at the base of the ventricles than at the apex (93). Wilson et al. (94) also concluded that the sum of the QRS and T areas was a measure of local variations in the duration of the excited state, that is, local variations in ventricular repolarization properties. In addition, Wilson and Johnston (92) noted that QRS and T loops of vectorcardiograms were inscribed in the same direction. This observation suggested that the sequence of recovery shares some features with activation sequence. More recently on the basis of analysis of the T loop of vectorcardiograms, Mashima et al. (51) suggested that activation and recovery sequences differed on the endocardial-epicardial axis, but that the two sequences were similar on the epicardial surface of the ventricles. Abildskov et al. (6) arrived at the same conclusion on the basis of theoretic derivations of the T loop. In those derivations the T loop was considered to consist of two parts, namely, the primary T loop and the secondary T loop. Conceptually, the secondary T loop is the T loop that would result if ventricular repolarization properties were uniform. It would be inscribed in the same direction as the QRS loop but would have a mean axis 180° from the mean axis of the QRS loop. In concept the primary T loop is the T loop that would result if the ventricles depolarized instantaneously and would reflect only ventricular repolarization properties. The addition of the primary and secondary T loops results in a T loop with a form due to both the effects of activation sequence and the effects of repolarization properties. This is equivalent to the recorded T loop. To obtain derived T loops that have the same orientation and inscription direction as normal recorded T loops, repolarization properties must be organized in such a way that 1) vectors of the primary T loop are of greater magnitude than those of the secondary T loop, 2) the primary T loop is located in the normal range of orientations of recorded T loops, and 3) the primary T loop has the same inscription direction as the QRS loop. Examples of additions of primary and secondary T loops are shown in Figure 1. A distribution of longer endocardial than epicardial repolarization properties layered with geometries similar to the geometry of activation wave fronts would result in that form of primary T loop. Horan et al. (43) have also divided the T wave into two imaginary parts and determined methods of choosing the most appropriate time alignment for subtracting the secondary T wave from recorded T waves to yield the primary T wave.

Analysis of body surface isopotential maps of normal ventricular recovery has also permitted inferences concerning features of ventricular recovery sequence (4). As shown in Figure 2, features of sequential QRS maps are repeated in all but the first few milliseconds of the first half of serial T maps recorded on normal subjects. This repetition of QRS features in sequential T maps is another body surface manifestation of similarities between activation and recovery sequences.

Cardiac Surface Electrograms. Analysis of the form of cardiac surface electrograms has also been used to infer features of ventricular repolarization properties. Burdon-Sanderson and Page (11, 12) were the first to recognize the long duration of the excited state in ventricular myocardium and further demonstrated that the duration of the excited state could be altered by injuring or heating the epicardial surface. Later Schütz (66, 67), Haas et al. (30) and Schaefer et al. (63) used the duration of monophasic action potentials to assess

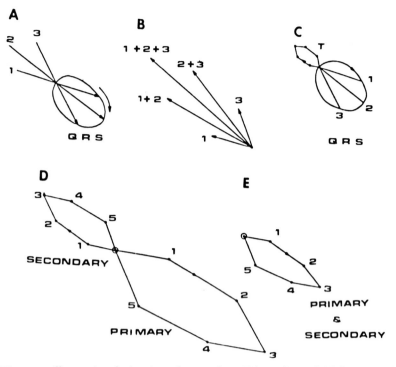

Fig. 1. Diagrams illustrating derivation of secondary T loop from QRS loop (*A–C*) and addition of secondary and primary T loops to give derived T loop with normal orientation and inscription direction (*D, E*). *A:* QRS loop with normal orientation and inscription directions; lines labeled 1, 2, and 3 indicate orientation of T vectors that have been directed opposite QRS vectors and in this step have been given magnitudes equal to QRS vector magnitudes. *B:* 2nd step in derivation, in which T vectors are added and then subtracted to represent increasing and then decreasing numbers of boundaries that exist during repolarization. *C:* last step in derivation of secondary T loop, which involves decreasing magnitude of the T vectors shown in *B* by a factor relating slope of action potential upstroke to slope of downstroke. QRS and secondary T loops both have clockwise inscription directions but have opposite orientations. *D:* secondary T loop and form of primary T loop that must be added to it to give T loop with a normal orientation and inscription direction. *E:* addition of this primary and secondary T loop.

ventricular repolarization and concluded that repolarization lasted longer in basal portions of the heart than at the apex. Pipberger et al. (58), however, found a base-to-apex sequence of recovery. These investigators took the time of the peak amplitude of T waves in multiple unipolar electrograms as their measure of recovery time. More recently Spach and Barr (74) recorded unipolar electrograms from more than 300 intramural and cardiac surface sites of dogs to evaluate ventricular recovery sequence. They displayed their data in the form of isopotential maps. They concluded that recovery proceeds from epicardium to endocardium and that recovery is completed later in the apical endocardial region than in the basal region of the left ventricle. In a later study these investigators found that the cardiac distribution of T potentials during ectopic ventricular drive was determined by the ventricular activation sequence (75). These findings suggest that differences in ventricular repolarization properties

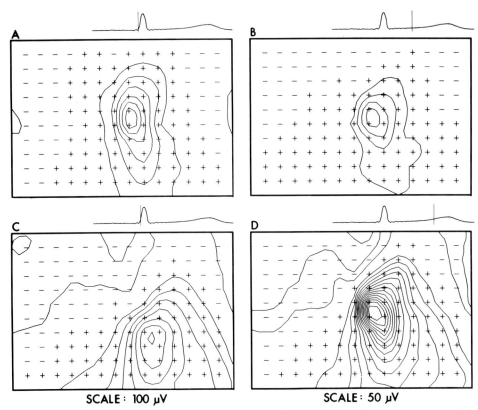

SCALE: 100 μV   SCALE: 50 μV

Fig. 2. Body surface isopotential maps of normal subject. Vertical lines in electrocardiograms shown above each map indicate time of map frames. Maps during early portions of the QRS are shown in *A* and *C*, and they have patterns similar to maps during early portions of the T (*B, D*). Map display represents torso as an unrolled cylinder split along posterior midline. The + and − signs indicate regions of positivity and negativity and also indicate electrode sites. Electrocardiograms recorded from each of these sites were referenced to a Wilson central terminal. *Upper edge* of map is at level of the supersternal notch, *lower edge* is at level of the umbilicus, and *lateral edges* are adjacent to the vertebral column. Electrode sites having same potential are connected. Isopotential lines during QRS are drawn in 100-μV increments and isopotential lines during T are drawn at 50-μV increments. [From Abildskov, Burgess, et al. (4) with permission of the American Heart Association.]

are greater than the time required for normal ventricular activation but are less than the time required for ventricular activation initiated from a ventricular site. That is, the time required to complete normal ventricular activation is less than normal differences in ventricular action potential durations. Recovery sequence is therefore largely determined by the differences in action potential duration, and T waves have the same polarities as QRS deflections. Following ectopic ventricular stimulation, however, the time required to complete ventricular activation exceeds the differences in ventricular action potential durations. Therefore, following ectopic activation, recovery sequence is largely dependent on ventricular activation sequence, and T waves have a polarity opposite to that of the QRS. Spach and Barr's studies differ from most other studies of ventricular repolarization in that they assessed the entire time course of repolarization rather than a single instant during the process such as the peak T amplitude, end of the T wave, or ventricular functional refractory period.

Another method for estimating ventricular repolarization properties and recovery sequence from unipolar cardiac surface electrograms has been developed in our laboratory (97). The method is based on the hypothesis that the time of the peak positive derivative during the T deflection in unipolar electrograms indicates local recovery time just as the time of the peak negative derivative during the QRS deflection indicates local activation time. Studies in which activation and recovery times measured from unipolar cardiac surface electrograms were compared with times measured from in vivo transmembrane action potentials recorded with floating glass micropipettes showed a high degree of correlation between measurements obtained with the two methods in normal and ischemic myocardium, and during various rates, activation sequences, and drug interventions. The ability to determine recovery times from unipolar electrograms will permit evaluation of ventricular recovery sequence during rapidly changing states.

Suction electrode recordings and, in a limited number of studies, recordings of in vivo transmembrane action potentials have also been used to assess the entire time course of ventricular repolarization. Techniques for recording injury potentials were first described by Schütz (65), later modified by Wiggers (89), and subsequently used by many investigators (7-9, 18, 20, 22, 26-29, 42, 45, 63, 68, 72, 78, 95). Wiggers (90) used suction electrograms to study the sequence of ventricular epicardial activation and noted delayed onset and shortening of monophasic action potentials at ventricular sites made ischemic by coronary ar⁺ery occlusion. Most early studies, however, emphasized the physical characteristics of suction electrode recordings and the physiological basis of the recordings rather than documentation of the cardiac state. Suction electrodes are constructed of hollow tubing with a central recording electrode. When negative pressure is applied to the tubing, tissue is sucked into it and damaged. It is generally accepted that cells injured by suction are partially depolarized and inexcitable, and the suction electrodes record potentials across the membranes of surrounding uninjured cells and partially depolarized injured cells (42). Hoffman et al. (42), Autenrieth et al. (8), and Churney and Ohshima (20) compared the forms of suction potential recordings to the form of transmembrane action potentials and concluded that, although the amplitudes of suction potentials were less than those of transmembrane action potentials, the suction recordings reflected the form of the downstroke of the transmembrane action potential with reasonable accuracy.

The sequence of ventricular recovery was studied with suction electrodes by Autenrieth et al. (8). Although they were successful in obtaining satisfactory recordings from only 5-14 sites in each dog, they were able to draw some conclusions about the distribution of ventricular repolarization properties. They found shorter monophasic action potentials on the posterior surface of the ventricles than on the anterior surface and shorter monophasic action potentials at the base than at the ventricular apex. The findings confirmed studies from our laboratory in which ventricular repolarization properties were evaluated with refractory period measurements (15). A more extensive study of the distribution of ventricular monophasic action potentials has recently been conducted in our laboratory (83); examples are shown in Figure 3. In that study suction electrograms were simultaneously recorded from four sites on the epicardial surface of the ventricles, and 32–43 sites were sampled in each dog. The durations of the monophasic action potentials were measured from the time of the rapid upstroke to the time of 90% return of the action potential downstroke to the base line. The sequence of 50% and 90% repolarization was also determined. To do this an arbitrary time in the cardiac cycle was chosen as a reference point, and the times the monophasic action potentials were 50% and 90% returned to base line were determined. These recovery times were dependent on both activation sequence and the duration of monophasic action potentials. Isochrone maps of 50% and 100% repolarization sequence were constructed, and the patterns of these maps were compared to each other and to

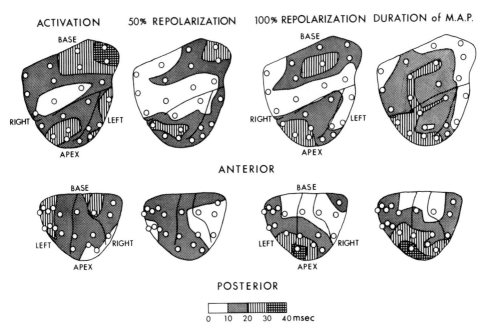

Fig. 3. Maps of epicardial activation sequence in dog, 50% repolarization sequence, 100% repolarization sequence, and pattern of distribution of monophasic action potential durations. Maps are displayed on diagrams of heart, and isochrones are separated by 10 ms. Maps of anterior surface of ventricles are shown in *upper portion* of figure, and maps of the posterior surface of ventricles are shown in *lower portion* of figure. Features of anterior ventricular activation sequence can be identified in both the 50% and 100% repolarization sequence maps of anterior surface of the ventricles. Features of pattern of distribution of monophasic action potential durations can also be identified in the 100% repolarization sequence map of anterior surface of ventricles. Map of 50% repolarization sequence of posterior surface of ventricles has little resemblance to either activation sequence map or pattern of distribution of monophasic action potential durations. However, map of 100% repolarization sequence of posterior surface of ventricles resembles pattern of distribution of monophasic action potential durations. Differences in the 50% and 100% repolarization sequence maps indicate configurations of action potential downstrokes are different in various portions of ventricles.

the pattern of isochrone maps of activation sequence. Maps of the patterns of distribution of monophasic action potentials were also constructed by joining sites at which the durations of monophasic action potentials were equal. The patterns of distribution of monophasic action potentials were also compared to activation sequence map patterns, and to the patterns of the 50% and 100% repolarization sequence maps. On the anterior surface of the ventricles, features of activation sequence were identified in both the 50% and 100% repolarization sequence maps. In addition the pattern of distribution of duration of monophasic action potentials could be identified in the 100% repolarization sequence maps. On the posterior surface of the ventricles the 50% repolarization sequence maps had little resemblance to either activation sequence maps or the pattern of distribution of monophasic action potential durations. There was, however, a striking similarity between the 100% repolarization sequence maps and the pattern of distribution of monophasic action potentials. The differences in the patterns of 50% and 100% repolarization sequence maps found indicate that there were differences in the configurations of transmembrane action potential downstrokes. Figure 4 shows two superimposed suction potentials that were recorded from a dog and demonstates differences in downstroke configurations. In the study of Autenrieth et al. (8) the configuration of the downstroke of suction potentials was the same at all sites sampled, but their sampling was not as extensive as ours. The differences in configuration of downstrokes of mon-ophasic action potentials that we found at various ventricular locations would be likely to affect measurements of the duration of refractoriness made with extrastimulus techniques and further complicate evaluation of repolarization properties with these measurements. Autenrieth et al. (9) have also used suction electrodes to study the effects of isoproterenol infusion on monophasic action potentials, and there have been a limited number of studies in which suction electrodes mounted on cardiac catheters were used in human subjects (45, 72). The latter studies show the feasibility and utility of the method in determining human ventricular repolarization properties. At the present time, however, the method has not been generally accepted, and extrastimulus techniques of measuring refractory periods are used almost exclusively to assess cardiac repolarization in human subjects.

   Although the downstrokes of monophasic action potentials recorded with suction electrodes seem to resemble the form of transmembrane action poten-

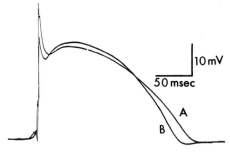

Fig. 4. Superimposed suction potential recordings from anterior epicardial surface of apex of left ventricle (*A*) and anterolateral portion of left ventricle (*B*). Configurations of suction potential downstrokes are different.

tials, additional records of in vivo transmembrane action potentials are desirable in defining the form of the action potential downstroke in both normal and abnormal conditions. A technique for recording transmembrane action potentials in intact hearts was described by Woodbury and Brady (96). The technique uses capillary glass microelectrodes suspended on fine wire and lowered on the heart in such a way that the electrodes "float" on the epicardial surface. Although the technique is ingenious, it still presents a great number of technical difficulties and has been utilized in only a limited number of studies to investigate the effects of myocardial ischemia on the transmembrane action potential (23, 44, 59, 62). There has been no systematic study of the duration and form of in vivo ventricular transmembrane action potentials. In an in vitro study on isolated Purkinje-papillary muscle and free wall of the right ventricle preparations, Moore et al. (57) found longer action potentials in endocardial than in epicardial muscle. This finding is in agreement with the suggestion of Wilson et al. (93) concerning the distribution of repolarization properties, which was based on an analysis of the T wave form and with studies of ventricular refractory periods (15, 85). Cohen et al. (21) also investigated action potential durations in isolated ventricular muscle preparations. In that investigation 2- to 4-mm-square and 1-mm-thick slices of tissue were used. The durations of transmembrane action potentials recorded from the base of the interventricular septum of sheep were compared to durations of transmembrane action potentials recorded from the epicardial surface of the apex. Recordings were obtained following a quiescent period and during steady-state drive. Following a quiescent period, basal endocardial transmembrane action potentials were shorter than the epicardial ones. During steady-state drive, however, apical epicardial action potentials were shorter than basal endocardial action potentials. The authors concluded that such a distribution of durations of transmembrane action potentials could account for the polarity of normal T waves. The study is complicated by the fact that endocardial-to-epicardial gradients of transmembrane action potential durations at the base and apex were not considered separately. That is, epicardial transmembrane action potentials at the apex were not compared to epicardial ones at the base, and endocardial transmembrane action potentials at the apex were not compared to the endocardial ones at the base. The study is further complicated by the fact that electrotonic interactions between Purkinje and muscle fibers present in large tissue preparations may not have been operating in the small sections of tissue used. Mendez et al. (52) showed that if a cut is made between the Purkinje–papillary muscle junction, Purkinje action potential durations shorten. Electrotonic effects on ventricular repolarization have also been suggested by studies in intact hearts in which extrastimulus techniques were used to measure refractory periods (2, 82).

Measurements of Repolarization Properties With Extrastimuli. By far the most extensive information about ventricular repolarization properties in intact hearts has been obtained with refractory period measurements. A relationship between the durations of ventricular refractory periods and monophasic action potential durations was first noted by Burdon-Sanderson and Page (11). The relationship has been generally accepted for normal ventricle (53, 73, 79, 91), and Weidmann (88) and Hoffman et al. (41) demonstrated a quantitative relation between the refractory period and the return of the ventricular transmembrane action potential to its resting level. However, there have been several reports demonstrating that, in the setting of myocardial ischemia, refractoriness can extend beyond the end of the transmembrane action potential (23, 25, 47). Such discrepancies between refractoriness and action potential duration may exist with other repolarization abnormalities as well.

In addition there is increasing evidence that ischemic myocardium responds differently to a wide variety of interventions than nonischemic myocardium. Propranolol prolongs refractory periods more in ischemic than in nonischemic tissue (46). Action potentials of ischemic His bundle fibers are longer at fast rates than at slow rates (47). Refractory periods of premature ventricular complexes at some coupling intervals are longer during control periods than during coronary occlusion (14). The refractory period of the beat following a premature ventricular complex, that is, the postextrasystolic depolarization, prolongs more during coronary occlusion than during control periods (13). Sympathetic stimulation prolongs the refractory period during one phase of coronary occlusion but shortens it in nonischemic tissue (38).

As a step in defining the distribution of mammalian ventricular repolarization properties, Reynolds and Vander Ark (60) measured refractory periods on the epicardial and underlying endocardial surfaces of ventricles of dogs and related the differences in endocardial and epicardial refractory periods to the polarity of T waves recorded with unipolar epicardial electrodes. These investigators rarely recorded positive T waves. In spite of their attempts to keep the hearts warm, there was probably cooling of the epicardial surface of the ventricles, resulting in prolongation of action potentials and increased refractory period duration. In the rare instances when positive T waves were present, however, epicardial refractory periods were shorter than endocardial refractory periods. When T waves were negative, epicardial refractory periods were longer than endocardial refractory periods. These investigators concluded that, although their experimental preparations were not normal, the endocardial-epicardial distribution of refractory periods accounted for the polarity of T waves in epicardial electrograms.

In a similar study van Dam and Durrer (85) used needle electrodes to measure the transmural distributions of refractory periods. They found that refractory periods were longest in endocardial layers of the ventricle, shortest in middle layers, and of intermediate duration in epicardial layers. A more detailed description of the distribution of ventricular refractory periods was subsequently reported from our laboratory (15). Unlike the findings of van Dam and Durrer, in our study refractory periods were shortest in epicardial layers of the ventricle and of intermediate duration in middle layers. The longest refractory periods were found in endocardial layers. In addition we found longer refractory periods at the apex than at the base in both the free wall of the left ventricle and the interventricular septum. We also found longer refractory periods on the left side of the septum than on the right side. The distribution of refractory periods seemed to be inversely related to normal activation sequence. Areas activated early had long refractory periods, and those activated late had short refractory periods. The data from this study are summarized in Figure 5.

Studies of ventricular refractory periods provide information concerning repolarization properties independent of activation sequence, but do not provide

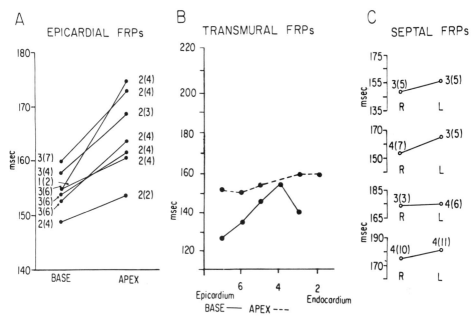

Fig. 5. Graphs showing distributions of ventricular functional refractory periods (FRPs).
A: graph showing that apical FRPs are longer than basal FRPS. Data were obtained from
averages of measurements in 7 dogs. Numbers beside data points indicate number of
sites included in averages, and numbers in parentheses indicate number of measurements
included in averages. B: graph of data from 1 dog illustrating that endocardial FRPs are
longer than epicardial FRPs at both ventricular base (*solid line*) and apex (*broken line*).
C: graphs of data from 4 dogs showing that FRPs on left of septum are longer than those
on right. Values shown are averages, and numbers beside data points indicate number of
sites and number of measurements included in averages.

information concerning the time of recovery following activation. To obtain
recovery sequence data, activation times at the test sites must be added to
refractory period measurements. Recovery sequence maps determined in this
way were reported by Abildskov (1) and were compared to activation sequence
maps. Examples are shown in Figure 6. On the epicardial surface of the
ventricles, recovery sequence was similar to normal activation sequence. However, on the endocardial-epicardial axis the recovery sequence was opposite the
sequence of activation. That is, the endocardium was activated before the
epicardium but recovered after the epicardium. In addition there were some
basal endocardial sites that recovered earlier than apical endocardial sites and
some basal endocardial sites that recovered later than apical endocardial sites.

Sufficient data about normal ventricular repolarization properties are now
available to explain body surface T wave form with reasonable accuracy.
Several models for relating ventricular repolarization properties to T wave form
have been developed, and these are reviewed in the next section.

Models of Ventricular Repolarization

An early model of repolarization was described by Macleod (49). That model
consisted of graphic derivation of T wave form from the graded stages of

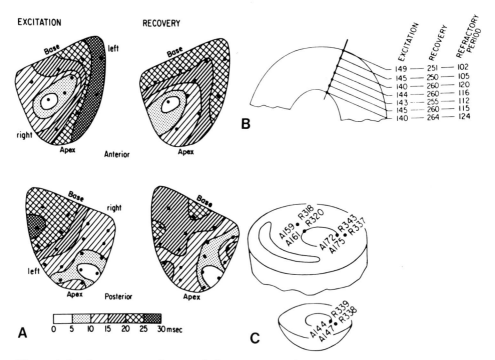

Fig. 6. *A:* isochrone maps of epicardial excitation and recovery sequences in dog. Maps of anterior surface of ventricles are shown in *upper* portion of panel, and those of posterior surface of ventricles are shown in *lower* portion of panel. Isochrones are shown at 5-ms intervals, and earliest activation and recovery times on epicardial surface are shown as zero. *B:* excitation times, recovery times, and refractory periods measured along endocardial-epicardial axis of one dog are shown. Activation proceeded from endocardium to epicardium. However, recovery sequence was from epicardium to endocardium because epicardial refractory periods were shorter than endocardial refractory periods. *C:* Activation (A) and recovery (R) times at basal and apical endocardial sites. Some basal endocardial sites recovered earlier and some later than apical endocardial sites.

recovery observed in muscle strips of frog atria. In Macleod's model, repolarization was considered to be uniform throughout the muscle strip. At the time of that report the form of cardiac transmembrane action potentials was not known. Later Churney et al. (19) and Hecht (40) developed models of T wave form based on the configuration of the downstroke of transmembrane action potentials and the temporal relationship of a limited number of action potentials to each other. In these models, as in Macleod's model, repolarization properties were considered to be uniform.

A satisfactory model of the relation of transmembrane action potential to T waves of the body surface electrocardiogram must consider 1) enough instants during repolarization to define transmembrane action potential downstroke configuration, 2) nonuniformity of transmembrane action potential durations, and 3) a sufficient number of action potentials to define the temporal relationship of repolarization in various portions of the ventricle. Progress in the development of such a model has been made. Harumi et al. (37) developed a theoretic model of the T wave in which the downstroke of a ventricular transmembrane action potential was divided into 54 time units. The difference in amplitude between the beginning and end of each time unit was taken as the change in potential during that moment of repolarization. The ventricles of dogs were divided into sections defined by Scher and Young (64) in their canine ventricular activation sequence maps, and action potential durations were assigned on the basis of van Dam and Durrer's (85) data on the distribution of canine ventricular repolarization properties. We used the model to derive QRS and T waves for normal dogs (37) and to predict T wave changes associated with thermally induced localized alterations of repolarization properties (16). We also used the model to analyze T wave abnormalities associated with myocardial infarction (50). Examples of derived ECGs of normal dogs and ECGs derived for states of acute and chronic ischemia are shown in Figure 7. In the derivations of normal ECGs, the amplitude of QRS vectors was defined by the lengths of lines necessary to close the activation wave fronts in the canine activation sequence maps of Scher and Young (64). The vectors were oriented from relatively negative depolarized areas toward relatively positive areas not yet activated and were made perpendicular to the lines closing activation wave fronts. Action potential durations were assigned to areas within each activation

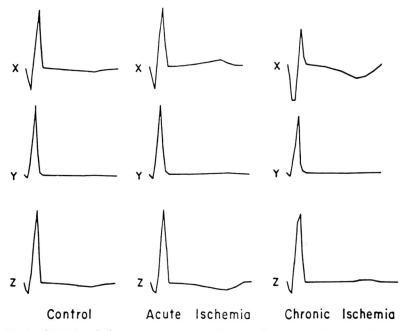

Control   Acute Ischemia   Chronic Ischemia

Fig. 7. Derived ECGs of dog representative of control state and states of acute and chronic ischemia. ECGs were derived using our theoretic model, Scher and Young's data on canine ventricular activation sequence, van Dam and Durrer's data on ventricular refractory period distributions, and our data on effects of acute and chronic ischemia on duration of functional refractory period. X, Y, and Z refer to orthogonal ECG leads. In horizontal lead, X, left-sided electrodes are positive with respect to right electrodes. In vertical lead, Y, leg electrode is positive with respect to neck electrode, and in the anteroposterior lead, Z, back electrode is positive with respect to anterior electrodes. [Adapted from Mandel, Burgess, et al. (50) with permission of the American Heart Association.]

wave front on the basis of the normal ventricular refractory period data of van Dam and Durrer (85). Instantaneous vector magnitudes were determined from the difference in action potential amplitudes in adjacent zones, and the T vectors were oriented from areas less completely repolarized to areas more completely repolarized. In the derivations of ECGs representative of acute and chronic ischemia, action potential durations were assigned to areas assumed to be ischemic on the basis of refractory period measurements. Action potentials of shorter than normal duration were assigned to simulate acute ischemia, and action potentials of longer than normal duration were assigned to simulate chronic ischemia. The model has been adapted (17) to incorporate the data of Durrer et al. (24) on human ventricular activation sequence and more complete data on the distribution of normal canine ventricular repolarization properties (15). In applications of the model to date, the heart has been considered to be in an infinite homogeneous conducting medium, and derived wave forms have been projected on vectorcardiographic leads.

In a related digital model of repolarization, Thiry and co-workers (80, 81) divided the ventricles into eleven segments and assigned action potentials with specific characteristics to each segment. In their model multiple dipole locations and orientations were fixed and dipole magnitudes were variable. T waves derived with this model corresponded to normal T configuration. Miller and Geselowitz (54, 55) also developed a computer simulation of the electrocardiogram.

Models such as these provide insights concerning the relation of the electrophysiological events of cardiac depolarization and repolarization to ECG wave form. They also provide insights concerning the limitations of the standard electrocardiogram for correctly inferring the electrophysiological state from recordings obtained with remote electrocardiographic leads. Models of course are simplifications, and even the most complex models do not completely describe the heart as a generator or the volume conductor properties of the body.

Spach and Barr (76) have taken a somewhat different approach with their analyses of cardiac surface ST-T potentials. They recorded cardiac surface potentials from the epicardial surface of dog hearts during drive of single or multiple ectopic sites. They then related the distribution of these potentials to predicted potentials that were computed from intracellular potential distributions based on action potential configuration and the geometric relations of the potentials to the epicardial recording sites. They found good agreement between the predicted and recorded potential distributions.

Inhomogeneity of Ventricular Repolarization and Arrhythmia Vulnerability

As discussed in the preceding sections, there is a relation between inhomogeneity of ventricular repolarization properties and the T wave of the body surface electrocardiogram. In addition to the relation between inhomogeneity of repolarization properties and T wave configuration, inhomogeneity of ventricular repolarization has been related to enhanced vulnerability of the ventricles to arrhythmias. In a series of papers, Han and co-workers (31–33, 35, 36) documented a relation between several interventions that increased dispersion of ventricular functional refractory periods and arrhythmia vulnerability induced by these interventions. The interventions tested included toxic doses of digitalis and quinidine, sympathetic stimulation, catecholamine infusion, myocardial ischemia, hypothermia, variation in heart rate, and premature ventricular depolarization. Some of the interventions such as sympathetic stimulation, catecholamine infusion, digitalis, premature depolarization, and myocardial ischemia shortened the average duration of refractory periods, but others such as hypothermia, quinidine, and bradycardia prolonged refractory periods. The significant relation between refractory periods and arrhythmia vulnerability was the degree of dispersion of refractory periods rather than refractory-period duration. Because inhomogeneity of ventricular refractory periods plays a role in both arrhythmia vulnerability and T wave form of the body surface ECG, useful prognostic information concerning arrhythmia vulnerability should be provided by the T waves of electrocardiograms of patients.

ECG Detection of Arrhythmia Vulnerability

Prolongation of the Q-T interval has been suggested as one ECG manifestation of increased inhomogeneity of ventricular refractory periods (34). A relation between prolonged Q-T intervals and enhanced arrhythmia vulnerability has been established for several clinical entities. A high incidence of sudden death in patients with the prolonged Q-T interval has been well documented, and this syndrome has been extensively reviewed (69, 86). Various drugs that prolong the Q-T interval are also associated with sudden death (61). During the last few years a relation between prolongation of the Q-T interval and sudden death in patients with coronary artery disease has also been noted (39, 70, 71). Haynes et al. (39) found that 37% of patients with ventricular fibrillation resuscitated outside of the hospital had prolonged Q-T intervals, whereas only 18% of patients with uncomplicated myocardial infarction had prolongation of the Q-T interval. Schwartz and Wolf (70, 71) recorded ECGs of 55 control subjects and 55 patients with recent myocardial infarction. Tracings were obtained at two-month intervals for up to seven years. Only 2% of the control subjects and 18% of surviving patients with myocardial infarction had prolongation of the Q-T interval. However, 57% of patients who had sudden deaths had prolonged Q-T intervals.

Because it is likely that there are conditions with increased dispersion of refractory periods without prolongation of the Q-T interval, other analyses of T wave form are desirable to detect those states. We have developed such an analysis in our laboratory (5, 84). The analysis is based on the hypothesis that normal ventricular repolarization and normal configuration of T waves is representative of a state of low arrhythmia vulnerability. It is also hypothesized that abnormal repolarization may not necessarily increase local dispersion of ventricular recovery times. The action potentials diagrammed in Figure 8, labeled 1, 2, and 3, represent endocardial, intramural, and epicardial action potentials, respectively. In the *upper panel* the action potentials have been assigned normal durations. The sequence of recovery is opposite that of activation, and there is a normal temporal dispersion between the action potential

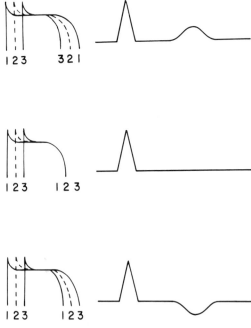

Fig. 8. Diagrams of transmembrane action potentials and electrocardiograms illustrating the concept that the distribution of repolarization properties may be abnormal without an increase in the dispersion of time of recovery. Diagrams at *top* show normal distribution of endocardial-to-epicardial repolarization properties and normal concordance of QRS and T deflections. Diagrams in the *middle* show abnormal distribution of action potentials with all action potentials ending at the same time and a flat T wave. In this example dispersion in time of recovery is less than normal. Diagrams at *bottom* show abnormal distribution of repolarization properties, and QRS and T deflections have opposite polarities. However, the amount of dispersion in time of recovery is no greater than the normal dispersion, illustrated in *top* diagram.

downstrokes. Such a distribution of action potentials would result in QRS and T deflections with the same polarity. In the other panels of Figure 8, the action potentials have abnormal durations. The action potentials in the *middle panel* are all completed at the same time, and the T wave is therefore isoelectric. This abnormality in distribution of action potential durations and abnormal T wave form, however, is associated with less temporal dispersion in action potential downstrokes than the dispersion associated with action potentials of normal duration. In the *lower panel* the sequences of recovery and activation are the same, and the QRS and T deflections would therefore have opposite polarities. The temporal dispersion of action potential downstrokes, however, is no greater than that associated with action potentials of normal duration.

The analysis requires recording electrocardiograms with a lead system sensitive to local cardiac events. We recorded electrocardiograms from 192 body surface sites or from 32 sites from which complete maps were computed (48). The potentials of the QRS and ST-T deflections in each lead were measured at 1-ms increments and integrated over the appropriate intervals. QRS, ST-T, and QRST isoarea maps were constructed by connecting sites at which the deflection areas were equal. A hypothetical series of QRST maps were then derived from an average of QRS isoarea maps obtained on normal subjects and fractions of average normal T maps. The fraction of the T map was allowed to vary from +1 to −1. These fractions of the T map represent states of repolarization without increased dispersion of ventricular refractory periods and were analogous to the two extremes of temporal dispersion of action potential downstrokes illustrated in Figure 8. The QRST isoarea map from the series that had the best match with the average normal QRST isoarea map was then subtracted from the map to be analyzed. The residual map, which we have called a vulnerability map, represented a map of increased dispersion of ventricular repolarization properties. The root mean square voltage of the vulnerability map was also calculated, and this we have called the vulnerability index. In animal experiments (84), states of enhanced vulnerability were induced by premature depolarizations, hypothermia, digitalis intoxication, and catecholamine infusion. In these experiments each dog's control map prior to the intervention was used to generate the hypothetical series of QRST maps against which the postintervention maps were compared. Examples of vulnerability maps of a dog obtained 4 and 40 min after intravenous injection of a toxic dose of ouabain are shown in Figure 9. The increasing density of lines in the two maps represents a state of increasing arrhythmia vulnerability. The animal developed ventricular fibrillation 1 min after the 40-min map was obtained. The root mean square values of vulnerability maps also increased with increasing prematurity of depolarizations, with degree of hypothermia, and during the first 3 min of catecholamine infusion. The analysis has also been applied to a study of a group of 70 normal subjects and 72 patients with old myocardial infarction (3). The vulnerability index of normal subjects averaged 78 mV·ms, and the vulnerability index of patients with old myocardial infarction averaged 148 mV·ms. The vulnerability index was more than 190 mV·ms in 15 of the myocardial infarction patients, and there was a 33% mortality rate in this group during the 2- to 3-yr follow-up period. Mortality rate during the same period was 8% in patients with vulnerability indices less than 130 mV·ms. The results of these studies suggest that vulnerability to arrhythmia can be detected by analysis of electrocardiograms recorded with lead systems sensitive to electrical activity in local cardiac regions. Additional studies of patients and experimental animals with states vulnerable to arrhythmia are being carried out to further document the validity of the analysis.

Conclusion

Considerable progress has been made in defining the normal distribution of ventricular repolarization properties and their relation to body surface T wave form and arrhythmia vulnerability. Most of the observations concerning normal ventricular repolarization have been based on measurements of refractory periods or time of recovery of excitability following activation. The findings have frequently confirmed early speculations concerning the distribution of repolarization properties that were based on analysis of ECG wave form. Studies of refractory periods and time of recovery of excitability, however, provide information concerning only one moment during the recovery process, and the measurements would be expected to be affected by variations in the

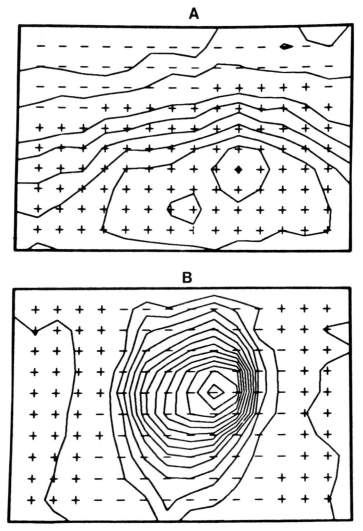

Fig. 9. Vulnerability maps of dog obtained 4 min (A) and 40 min (B) after injection of lethal dose of ouabain. Format of maps is same as that used for isopotential maps in Fig. 2. Increasing density of lines in maps indicates increasing dispersion of repolarization. Dog developed ventricular fibrillation 1 min after map shown in B was recorded.

slope of the action potential downstroke. In addition refractory periods can be measured at only one site at a time, and evaluation of changes in repolarization properties at multiple sites during rapidly changing states is therefore difficult with this method. To advance understanding of ventricular repolarization and its relation to T wave form and arrhythmia vulnerability, further studies of the effects of abnormal cardiac states on the distribution of repolarization properties are needed. Ideally these studies will include not only measurement of refractory periods but new methods of evaluating the entire time course of repolarization and recording methods for evaluating repolarization properties during rapidly changing states. Use of cardiac surface isopotential maps as proposed by Spach and Barr (74, 75), evaluation of the time of peak T wave amplitude in bipolar leads proposed by Spear et al. (77) and Wyatt et al. (97), or peak positive derivative of the T wave in unipolar electrograms proposed by Wyatt (98) are methods especially promising for these purposes. As additional information concerning the configuration of transmembrane action potential downstroke in normal and disease states becomes available, both the diagnostic and prognostic utility of the electrocardiogram should be improved.

Acknowledgments

This research was supported, in part, by Program Project Grant HL-13480 and Research Grant HL-12611 from the National Institutes of Health, Award 76-760 from the American Heart Association, and the Richard A. and Nora Eccles Harrison Fund for Cardiovascular Research.

References

1. ABILDSKOV, J. A. The sequence of normal recovery of excitability in the dog heart. *Circulation* 52: 442–446, 1975.
2. ABILDSKOV, J. A. Effects of activation sequence on the local recovery of ventricular excitability in the dog. *Circ. Res.* 38: 240–243, 1976.
3. ABILDSKOV, J. A., M. J. BURGESS, I. ERSHLER, R. LUX, AND P. URIE. Electrocardiographic recognition of states at high risk of ventricular arrhythmias (Abstract). *Circulation* 58: II-153, 1978.
4. ABILDSKOV, J. A., M. J. BURGESS, R. L. LUX, R. WYATT, AND G. M. VINCENT. The expression of normal ventricular repolarization in the body surface distribution of T potentials. *Circulation* 54: 901–905, 1976.
5. ABILDSKOV, J. A., M. J. BURGESS, P. URIE, R. L. LUX, AND R. F. WYATT. The unidentified information content of the electrocardiogram. *Circ. Res.* 40: 3–7, 1977.
6. ABILDSKOV, J. A., K. MILLAR, M. J. BURGESS, AND L. S. GREEN. Characteristics of ventricular recovery as defined by the vectorcardiographic T loop. *Am. J. Cardiol.* 28: 670–674, 1971.
7. ASHMAN, R., W. S. WILDE, AND N. WOODY. The positive phase of the "injury action potential." *Am. J. Physiol.* 129: 301–302, 1940.
8. AUTENRIETH, G., B. SURAWICZ, AND C. S. KUO. Sequence of repolarization on the ventricular surface in the dog. *Am. Heart J.* 89: 463–469, 1975.
9. AUTENRIETH, G., B. SURAWICZ, C. S. KUO, AND M. ARITA. Primary T wave abnormalities caused by uniform and regional shortening of ventricular monophasic action potential in dog. *Circulation* 51: 668–676, 1975.
10. BAYLISS, W. M., AND E. H. STARLING. On the electromotive phenomena of the mammalian heart. *Int. Monatschr. Anat. Physiol.* 9: 256–281, 1892.
11. BURDON-SANDERSON, J. S., AND F. J. M. PAGE. On the time relations of the excitatory process in the ventricle of the heart of the frog. *J. Physiol. London* 2: 384–435, 1880.
12. BURDON-SANDERSON, J. S., AND F. J. M. PAGE. On the electrical phenomena of the excitatory process in the heart of the frog and of the tortoise, as investigated photographically. *J. Physiol. London* 4: 327–338, 1883.
13. BURGESS, M. J. Refractoriness of postextrasystolic depolarizations of ischemic and non-ischemic myocardium (Abstract). *Clin. Res.* 28-2: 468A, 1980.
14. BURGESS, M. J., AND J. COYLE. The effect of coupling interval on refractory periods in ischemia (Abstract). *Circulation* 56: III-18, 1977.
15. BURGESS, M. J., L. S. GREEN, K. MILLAR, R. WYATT, AND J. A. ABILDSKOV. The sequence of normal ventricular recovery. *Am. Heart J.* 84: 660–669, 1972.
16. BURGESS, M. J., K. HARUMI, AND J. A. ABILDSKOV. Application of a theoretic model to experimentally induced T-wave abnormalities. *Circulation* 34: 669–678, 1966.
17. BURGESS, M. J., AND R. L. LUX. Physiologic basis of the T wave. In: *Advances in Electrocardiography* (2nd ed.), edited by R. C. Schlant and J. W. Hurst, New York: Grune, 1976, p. 327–337.
18. CALABRUSI, M., AND A. J. GEIGER. Potential changes in injured cardiac muscle. *Am. J. Physiol.* 137: 440–446, 1942.
19. CHURNEY, L., R. ASHMAN, AND E. BYER. Electrogram of turtle heart immersed in a volume conductor. *Am. J. Physiol.* 154: 214–250, 1948.

20. CHURNEY, L., AND H. OHSHIMA. Improved suction electrode for recording from the dog heart in situ. *J. Appl. Physiol.* 19: 793–798, 1964.
21. COHEN, I., W. GILES, AND D. NOBEL. Cellular basis for the T wave of the electrocardiogram. *Nature London* 262: 657–661, 1976.
22. CRANEFIELD, P. F., J. A. E. EYSTER, AND W. E. GILSON. Electrical characteristics of injury potentials. *Am. J. Physiol.* 167: 450–456, 1951.
23. DOWNAR, E., M. J. JANSE, AND D. DURRER. The effect of "ischemic" blood on transmembrane potentials of normal porcine ventricular myocardium. *Circulation* 55: 455–462, 1977.
24. DURRER, D., R. TH. VAN DAM, G. E. FREUD, M. J. JANSE, F. L. MEIJLER, AND R. C. ARZBAECHER. Total excitation of the isolated human heart. *Circulation* 41: 899–912, 1970.
25. EL-SHERIF, N., B. J. SCHERLAG, R. LAZZARA, AND P. SAMET. Pathophysiology of tachycardia and bradycardia-dependent block in the canine proximal His-Purkinje system after acute ischemia. *Am. J. Cardiol.* 33: 529–540, 1974.
26. EYSTER, J. A. E., AND W. E. GILSON. The development and contour of cardiac injury potential. *Am. J. Physiol.* 145: 507–520, 1946.
27. EYSTER, J. A. E., AND W. E. GILSON. Electrical characteristics of injuries to heart muscle. *Am. J. Physiol.* 150: 572–579, 1947.
28. EYSTER, J. A. E., AND W. J. MEEK. Cardiac injury potentials. *Am. J. Physiol.* 138: 166–174, 1942.
29. EYSTER, J. A. E., W. J. MEEK, H. GOLDBERG, AND W. E. GILSON. Potential changes in an injured region of cardiac muscle. *Am. J. Physiol.* 124: 717–728, 1938.
30. HAAS, H. G., A. BLÖMER, M. LEY, AND H. SCHAEFER. Experimentelle Untersuchungen am Hundeherzen zum Problem des Ventrikelgradienten. *Cardiologia* 37: 66, 1960.
31. HAN, J., J. DE TRAGLIA, D. MILLET, AND G. K. MOE. Incidence of ectopic beats as a function of basic rate in the ventricle. *Am. Heart J.* 72: 632–639, 1966.
32. HAN, J., P. GARCIA DE JALON, AND G. K. MOE. Adrenergic effects on ventricular vulnerability. *Circ. Res.* 14: 516–524, 1964.
33. HAN, J., P. GARCIA DE JALON, AND G. K. MOE. Fibrillation threshold of premature ventricular responses. *Circ. Res.* 18: 18–25, 1966.
34. HAN, J., AND B. S. GOEL. Electrophysiologic precursors of ventricular tachyarrhythmias. *Arch. Intern. Med.* 129: 749–755, 1972.
35. HAN, J., D. MILLET, B. CHIZZONITTI, AND G. K. MOE. Temporal dispersion of recovery of excitability in atrium and ventricle as a function of heart rate. *Am. Heart J.* 71: 481–487, 1966.
36. HAN, J., AND G. K. MOE. Nonuniform recovery of excitability in ventricular muscle. *Circ. Res.* 14: 44–60, 1964.
37. HARUMI, K., M. J. BURGESS, AND J. A. ABILDSKOV. A theoretic model of the T wave. *Circulation* 34: 657–668, 1966.
38. HAWS, C. W., AND M. J. BURGESS. Effects of sympathetic stimulation and ablation on refractoriness of ischemic tissue. (Abstract). *Clin. Res.* 27-2: 438A, 1979.
39. HAYNES, Q. E., A. P. HALLSTROM, AND L. A. COBB. Repolarization abnormalities in sudden cardiac death syndrome (Abstract). *Clin. Res.* 25: 143A, 1977.
40. HECHT, H. Some observations and theories concerning the electrical behavior of heart muscle. *Am. J. Med.* 30: 720–746, 1961.
41. HOFFMAN, B. F., C. Y. KAO, AND E. E. SUCKLING. Refractoriness in cardiac muscle. *Am. J. Physiol.* 190: 473–482, 1957.
42. HOFFMAN, B. F., P. F. CRANEFIELD, E. LEPESCHKIN, B. SURAWICZ, AND H. C. HERRLICH. Comparison of cardiac monophasic action potentials recorded by intracelluar and suction electrodes. *Am. J. Physiol.* 196: 1297–1301, 1959.
43. HORAN, L. G., R. C. HAND, J. C. JOHNSON, M. SRIDHARAN, T. B. RANKIN, AND N. C. FLOWERS. A theoretic examination of ventricular repolarization and the secondary T wave. *Circ. Res.* 42: 750–756, 1978.
44. KLÉBER, A. G., M. J. JANSE, F. J. L. VAN CAPELLE, AND D. DURRER. Mechanism and time course of S-T and T-Q segment changes during acute regional ischemia in the pig heart determined by extracellular and intracellular recordings. *Circ. Res.* 42: 603–613, 1978.
45. KORSGREN, M., E. LESKINEN, U. SJÖSTRAND, AND E. VARNAUSKAS. Intracardiac recording of monophasic action potentials in the human heart. *Scand. J. Clin. Lab. Invest.* 18: 561–564, 1966.
46. KUPERSMITH, J., H. SHIANG, R. L. LITWAK, AND M. V. HERMAN. Electrophysiological and antiarrhythmic effects of propranolol in canine acute myocardial ischemia. *Circ. Res.* 38: 302–307, 1976.

47. LAZZARA, R., N. EL-SHERIF, AND B. J. SCHERLAG. Disorders of cellular electrophysiology produced by ischemia of the canine His bundle. *Circ. Res.* 36: 444–454, 1975.

48. LUX, R. L., C. R. SMITH, R. F. WYATT, AND J. A. ABILDSKOV. Limited lead selection for estimation of body surface potential maps in electrocardiography. *IEEE Trans. Biomed. Eng.* BME-25: 270–275, 1978.

49. MACLEOD, A. G. The electrocardiogram of cardiac muscle, an analysis which explains the regression or T wave deflection. *Am. Heart J.* 15: 165–186, 1938.

50. MANDEL, W. J., M. J. BURGESS, J. NEVILLE, AND J. A. ABILDSKOV. Analysis of T wave abnormalities associated with myocardial infarction using a theoretic model. *Circulation* 38: 178–188, 1968.

51. MASHIMA, S., L. FU, AND K. FUKUSHIMA. The ventricular gradient and the vectorcardiographic T loop in left ventricular hypertrophy. *J. Electrocardiol.* 2: 55–62, 1969.

52. MENDEZ, C., W. J. MUELLER, J. MEREDITH, AND G. K. MOE. Interaction on transmembrane potentials in canine Purkinje fibers and at Purkinje fiber-muscle junctions. *Circ. Res.* 24: 361–372, 1969.

53. MINES, G. R. On dynamic equilibrium in the heart. *J. Physiol. London* 46: 349–383, 1913.

54. MILLER, W. T., AND D. B. GESELOWITZ. Simulation studies of the electrocardiogram. I. The normal heart. *Circ. Res.* 43: 301–314, 1978.

55. MILLER, W. T., AND D. B. GESELOWITZ. Simulation studies of the electrocardiogram. II. Ischemia and infarction. *Circ. Res.* 43: 315–323, 1978.

56. MINES, G. R. On functional analysis by the action of electrolytes. *J. Physiol. London* 46: 188–235, 1913.

57. MOORE, E. N., J. B. PRESTON, AND G. K. MOE. Durations of transmembrane action potentials and functional refractory periods of canine false tendon and ventricular myocardium. *Circ. Res.* 17: 259–273, 1965.

58. PIPBERGER, H., L. SCHWARTZ, R. MASSUMI, AND M. PRINZMETAL. Studies on the nature of the repolarization process. XIX. Studies on the mechanism of ventricular activity. *Am. Heart J.* 53: 100–124, 1957.

59. PRINZMETAL, M., H. TOYOSHIMA, A. EKMEKCI, Y. MIZUNO, AND T. NAGAYA. Myocardial ischemia. Nature of ischemic electrocardiographic patterns in the mammalian ventricles as determined by intracellular electrocardiographic and metabolic changes. *Am. J. Cardiol.* 8: 493–503, 1961.

60. REYNOLDS, E. W., AND C. R. VANDER ARK. An experimental study on the origin of T-waves based on determinations of effective refractory period from epicardial and endocardial aspects of the ventricle. *Circ. Res.* 8: 943–949, 1959.

61. REYNOLDS, E. W., AND C. R. VANDER ARK. Quinidine syncope and the delayed repolarization syndromes. *Mod. Concepts Cardiovasc. Dis.* 45: 117–122, 1976.

62. SAMSON, W. E., AND A. M. SCHER. Mechanism of ST-segment alteration during acute myocardial injury. *Circ. Res.* 8: 780–787, 1960.

63. SCHAEFER, H., A. PENA, AND P. SCHOLMERICH. Der monophasische aktions-strom vom Spitze und Basis des Warmbluterherzens und die Theorie der T-Welle des Ekg. *Pfluegers Arch.* 246: 728–745, 1943.

64. SCHER, A. M., AND A. C. YOUNG. Ventricular depolarization and the genesis of QRS. *Ann. NY Acad. Sci.* 65: 768–778, 1957.

65. SCHÜTZ, E. Einphasische Aktionsstrome vom in situ durchbluteten Saugertierherzen. *Z. Biol.* 92: 441–452, 1932.

66. SCHÜTZ, E. Elektrophysiologic des Herzens bei einphasischer Albeitung. *Ergeb. Physiol. Biol. Chem. Exp. Pharmakol.* 38: 493–620, 1936.

67. SCHÜTZ, E. Der monophasische Aktionsstrom. *Vehr. Dtsch. Ges. Kreislaufforsch.* 12: 15–43, 1939.

68. SCHÜTZ, E., AND H. LEHNE. Ein Grundexperiment zur Duetung de monophasischem Aktionsstroma Bleitung. *Z. Ges. Exp. Med.* 110: 137–142, 1942.

69. SCHWARTZ, P. J., M. PERITI, AND A. MALLIANI. The long QT syndrome. *Am. Heart J.* 89: 378–390, 1975.

70. SCHWARTZ, P. J., AND S. WOLF. QT prolongation as predictor of sudden death in patients with myocardial infarction. *Proc. Eur. Conf. Cardiol. 7th, Amsterdam* 1: 53, 1976.

71. SCHWARTZ, P. J., AND S. WOLF. QT interval prolongation as predictor of sudden death in patients with myocardial infarction. *Circulation* 57: 1074–1077, 1978.

72. SHABETAI, R., B. SURAWICZ, AND W. HAMMIL. Monophasic action potentials in man. *Circulation* 38: 341–352, 1968.

73. SIKAND, R. S., L. H. NAHUM, H. LEVINE, AND H. GELLER. Excitability of the intact dog heart. *Yale J. Biol. Med.* 34: 366–380, 1952.

74. SPACH, M., AND R. C. BARR. Ventricular intramural and epicardial potential distributions during ventricular activation and repolarization in the intact heart. *Circ. Res.* 37: 243–257, 1975.

75. SPACH, M., AND R. C. BARR. Analysis of ventricular activation and repolarization from intramural and epicardial potential distributions for ectopic beats in the intact dog. *Circ. Res.* 37: 830–843, 1975.

76. SPACH, M. S., AND R. C. BARR. Origin of epicardial ST-T wave potentials in the intact dog. *Circ. Res.* 39: 475–487, 1976.

77. SPEAR, J. F., E. N. MOORE, AND L. N. HOROWITZ. Effect of current pulses delivered during the ventricular vulnerable period upon the ventricular fibrillation threshold. *Am. J. Cardiol.* 32: 814–822, 1973.

78. SUGARMAN, H., L. N. KATZ, A. SANDERS, AND K. JOCHIM. Observations of the genesis of the electrical currents established by injury to the heart. *Am. J. Physiol.* 130: 130–143, 1940.

79. TAIT, J. The relation between refractory phase and electrical change. *Proc. Physiol. Soc. London,* 1910, p. 37.

80. THIRY, P., AND R. M. ROSENBERG. On electrophysiological activity of the normal heart. *J. Franklin Inst. (Mathematical Models of Biological Systems)* 297: 377–396, 1974.

81. THIRY, P. S., R. M. ROSENBERG, AND J. A. ABBOTT. A mechanism for the electrocardiogram response to left ventricular hypertrophy and acute ischemia. *Circ. Res.* 36: 92–104, 1975.

82. TOYOSHIMA H., AND M. J. BURGESS. Electrotonic interaction during canine ventricular repolarization. *Circ. Res.* 43: 348–356, 1978.

83. TOYOSHIMA, H., R. L. LUX, R. F. WYATT, M. J. BURGESS, AND J. A. ABILDSKOV. Sequences of early and late phases of repolarization on dog ventricular epicardium. *J. Electrocardiol.* 14: 143–152, 1981.

84. URIE, P., M. J. BURGESS, R. L. LUX, R. F. WYATT, AND J. A. ABILDSKOV. The electrocardiographic recognition of cardiac states at high risk of ventricular arrhythmias: an experimental study in dogs. *Circ. Res.* 42: 350–358, 1978.

85. VAN DAM, R. TH., AND D. DURRER. Experimental study on the intramural distribution of the excitability cycle and on the form of the epicardial T wave in the dog heart in situ. *Am. Heart J.* 61: 537–542, 1961.

86. VINCENT, G. M., J. A. ABILDSKOV, AND M. J. BURGESS. QT interval syndromes. *Prog. Cardiovasc. Dis.* 16: 523–530, 1974.

87. WALLER, A. D. A demonstration on man of electromotive changes accompanying the heart's beat. *J. Physiol. London* 8: 229–234, 1887.

88. WEIDMANN, S. The effect of the cardiac membrane potential on the rapid availability of the sodium carrying system. *J. Physiol. London* 127: 213–224, 1955.

89. WIGGERS, H. C. Pure monophasic action potentials and their empolyment in studies of ventricular surface negativity. *Proc. Soc. Exp. Biol. Med.* 34: 337–340, 1936.

90. WIGGERS, H. C. The sequence of ventricular surface excitation determined by registration of monophasic action potentials. *Am. J. Physiol.* 118: 333–344, 1937.

91. WILSON, F. N., AND G. R. HERMANN. An experimental study of incomplete bundle branch block and the refractory period of the heart of the dog. *Heart* 8: 229–296, 1921.

92. WILSON, F. N., AND F. D. JOHNSTON. The vectorcardiogram. *Am. Heart J.* 16: 14–28, 1938.

93. WILSON, F. N., A. G. MACLEOD, AND P. S. BARKER. The T deflection of the electrocardiogram. *Trans. Assoc. Am. Physicians* 46: 29–38, 1931.

94. WILSON, F. N., A. G. MACLEOD, P. S. BARKER, AND F. D. JOHNSTON. The determination and the significance of the areas of the ventricular deflections of the electrocardiogram. *Am. Heart J.* 10: 44–61, 1934.

95. WILSON, F. N., A. G. MACLEOD, F. D. JOHNSTON, AND I. G. W. HILL. Monophasic electrical response produced by the contraction of injured heart muscle. *Proc. Soc. Exp. Biol. Med.* 30: 797–799, 1933.

96. WOODBURY, J. W., AND A. J. BRADY. Intracellular recording from moving tissue with a flexibly mounted ultra microelectrode. *Science* 123: 100–101, 1956.

97. WYATT, R. F. Comparison of estimates of activation and recovery times from bipolar and unipolar electrograms to in vivo transmembrane action potential durations. *Proc. IEEE/Eng. Med. Biol. Soc. 2nd Ann. Conf.,* Washington, DC, Sept., 1980. In press.

98. WYATT, R. F., J. A. ABILDSKOV, AND M. J. BURGESS. Measurement of ventricular recovery properties and repolarization sequence from bipolar electrograms (Abstract). *Circulation* 58: II-46, 1978.

CHAPTER 9

Brain Stem Mechanisms and Substrates Involved in Generation of Sympathetic Nerve Discharge

GERARD L. GEBBER
Department of Pharmacology and Toxicology,
Michigan State University, East Lansing,
Michigan

Historical Overview	**Slower Rhythms**
Cardiac-Related Rhythm	**Brain Stem Sympathetic Neurons**
Respiratory-Related Rhythm	

In this chapter the current state of knowledge concerning central mechanisms and substrates responsible for the background discharges in sympathetic nerves is discussed. It is apparent from recent investigations that the classic concept of a randomly discharging and diffusely organized brain stem network onto which rhythms (cardiac- and respiratory-related) are imposed by extrinsic inputs has not passed the test of time. Rather, brain stem networks that govern the discharges of sympathetic nerves are inherently capable of rhythm generation. Sympathetic nerve rhythms inherent to the central nervous system imply the existence of neuronal circuits that are capable of oscillatory activity. The importance of such oscillating circuits is that they provide a mechanism for the coordination of the activity of populations of sympathetic neurons in the absence of periodic input from sources extrinsic to the central nervous system. Indeed the thesis is developed that, rather than creating rhythms in sympathetic nerve discharge, the function of periodic input from extrinsic sources such as the baroreceptors is to entrain rhythms of central origin. Finally recent work on the identification and classification of brain stem neuronal types that comprise those circuits responsible for sympathetic nerve discharge is reviewed.

Historical Overview

A basic problem in research on the neural control of the heart and circulation concerns the origin of the background discharges in sympathetic nerves. This problem was recognized when Bernard (10) demonstrated that transection of the cervical spinal cord led to a pronounced fall in blood pressure that was sustained throughout the course of an acute experiment. The implications of this experiment are clear. First, there is a neurogenic component for the support of resting blood pressure that arises from the background discharges in sympathetic nerves. Second, the background discharges in sympathetic nerves primarily are the consequence of activity generated in neural networks located above the level of spinal transection, i.e., the brain.

In the early 1870s Dittmar (23) and Owsjannikow (59) defined those regions of the brain responsible for the background discharges in sympathetic nerves. Their approach was to study the effects on blood pressure produced by serial transections of the brain stem in the rabbit. Transection of the neuraxis above the caudal one-third of the pons produced little change in blood pressure. Thus it was assumed that the forebrain was not intimately involved in maintaining blood pressure, at least in the anesthetized animal. Transections made more caudally revealed that the background discharges in sympathetic nerves arise primarily from networks located in the caudal one-third of the pons and rostral two-thirds of the medulla (i.e., between A and C in Fig. 1). This point was demonstrated directly with recordings made from the cervical and inferior cardiac sympathetic nerves of the cat in a later study by Alexander (3).

As a consequence of the studies described above, Ranson and Billingsley (63) began an investigation of the location of cardiovascular reactive sites in the brain stem of the cat. They found two discrete areas on the dorsal surface of the medulla (i.e., floor of fourth ventricle) from which blood pressure could be changed with electrical stimulation. A decrease in blood pressure (later shown to be due to inhibition of sympathetic nerve traffic) could be elicited from a medial medullary point near the obex. An increase in blood pressure was produced by stimulation of an area somewhat lateral and rostral to the depressor site. Although Ranson and Billingsley were careful not to attribute their results to the activation of functionally discrete and anatomically circumscribed "centers," others were not so cautious in their interpretations. Thus the misconception of discrete centers for generation of the background discharges in sympathetic nerves arose. The pendulum swung to the opposite extreme in 1939 when Wang and Ranson (74) reexplored the brain stem for cardiovascular reactive sites. The depths as well as the dorsal surface of the brain stem were electrically stimulated in this study. As shown in Figure 1B, the pressor and depressor points found earlier on the dorsal surface of the medulla by Ranson and Billingsley (63) were not anatomically discrete centers but rather the apexes of two triangles that extended almost to the ventral surface of the brain stem through the reticular formation. Decreases in blood pressure were produced most often by stimulation of the medial reticular formation. Increases in blood pressure usually were elicited by stimulation of the periventricular gray and lateral reticular formation. In marked contrast to current views (12, 66, 67) the reticular formation was considered to be diffusely organized by anatomists and physiologists in the 1930s and 1940s. As a consequence the results of Wang and Ranson were interpreted as indicating that the background discharges in sympathetic nerves were randomly generated within a diffusely interconnected network of brain stem reticular neurons. This concept, which still appears in many current textbooks, is the subject of challenge in this chapter.

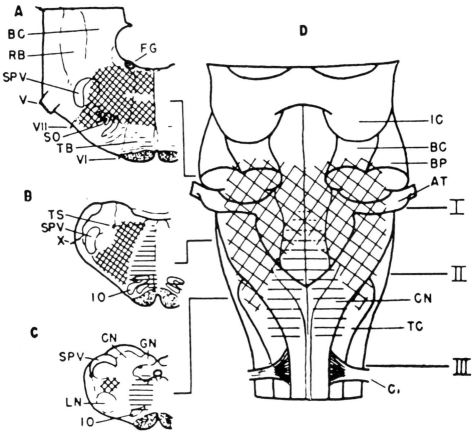

Fig. 1. Pressor and depressor regions of brain stem of cat. Pressor region indicated by *cross-hatching*, depressor region by *horizontal ruling*. A–C: frontal sections through medulla at levels indicated by guide lines to D. D: pressor and depressor regions projected onto dorsal surface of brain stem. I, II, and III in D are levels of transection discussed by Alexander (3). AT, auditory tubercle; BC, brachium conjunctiva; BP, brachium pontis; C_1, first cervical nerve; CN, cuneate nucleus; FG, facial genu; GN, gracilis nucleus; IC, inferior colliculus; IO, inferior olivary nucleus; LN, lateral reticular nucleus; RB, restiform body; SO, superior olivary nucleus; SPV, spinal trigeminal tract; TB, trapezoid body; TC, tuberculum cinereum; TS, tractus solitarius; V, VI, VII, corresponding cranial nerves. [From Alexander (3).]

The background discharges in sympathetic nerve bundles were first characterized by Adrian et al. (2) and Bronk et al. (13) in the 1930s. Most commonly the discharges of populations of pre- or postganglionic sympathetic neurons are synchronized into bursts (i.e., slow waves as recorded with a preamplifier band pass of 1–1,000 Hz) locked in a 1:1 relation to the cardiac cycle (Fig. 2). In addition the amplitude of the cardiac-related slow waves in sympathetic nerve discharge (SND) waxes and wanes with the period of the respiratory cycle. In vagotomized animals the amplitude of the cardiac-related slow waves usually is greatest during the inspiratory phase of the phrenic nerve cycle (Fig. 2). Thus SND contains periodicities (i.e., rhythms) related to both the respiratory and cardiac cycles. As is discussed subsequently, a more rapid periodicity (10 Hz)

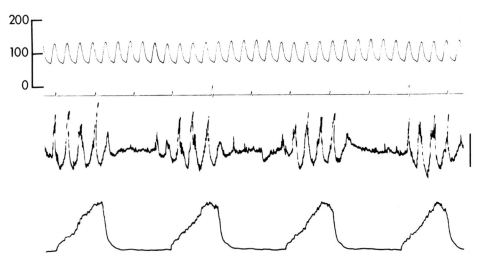

Fig. 2. Rhythmic components in sympathetic nerve discharge of a vagotomized cat. *Top trace* is blood pressure (mmHg). *Middle trace* shows background discharges of external carotid postganglionic sympathetic nerve (negativity recorded as an upward deflection in this and in subsequent figures; preamplifier band pass was 1–1000 Hz). *Bottom trace* shows RC integrated (time constant 0.05 s) phrenic nerve discharge (inspiration recorded as an upward deflection). Time base (below blood pressure) 1 s/division; vertical calibration 40 μV and applies to SND. [From Barman and Gebber (5).]

can be observed in place of or in combination with the cardiac-related rhythm (17, 18, 36–38, 40, 54).

The early electrophysiological experiments led to a rather simple view of the origin of the background discharges in sympathetic nerves. It was assumed that factors such as the local chemical environment (pH, P_{CO_2}, P_{O_2}) and converging inputs from various sensory pathways led to the random generation of discharges within a diffuse brain stem reticular network governing the discharges of preganglionic sympathetic neurons. Furthermore the randomly generated discharges of populations of brain stem reticular neurons were believed to be synchronized by specialized inputs so as to produce the cardiac- and respiratory-related rhythms in SND. The cardiac-related rhythm was considered to result as a simple consequence of the baroreceptor reflexes (2, 18, 41, 43). That is, increased baroreceptor nerve discharge during systole was thought to lead to a delayed central inhibition of SND, whereas the removal of inhibition during diastole was presumed to elicit a reflex increase in SND. It was also assumed (and still is by many investigators) that the brain stem respiratory oscillator imposes its rhythm on the reticular network responsible for the background discharges in sympathetic nerves (18, 51, 60). Thus the cardiac- and respiratory-related rhythms in SND traditionally have been attributed to sources extrinsic to the brain stem generator.

The first important departure from the traditional view for the generation of the background discharges in sympathetic nerves was made by Cohen and Gootman (17, 18) and Green and Heffron (40). These investigators observed in the discharges of sympathetic nerve bundles a 10-Hz rhythm that most often was not related to the phases of the cardiac cycle. An example of 10-Hz activity in splanchnic SND of the vagotomized cat is shown in Figure 3. This work strongly suggests that central networks controlling the discharges of preganglionic sympathetic neurons (i.e., central sympathetic networks) are inherently capable of rhythm generation. Sympathetic nerve rhythms intrinsic to the brain stem and/or spinal cord imply the existence of complexly organized central networks containing either pacemaker cells or oscillating circuits. Thus, as was first suggested by Cohen and Gootman (18), certain patterns of rhythmic activity might be representative of the fundamental organization of those central circuits responsible for the background discharges in sympathetic nerves.

Subsequent work by McCall and Gebber (54) demonstrated a 10-Hz periodicity in autocorrelograms of renal SND during asphyxia in some high spinal cats. Although Gootman and Cohen (38) have not observed a 10-Hz rhythm in SND of spinal cats, McCall and Gebber concluded that this rhythm is generated in a spinal sympathetic network. The question remained open, however, whether brain stem sympathetic networks are inherently capable of rhythm generation.

Cardiac-Related Rhythm

In a series of papers published by our laboratory (27–29, 34, 72) we demonstrated that the cardiac-related rhythm in the discharges of pre- or postganglionic sympathetic nerves does not result as a simple consequence of the baroreceptor reflexes. This contention was based on the following observations. First, a rhythm with a frequency (2–6 Hz) close to that of heart rate persisted in

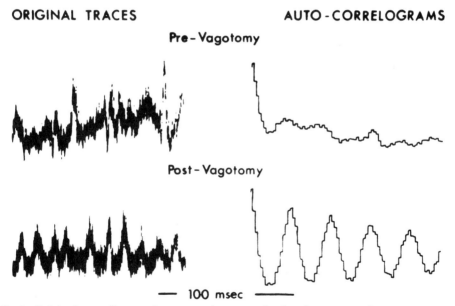

ORIGINAL TRACES   **AUTO-CORRELOGRAMS**

Pre-Vagotomy

Post-Vagotomy

— 100 msec ———

Fig. 3. Original recordings and autocorrelograms of splanchnic sympathetic nerve activity before and after vagotomy in the cat. Autocorrelograms were derived from sample runs containing approximately 500 cardiac cycles. Address bin, 6 ms. Note prominent 10-Hz periodicity in autocorrelogram of SND after vagotomy. [From Cohen and Gootman (18).]

SND of anesthetized cats after complete baroreceptor denervation (Fig. 4). The 2- to 6-Hz rhythm could not be attributed to cardiac-related activity in extra-baroreceptor cardiovascular afferents because the phase relations between SND and the cardiac cycle were completely disrupted after bilateral section of the carotid sinus, aortic depressor, and vagus nerves. This point is demonstrated in Figure 4*IIB*, where it can be seen that the R wave–triggered average of SND

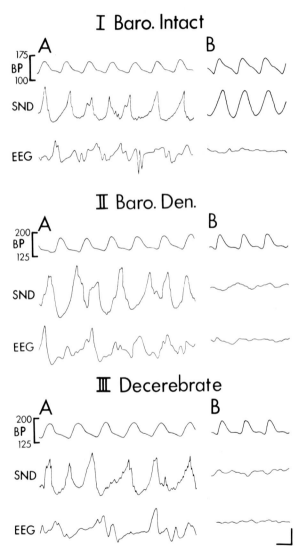

Fig. 4. Activity patterns in splanchnic SND and EEG before and after baroreceptor denervation and after decerebration. *I:* baroreceptor reflexes intact. *A:* oscilloscopic records of blood pressure in mmHg (*top*), splanchnic SND (*middle*), and parietal-frontal EEG (*bottom*). *B:* R wave–triggered computer averaged records (64 trials). Sequence of traces same as in *A*. Address bin was 1 ms. *IIA, B:* same, but after baroreceptor denervation. *IIIA, B:* same, but after midcollicular decerebration. Horizontal calibration 200 ms; vertical calibration 100 μV for SND and 50 μV for the EEG. [From Barman and Gebber (6).]

approached a straight line after baroreceptor denervation. Second, dramatic shifts in the phase relations between carotid sinus baroreceptor nerve activity and SND could be produced by slowing heart rate (with efferent vagal stimulation) in cats with sectioned aortic depressor and vagus nerves (Fig. 5). This observation failed to support the traditional view for the genesis of the cardiac-related rhythm. That is, the phase relations between carotid sinus nerve activity and SND should not have changed when heart rate was slowed if the rhythm indeed resulted as a simple consequence of the baroreceptor reflexes. Third, a single shock applied to the carotid sinus nerve or to intramedullary components of the baroreceptor reflex arc early (50 ms after R wave) in the cardiac cycle extinguished a complete cardiac-related slow wave of SND. In contrast, stimuli

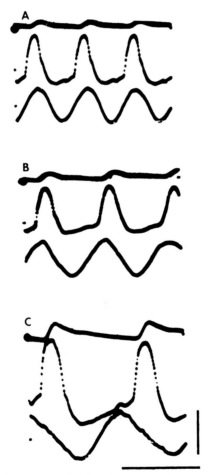

Fig. 5. Phase relations between carotid sinus baroreceptor nerve activity and renal SND at different heart rates in a cat with sectioned aortic depressor and vagus nerves. Records are sum of 64 R wave–triggered trials. Address bin was 1 ms. *Top traces* in each panel show arterial pulse. *Middle traces* show carotid sinus baroreceptor nerve discharge. *Bottom traces* show renal SND. A: control heart rate. B: during slowing of heart rate produced by stimulation of peripheral end of cut right vagus nerve (3 Hz). C: during slowing produced by vagus nerve stimulation (5 Hz). Horizontal calibration 500 ms; vertical calibration 534 μV. [From Gebber (27).]

applied later in the cardiac cycle had no effect on sympathetic nerve slow waves. These observations also failed to support the view that the cardiac-related rhythm results as the consequence of a simple inverse relationship between activity in baroreceptor and sympathetic nerves inasmuch as the slow wave in SND exhibited "all-or-none" characteristics.

The experiments described above led us to conclude that the 2- to 6-Hz rhythm in SND is generated within the central nervous system and then entrained in a 1:1 relation to the cardiac cycle by the baroreceptor reflexes. Subsequent work performed by Barman and Gebber (6) and Gebber and Barman (31) revealed that the rhythm in SND is irregular in form after baroreceptor denervation. That is, the frequency of sympathetic nerve–slow wave occurrence varies from moment to moment between 2 and 6 Hz. Thus, entrainment of sympathetic nerve slow waves in a 1:1 relation to the cardiac cycle by the baroreceptor reflexes serves the purpose of stabilizing the rhythm of central origin. In addition Barman and Gebber (6) demonstrated that the irregular 2 to 6 Hz rhythm in SND is transformed into a regular rhythm (i.e., constant inter-slow wave intervals) during short periods (20–30 s) of asphyxia in the barore-ceptor-denervated cat. Importantly, the rhythm during asphyxia approached the upper limit of the 2- to 6-Hz frequency band (i.e., a value approximately double that of heart rate). Normally the 1:1 relationship between slow waves in SND and the cardiac cycle is maintained during short periods of asphyxia in cats with intact baroreceptor reflexes. Thus entrainment of the 2- to 6-Hz rhythm by the baroreceptor reflexes also prevents large increases in the frequency of occurrence of sympathetic nerve slow waves during stressful situations.

Additional experimentation in our laboratory indicated that the 2- to 6-Hz rhythm in SND is generated in the brain stem. First, although SND and EEG activity contained common frequency components in the range between 2 and 6 Hz after baroreceptor denervation (as demonstrated with spectral and crosscorrelation analyses), we (6, 31) found that midcollicular decerebration did not eliminate the 2- to 6-Hz rhythm in SND (Fig. 4III). This observation does not support the previous suggestion of Camerer et al. (14) that the 2- to 6-Hz rhythm in SND is dependent on the integrity of interconnections between the forebrain and brain stem. Second, McCall and Gebber (54) failed to observe 2- to 6-Hz slow waves in residual SND of the acutely spinalized (C₁ transected) cat either in the normocapnic state or during asphyxia. Thus our experiments seem to rule out the spinal cord and forebrain as sites of origin of the 2- to 6-Hz rhythm in SND.

Gootman and Cohen (38) recently observed irregular oscillations in the 2- to 3-Hz range in SND of high spinal cats during asphyxia or after intravenous strychnine. In contrast to their results in cats with an intact neuraxis, however, there was no correlation on a short time scale (100–500 ms) between irregularly occurring oscillations recorded simultaneously from different sympathetic nerves. Thus Gootman and Cohen (38) concluded that even though oscillations can be generated at the spinal level the normally occurring synchrony between the discharges in different sympathetic nerves is a property of the brain stem. As is discussed subsequently, synchronization on a short time scale between 2 and 6 Hz slow waves in different sympathetic nerves is the rule in decerebrate baroreceptor denervated cats (30). This observation further supports the contention that the 2- to 6-Hz rhythm reported in our studies is generated in the brain stem.

As already mentioned, the 2- to 6-Hz rhythm in SND is irregular in form (i.e., variable inter–slow wave intervals) in the normocapnic baroreceptor denervated cat. Gebber and Barman (30, 32) have tested a number of models that might explain the 2- to 6-Hz rhythm and thus provide information about the intrinsic organization of the brain stem network responsible for the background discharges in sympathetic nerves (i.e., brain stem sympathetic network). One possibility considered was that the rhythm is generated by a solitary brain stem oscillator that is unstable in the absence of baroreceptor entrainment. In this case slow waves in different sympathetic nerves would be locked in a 1:1 relation after as well as before baroreceptor denervation. In addition shifts in the phase relations between slow waves in different sympathetic nerves would fall within the range of conduction times in pathways from the oscillator to the peripheral nerves. A second possibility is that the brain stem generator is comprised of a number of oscillators, each of which drives a separate group of sympathetic neurons. If the oscillators are not coupled and each runs at a different frequency that can vary between 2 and 6 Hz, then slow waves in different sympathetic nerves would not be related in a 1:1 fashion after baroreceptor denervation. Alternatively, the brain stem oscillators might be directly coupled to each other and/or receive common inputs from extrabaroreceptor sources. If the lead oscillator remains the same in each cycle, then the relationships between activity in different sympathetic nerves would be the same as those listed for the single unstable oscillator hypothesis. If the lead oscillator changes from cycle to cycle, however, then shifts in the phase relations between slow waves in different sympathetic nerves might exceed the range of conduction times in pathways from the coupled oscillators to the peripheral nerves. Shifts in the phase relations between slow waves in different sympathetic nerves would result from changes in the sequence of activation of the coupled oscillators. As described below, the data obtained by Gebber and Barman (30, 32) most strongly support the model of coupled oscillators with a shifting lead.

Although baroreceptor denervation transformed the cardiac-related rhythm in SND into an irregular 2- to 6-Hz rhythm, slow waves in postganglionic nerves that exit from different ganglia remained related in a 1:1 fashion. The data presented in Figure 6 are representative of those derived from recordings of activity from different sets of sympathetic nerves including combinations of the external carotid, inferior cardiac, and renal postganglionic nerves. The irregular 2- to 6-Hz rhythm in SND after baroreceptor denervation can be discerned from the inconstant inter–slow wave intervals in each nerve (Fig. 6IIA), the multiple peaks in the power spectra (Fig. 6IIC), and essentially flat autocorrelograms (Fig. 6IID). Persistence of the 1:1 relationship between slow waves in different sympathetic nerves after baroreceptor denervation is indicated by the oscilloscopic traces (Fig. 6IIA) and by the peak at zero lag in the crosscorrelogram of activity in both nerves (Fig. 6IID). Importantly, marked shifts in the phase relations between slow waves in different sympathetic nerves were observed both before and after baroreceptor denervation. This point is demonstrated by the histograms in Figure 7, which show distributions of the intervals between

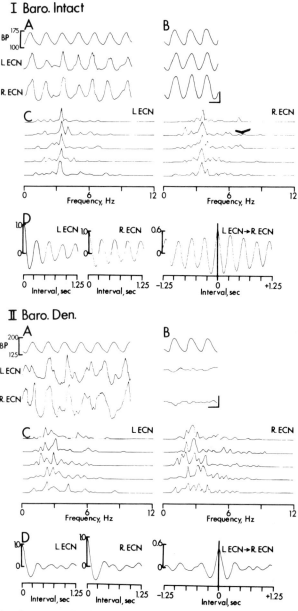

Fig. 6. Relationship between discharges of left and right external carotid postganglionic sympathetic nerves (ECNs) before and after baroreceptor denervation. *I*: baroreceptor reflexes intact. *A*: oscilloscopic records of blood pressure in mmHg (*top*), left ECN discharge (*middle*), and right ECN discharge (*bottom*). *B*: R wave–triggered computer averaged records (64 trials); address bin 1 ms. Sequence of traces same as in *A*. *C*: power spectra for left and right ECN discharges. *Traces* (from *bottom* to *top*) are consecutive spectra; each based on 5.1 s of data. Resolution was 0.05 Hz. Vertical range of computer same for all traces. *D*: traces show left ECN autocorrelogram, right ECN autocorrelogram, and left ECN → right ECN crosscorrelogram. Left ECN → right ECN means that activity in right nerve lags that in left to right of zero time and leads to left of zero time. Analysis time for correlograms 30.7 s; address bin 10 ms. *IIA–D*: same, but after baroreceptor denervation (i.e., bilateral section of carotid sinus, aortic depressor, and vagus nerves). Horizontal calibration 200 ms for *A* and *B*; vertical calibration 50 μV for *left* ECN and 100 μV for right in *A* and *B*. [From Gebber and Barman (30).]

212

the peaks of corresponding slow waves in the external carotid postganglionic branches of the left and right superior cervical ganglia. The peak of the slow wave in the right external carotid nerve could precede or lag that in the left nerve by as much as 30 ms before baroreceptor denervation (Fig. 7IA). Thus the range over which the phase relations shifted was 60 ms. As shown in Figure 7IIA, baroreceptor denervation significantly increased the range over which the phase relations between slow waves in the two nerves shifted. The range of shifts in phase relations between spontaneously occurring slow waves (135 ± 15 ms in 13 experiments) after baroreceptor denervation was more than two-fold greater than the range (50 ms) of conduction times in excitatory pathways

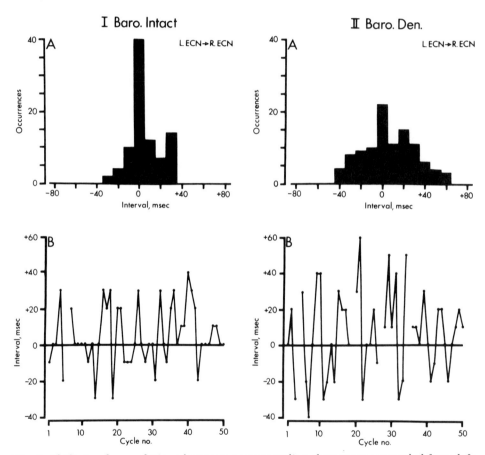

Fig. 7. Shifts in phase relations between corresponding slow waves recorded from left and right ECNs before and after baroreceptor denervation. Data are from same animal as in Fig. 6. *I*: baroreceptor reflexes intact. *A*: histogram shows distribution of intervals between peaks of corresponding slow waves in left and right ECNs and is based on 100 cycles of SND. Number of occurrences is plotted against interval (ms). Positive intervals show instances when peak of slow wave in right ECN lagged that in left ECN (left ECN → right ECN). Negative intervals show instances when peak of slow wave in right ECN led that in left ECN. *Bars* are 10 ms in width centered around any given interval. *B*: interval between peaks of slow waves (left ECN → right ECN) is plotted for 50 consecutive cycles. Breaks in plots are indicative of occurrence of a local slow wave in one of the two nerves. *IIA,B*: same, but after baroreceptor denervation. [From Gebber and Barman (30).]

from the brain stem to the external carotid postganglionic nerve reported by Gebber et al. (35). This study showed that the onset latencies of potentials evoked in the external carotid nerve by stimuli applied to widely separated sympathoexcitatory sites in the lower brain stem ranged from 50 to 100 ms. Figure 7 also shows plots of the intervals between the peaks of corresponding slow waves in the two nerves for 50 consecutive cycles before (*IB*) and after (*IIB*) baroreceptor denervation. These data demonstrate that there was no recurring pattern in the shifts in phase relations between slow waves in the left and right external carotid nerves. The breaks in the plots (Fig. 7*IB, IIB*) indicate that the 1:1 relationship between slow waves in both nerves was occasionally disrupted. That is, local slow waves appeared in one or the other nerve.

The results described above led Gebber and Barman (30, 32) to propose the model for the generation of the 2- to 6-Hz rhythm in SND that is depicted in Figure 8. Shifts in the phase relations between slow waves in postganglionic sympathetic nerves that exit from different ganglia suggested that the driving inputs to these nerves (S_1 and S_2) arise from separate pools of brain stem neurons. Each pool of brain stem neurons (O_1 and O_2) is presumed to form a circuit capable of oscillation in the 2- to 6-Hz range. That the brain stem oscillators are directly coupled to each other and/or receive common inputs from extrabaroreceptor sources (tonic inputs, T) was suggested by two observations. First, the 1:1 relationship between slow waves in postganglionic nerves that exit from different ganglia was maintained after baroreceptor denervation. Second, the value of the crosscorrelation function relating the activity of different sympathetic nerves near zero lag was only slightly reduced by baroreceptor denervation (see Fig. 6*ID, IID*). Importantly the range of shifts in phase relations between slow waves in different sympathetic nerves after baroreceptor denervation was more than two-fold greater than the range of conduction times in excitatory pathways from the brain stem to the postganglionic nerves. This observation further suggested to Gebber and Barman (30, 32) that the leading focus of brain stem activity (i.e., lead oscillator) and consequently the sequence of activation of coupled oscillators changes from cycle to cycle. The shifts in

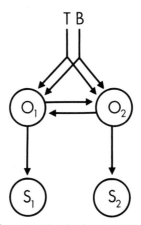

Fig. 8. Model for generation of 2- to 6-Hz rhythm in SND. B, baroreceptor input; O_1, O_2, brain stem oscillators; S_1, S_2, groups of sympathetic neurons; T, tonic extrabaroreceptor inputs to brain stem oscillators. See text for description. [From Gebber and Barman (30).]

phase relations between slow waves failed to show a recurring pattern. Thus the choice of the lead oscillator does not appear to be related specifically to the phases of slower sympathetic nerve rhythms such as the respiratory-related periodicity. Rather the choice of the lead oscillator from cycle to cycle of SND apparently is governed by stochastic processes. These processes, however, are limited to an extent by baroreceptor input. In this regard the range of shifts in phase relations between slow waves recorded in different sympathetic nerves was found to be significantly greater after baroreceptor denervation.

The contention that shifts in the phase relations between slow waves in different sympathetic nerves result primarily from changes in the leading focus of brain stem activity and consequently the sequence of activation of coupled oscillators was further supported by results obtained in experiments with electrical stimulation of brain stem pressor sites. Gebber and Barman (30, 32) demonstrated that the phase relations between potentials evoked in different sympathetic nerves by single shocks applied repetitively (2–6 Hz) to the same pressor site in the pontine or medullary reticular formation were significantly less variable than those between spontaneously occuring slow waves of similar amplitude. Although such experiments did not take into account the element of synchrony of evoked potentials, they nevertheless suggested that shifts in the phase relations between discharges recorded in different sympathetic nerves can be minimized by controlling the leading focus of activity in the brain stem.

The model proposed by Gebber and Barman (30, 32) would explain the inconstant inter-slow wave intervals in any particular nerve in the following manner. The variability of sympathetic nerve-slow wave occurrence between 2 and 6 Hz would result as the consequence of changes in the sequence of activation of coupled brain stem oscillators. That is, the variability of inter-slow wave intervals in any particular postganglionic nerve would arise from changes in the timing of activation of the appropriate pool of brain stem neurons relative to the leading focus of activity in each cycle of SND. Local slow waves occasionally observed in one of a pair of postganglionic nerves that exit from different ganglia might be indicative of the failure of the brain stem oscillator controlling the second nerve to follow the leading focus in a given cycle.

Respiratory-Related Rhythm

As already noted, the background discharges of sympathetic nerves often exhibit a slow rhythmic component with the period of the respiratory cycle. The slow rhythm in SND in part arises from the inhibitory influence of pulmonary inflation afferents in the vagus nerve (19). However, a strong respiratory-related periodicity persists in SND after bilateral vagotomy in paralyzed, pneumothoracotomized, and artificially ventilated cats (Fig. 2). This rhythm generally is assumed to be extrinsically imposed on central sympathetic networks by elements of the brain stem respiratory oscillator (18, 51, 60). The interaction is thought to occur in the brain stem because the discharges of some brain stem units related in time either to the R wave of the ECG (presumably through baroreceptor phasing mechanisms) or to SND also exhibit a respiratory-related periodicity (39, 46, 71). A number of observations made in our laboratory (5, 29), however, contradict the view that the brain stem respiratory oscillator directly imposes its rhythm on central sympathetic networks. Rather our results support the hypothesis that the slow periodic components in sympathetic and phrenic nerve discharges of vagotomized cats are generated by independent brain stem oscillators normally coupled to each other.

First, changes in respiratory rate (as monitored from the phrenic nerve in paralyzed, vagotomized, and artificially ventilated cats) were accompanied by dramatic shifts in the phase relations between phrenic and sympathetic nerve activity. An example is shown in Figure 9. The pattern of phase relations between phrenic and sympathetic nerve activity could be described as inspiratory when respiratory rate was 31 cycles/min (Fig. 9A). That is, SND began to increase near the beginning of inspiration, reached a maximum near the peak of inspiration, and then decayed in time with phrenic nerve activity. This relationship was changed to an expiratory-inspiratory phase-spanning pattern when respiratory rate spontaneously decreased to 24 cycles/min (Fig. 9B). At this time, SND began to increase in early expiration and reached a maximum

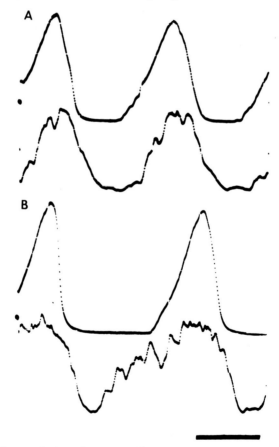

Fig. 9. Shift in phase relations between phrenic nerve activity and external carotid postganglionic SND accompanying decrease in respiratory rate in vagotomized cat. Each panel shows computer-summed records (32 trials) of RC integrated (time constant 0.05 s) phrenic (*top traces*) and sympathetic (*bottom traces*) nerve activity. Sweep of computer was triggered by timing pulse derived near beginning of inspiratory phase of phrenic nerve discharge cycle. Respiratory rate 31 cycles/min in *A* and 24 cycles/min in *B*; horizontal calibration 1 s. [From Barman and Gebber (5).]

during inspiration. These results make it difficult to accept the notion that the respiratory-related periodicity in SND is directly imposed on central sympathetic networks by elements of the brain stem respiratory oscillator. If such was the case, then the phase relations between phrenic and sympathetic nerve discharges should have been independent of respiratory rate.

The independent oscillator hypothesis was further supported by the observation that the slow periodic component in SND persisted in 11 of 18 experiments when the rhythmic discharge pattern of the phrenic nerve disappeared during hyperventilation (5). A typical experiment with this result is shown in Figure 10A. Hyperventilation to the point of phrenic nerve quiescence was accomplished by increasing the respirator pump rate. As reported by others (51, 60), the slow rhythms in phrenic and sympathetic nerve discharges disappeared in parallel in the remaining experiments (Fig. 10B).

Persistence of the slow rhythmic component in SND during hyperventilation is pertinent when viewed in the light of experiments performed by Cohen (15). He reported that hypocapnia in the vagotomized cat led to disappearance of the rhythmic discharges of brain stem respiratory neurons in parallel with those of the phrenic nerves. Thus phrenic nerve quiescence produced by hyperventilation in our experiments presumably was associated with disappearance of the rhythmic discharges in those mutually inhibitory neuronal pairs that are believed to comprise the brain stem respiratory oscillator (16, 21, 44, 45). If this assumption is accepted, then it would be impossible to attribute the slow

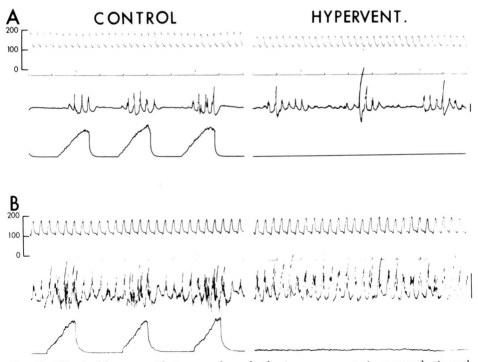

Fig. 10. Effect of hyperventilation on slow rhythmic components in sympathetic and phrenic nerve discharges in two vagotomized cats (A and B). Sequence of *traces* in each *panel* is as described in Fig. 2. Vertical calibrations 40 μV; time base 1 s/division. [From Barman and Gebber (5).]

rhythmic component in SND directly to inputs from the respiratory oscillator. Rather the slow rhythmic components in sympathetic and phrenic nerve discharges would have been generated by independent brain stem oscillators normally coupled to each other. It would also follow that the neuronal types that constitute the slow sympathetic oscillator are less apt to lose their rhythmic discharge patterns during hyperventilation than are those neurons that comprise the brain stem respiratory oscillator.

Slower Rhythms

Oscillations of blood pressure with repetition rates of a few cycles per minute have been observed in experimental animals under various conditions (e.g., hemorrhage). These oscillations are much slower than the respiratory rhythm and apparently are the result of analogous oscillations of SND (24, 48, 49, 61). Such low-frequency blood pressure oscillations have been referred to as third-order, vasomotor, or Mayer waves. A number of investigators (4, 42) have attempted to establish a relationship between the appearance of low-frequency rhythms in chemoreceptor or baroreceptor nerve discharge and Mayer waves. The recent experiments of Preiss and Polosa (61), however, have dispelled the notion that Mayer waves are the direct consequence of cyclic changes in sensory feedback. They demonstrated that carotid and aortic baroreceptor and chemoreceptor deafferentation did not prevent the appearance of Mayer waves in response to hemorrhage. Moreover they found that the rhythm in SND that accompanied Mayer waves was not eliminated when the third-order blood pressure waves were abolished by α-adrenergic blockade or arterial pressure stabilization. This series of experiments, an example of which is shown in Figure 11, precluded sensory feedback of extrasinoaortic nerve origin as a primary factor in the generation of Mayer waves. Thus the results presented by Preiss and Polosa (61) strongly support the view that a central oscillator is responsible for Mayer waves. Mayer waves and analogous rhythms in SND have been observed in spinal (24, 48) as well as in intact animals. This observation suggests that the pattern generator for these very slow rhythms is located in the spinal cord.

Brain Stem Sympathetic Neurons

Models that may explain the genesis of the 2- to 6-Hz rhythm and thus provide information about the intrinsic organization of the brain stem network responsible for the background discharges in sympathetic nerves ultimately must be based on data collected from single neurons. First, those neuronal types that comprise the brain stem 2- to 6-Hz oscillator must be identified. Second, the synaptic interactions among such neurons have to be defined.

The majority of studies directed to the problem of identifying brain stem sympathetic neurons represent attempts to locate single neurons whose discharges are influenced by the baroreceptor reflexes. Brain stem neurons have been located that change their basal discharge rate in response to electrically or pressure-induced activation of baroreceptor nerves (1, 11, 47, 52, 56, 57, 68, 75) and to mechanically or drug-induced alterations in blood pressure (46, 50, 62, 65). In addition post–R wave interval analysis has been used to locate brain

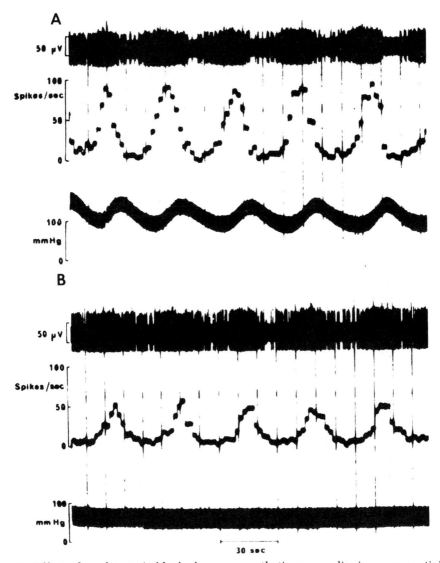

Fig. 11. Effect of α-adrenergic blockade on sympathetic preganglionic neuron activity during an episode of Mayer waves. From *top* to *bottom*: sympathetic cervical neurogram; integrated sympathetic cervical neurogram; systemic arterial pressure. A: control. B: after intravenous administration of 200 μg/kg phentolamine mesylate. Notice persistence of Mayer rhythm in sympathetic nerve activity after elimination of systemic arterial pressure oscillations. [From Preiss and Polosa (61).]

stem neurons whose background discharges exhibit a cardiac-related periodicity, presumably via baroreceptor phasing mechanisms (7, 26, 39, 55, 69, 70). Baroreceptor interneurons within the nucleus of the tractus solitarius have been successfully identified with these approaches (11, 47, 52, 55, 56, 68, 70). However, the usefulness of such studies in identifying neurons in efferent networks that specifically govern the discharges of sympathetic nerves is questionable. As discussed by Barman and Gebber (7, 8), brain stem systems that control nonautonomic function [e.g., respiratory (64), somatomotor (22), cortical (9, 53)] share baroreceptor input with the sympathetic network.

A more promising approach to the problem of identifying brain stem sympathetic neurons has recently been employed by Gootman et al. (39). These workers used the method of crosscorrelation to establish a relationship between brain stem unit activity and splanchnic SND in the cat. Crosscorrelation analysis is a statistical method that allows for a decision as to whether two neuronal elements are interconnected or receive common input (58). The crosscorrelation function between elements A and B is a measure of the expected activity of B relative to the firing times of A. This method has been used extensively to analyze neuronal connections in the thalamus and cortex (25, 73) and in respiratory networks (16, 20). An example of a brain stem unit → splanchnic SND crosscorrelogram from the paper of Gootman et al. (39) is shown in Figure 12. The crosscorrelogram (*top trace*; spike → spl) indicates that peak activity in

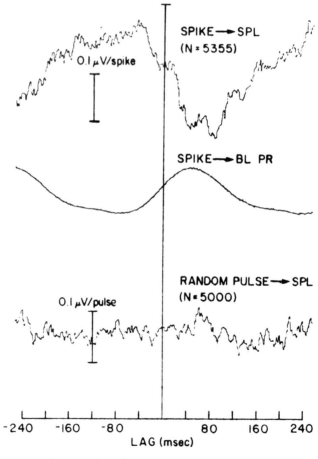

Fig. 12. *Top*: crosscorrelogram describing relationship between spontaneous discharges of a brain stem unit and splanchnic SND. Mean firing frequency of unit was 13.4 Hz. *Middle*: crosscorrelogram showing relationship between discharges of same unit and blood pressure wave. *Bottom*: crosscorrelogram between random pulses and splanchnic SND from same recording. Address bin 2 ms. [From Gootman et al. (39).]

the splanchnic nerve preceded the discharge of the brain stem unit by approximately 40 ms and that minimum splanchnic nerve activity followed spike
occurrence by approximately 80 ms. This relationship suggests that the discharge of the brain stem unit led to a decrease in splanchnic SND. This
interpretation, however, must be viewed cautiously, because the experiments
of Gootman et al. (39) were performed in baroreceptor-innervated cats. Indeed,
the unit → blood pressure crosscorrelogram in Figure 12 (*middle trace*; spike
→ bl pr) shows that the brain stem unit decreased its discharge rate during the
rising phase of systole. Thus the brain stem unit in question may not have been
part of a sympathetic network but rather a component of some other system
that was coupled to the sympathetic network by common inputs from the
baroreceptors. For this reason we have performed crosscorrelation analysis of
brain stem unit activity and SND in baroreceptor denervated cats using the
method of midsignal spike-triggered averaging.

Thirty percent of the medullary reticular neurons sampled in baroreceptor-
denervated cats by Gebber and Barman (33) exhibited discharges that were
temporally related to the 2- to 6-Hz rhythm in inferior cardiac SND. These
neurons were located primarily in nucleus reticularis parvocellularis and nucleus reticularis ventralis. The relationship between the discharges of one of
these neurons and inferior cardiac nerve activity is shown in Figure 13. The
average of SND in Figure 13A was constructed from data that were acquired
prior to and after the discharge of the brain stem unit. Unit spike occurrence is

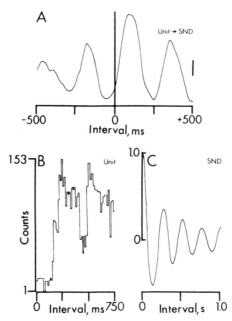

Fig. 13. Relationship between brain stem unit activity and inferior cardiac SND in a
baroreceptor denervated cat. *A*: midsignal spike-triggered average of inferior cardiac
SND (1,200 trials). Unit spike occurrence is at time 0; address bin 1 ms; vertical calibration
5 μV. *B*: autocorrelogram of unit discharges. Address bin 12 ms; analysis based on 2432
spikes. *C*: autocorrelogram of inferior cardiac SND. Analysis time 40 s; address bin 10
ms. [Adapted from Figs. 2 and 3 in Gebber and Barman (33).]

at time 0 in the average of SND. The portion of the average to the left of time 0 is for inferior cardiac nerve activity that preceded unit spike occurrence, while that to the right of time 0 is for SND that followed the discharge of the brain stem neuron. The average of SND contains a rhythm in the 2- to 6-Hz range in activity that both preceded and followed unit spike occurrence. The period of this rhythm was the same as that appearing in the autocorrelograms of the discharges of the brain stem unit (Fig. 13B) and inferior cardiac nerve (Fig. 13C). Identical periodicities in the autocorrelograms and spike-triggered average clearly indicate that the 2- to 6-Hz rhythm in unit and inferior cardiac nerve activity were locked to each other. Thus it is likely that such neurons either were contained in or received inputs from the brain stem oscillator responsible for the 2- to 6-Hz rhythm in SND of the baroreceptor denervated cat.

Gebber and Barman (33) also noted that the discharges of different medullary reticular neurons could be locked to points on the 2- to 6-Hz slow wave in SND that were nearly 180° out of phase. Examples are shown in Figures 13 and 14. Unit spike occurrence coincided with a point close to the onset of the rising phase of the 2- to 6-Hz slow wave in Figure 13 and with a point near the start of the falling phase in Figure 14. These data support the view that the 2- to 6-Hz rhythm in SND results as the consequence of alternations in the discharges of two separate groups of brain stem neurons. Future studies with intracellular recordings and unit → unit crosscorrelation analysis will be required to define the synaptic interrelations of such neuronal pairs.

Figure 15 shows an example of the relationship between the discharges of another type of brain stem neuron and the inferior cardiac nerve. The portion of the average of SND that preceded the unit spike occurrence (i.e., to the left of time 0) was flat. This observation indicates that this brain stem neuron was not influenced by preceding activity in the oscillator responsible for the 2- to 6-

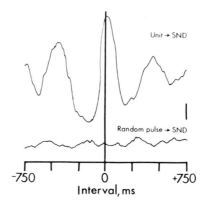

Fig. 14. Synchronization of brain stem unit discharges to a point near beginning of falling phase of the 2- to 6-Hz slow wave in inferior cardiac SND of baroreceptor denervated cat. *Top trace* is midsignal spike-triggered average of inferior cardiac SND; unit spike occurrence at time 0. *Bottom trace* is average of SND constructed with triggers derived from a random pulse train (i.e., "dummy" average). Deflections in the spike-triggered average that exceeded those in the dummy average by at least a factor of 2 were considered to represent changes in SND temporally related to discharges of brain stem unit. Number of trials for each average was 860; address bin 1.5 ms; vertical calibration 5 μV.

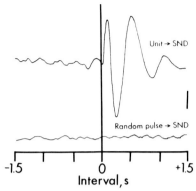

Fig. 15. Resetting of 2- to 6-Hz rhythm in inferior cardiac SND by discharges of brain stem unit in a baroreceptor denervated cat. *Top trace* is midsignal spike-triggered average of SND. *Bottom trace* is average of SND constructed with triggers derived from random pulse train. Number of trials for each average was 769; address bin 3 ms; vertical calibration 20 μV.

Hz rhythm in SND. Thus this neuron neither received inputs from nor was contained in the brain stem sympathetic oscillator. Rather this neuron appeared to provide input to the oscillator as indicated by the appearance of a prominent 2- to 6-Hz rhythm in the portion of the average of SND that followed unit spike occurrence (i.e., to the right of time 0). Furthermore, replication of the rhythm in the discharges of the inferior cardiac nerve to the right but not the left of time 0 in the midsignal spike-triggered average of SND indicates that the discharges of such neurons reset the brain stem oscillator. Gebber and Barman (33) noted that neurons of this type were intermingled with those classified as components of the oscillator.

Although Gebber and Barman (33) were successful in identifying brain stem units with activity related to that in the inferior cardiac nerve of baroreceptor denervated cats, it is impossible to state at this time whether all of these neurons were contained within networks that specifically govern SND. It cannot be denied that some of these neurons may have been contained in nonsympathetic networks coupled to sympathetic circuits via local connections. Nevertheless Gebber and Barman (33) noted that electrical stimuli applied through the microelectrode at sites of sympathetic nerve-related unit activity usually elicited changes in inferior cardiac SND similar to those observed in the corresponding spike-triggered average. Similarities in corresponding spike-triggered and post-stimulus averages of inferior cardiac nerve activity favor the view that, in the majority of instances, Gebber and Barman (33) were dealing with neurons that specifically governed SND.

Acknowledgments

The research from the author's laboratory contained within this chapter was supported by Public Health Service Grant HL-13187.

References

1. ADAIR, J. R., AND J. W. MANNING. Hypothalamic modulation of baroreceptor afferent unit activity. *Am. J. Physiol.* 229: 1357–1364, 1975.
2. ADRIAN, E. D., D. W. BRONK, AND G. PHILLIPS. Discharges in mammalian sympathetic nerves. *J. Physiol. London* 74: 115–133, 1932.

3. ALEXANDER, R. S. Tonic and reflex functions of medullary sympathetic cardiovascular centers. *J. Neurophysiol.* 9: 205–217, 1946.
4. ANDERSSON, B., R. A. KENNEY, AND E. NEIL. The role of chemoreceptors of the carotid and aortic regions in the production of Mayer waves. *Acta Physiol. Scand.* 20: 203–220, 1950.
5. BARMAN, S. M., AND G. L. GEBBER. Basis for synchronization of sympathetic and phrenic nerve discharges. *Am. J. Physiol.* 231: 1601–1607, 1976.
6. BARMAN, S. M., AND G. L. GEBBER. Sympathetic nerve rhythm of brain stem origin. *Am. J. Physiol.* 239 (*Regulatory Integrative Comp. Physiol.* 8): R42–R47, 1980.
7. BARMAN, S. M., AND G. L. GEBBER. Brain stem neuronal types with activity patterns related to sympathetic nerve discharge. *Am. J. Physiol.* 240 (*Regulatory Integrative Comp. Physiol.* 9): R335–R347, 1981.
8. BARMAN, S. M., AND G. L. GEBBER. Problems associated with the identification of brain stem neurons responsible for sympathetic nerve discharge. *J. Autonomic Nerv. Syst.* 3: 369–377, 1981.
9. BAUST, W., AND H. HEINEMANN. The role of the baroreceptors and of blood pressure in the regulation of sleep and wakefulness. *Exp. Brain Res.* 3: 12–24, 1967.
10. BERNARD, C. *Lecons sur la Physiologie et la Pathologie du Systeme Nerveux.* Paris: Bailliere, vol. 1, 1863.
11. BISCOE, T. J., AND S. R. SAMPSON. Responses of cells in the brain stem of the cat to stimulation of the sinus, glossopharyngeal, aortic and superior laryngeal nerves. *J. Physiol. London* 209: 359–373, 1970.
12. BRODAL, A. *The Reticular Formation of the Brain Stem. Anatomical Aspects and Functional Correlations.* London: Oliver and Boyd, 1957.
13. BRONK, D. W., L. K. FERGUSON, R. MARGARIA, AND D. Y. SOLANDT. The activity of the cardiac sympathetic centers. *Am. J. Physiol.* 117: 237–249, 1936.
14. CAMERER, H., M. STROH-WERZ, B. KRIENKE, AND P. LANGHORST. Postganglionic sympathetic activity with correlation to heart rhythm and central cortical rhythms. *Pfluegers Arch.* 370: 221–225, 1977.
15. COHEN, M. I. Discharge patterns of brain-stem respiratory neurons in relation to carbon dioxide tension. *J. Neurophysiol.* 31: 142–165, 1968.
16. COHEN, M. I. Neurogenesis of respiratory rhythm in the mammal. *Physiol. Rev.* 59: 1105–1173, 1979.
17. COHEN, M. I., AND P. M. GOOTMAN. Spontaneous and evoked oscillations in respiratory and sympathetic discharge. *Brain Res* 16: 265–268, 1969.
18. COHEN, M. I., AND P. M. GOOTMAN. Periodicities in efferent discharges of splanchnic nerve of the cat. *Am. J. Physiol.* 218: 1092–1101, 1970.
19. COHEN, M. I., P. M. GOOTMAN, AND J. L. FELDMAN. Inhibition of sympathetic discharge by lung inflation. In: *Arterial Baroreceptors and Hypertension*, edited by P. Sleight. New York: Oxford Univ. Press, 1980, p. 161–167.
20. COHEN, M. I., M. F. PIERCEY, P. M. GOOTMAN, AND P. WOLOTSKY. Synaptic connections between medullary and phrenic motoneurons as revealed by cross-correlation. *Brain Res.* 81: 319–324, 1974.
21. COHEN, M. I., M. F. PIERCEY, P. M. GOOTMAN, AND P. WOLOTSKY. Respiratory rhythmicity in the cat. *Federation Proc.* 35: 1967–1974, 1976.
22. COOTE, J. H., AND V. H. MACLEOD. Evidence for the involvement in the baroreceptor reflex of a descending inhibitory pathway. *J. Physiol. London* 241: 477–496, 1974.
23. DITTMAR, C. Uber die Lage des sogenannten Gefasscentrums der Medulla oblongata. *Ber. Verh. Saechs. Wiss. Leipzig Math. Phys. Kl.* 25: 449–479, 1873.
24. FERNANDEZ DE MOLINA, A., AND E. R. PERL. Sympathetic activity and the systemic circulation in the spinal cat. *J. Physiol. London* 181: 82–102, 1965.
25. FROST, J. D., AND Z. ELAZAR. Three-dimensional selective amplitude histograms: a statistical approach to EEG-single neuron relationships. *Electroencephalogr. Clin. Neurophysiol.* 25: 499–503, 1968.
26. GEBBER, G. L. The probabilistic behavior of central 'vasomotor' neurons. *Brain Res.* 96: 142–146, 1975.
27. GEBBER, G. L. Basis for phase relations between baroreceptor and sympathetic nervous discharge. *Am. J. Physiol.* 230: 263–270, 1976.
28. GEBBER, G. L. Central oscillators responsible for sympathetic nerve discharge. *Am. J. Physiol.* 239 (*Heart Circ. Physiol.* 8): H143–H155, 1980.

29. GEBBER, G. L., AND S. M. BARMAN. Brain stem vasomotor circuits involved in the genesis and entrainment of sympathetic nervous rhythms. In: *Progress in Brain Research. Hypertension and Brain Mechanisms,* edited by W. De Jong, A. P. Provoost, and A. P. Shapiro. Amsterdam: Elsevier, 1977, vol. 47, p. 61-75.
30. GEBBER, G. L., AND S. M. BARMAN. Basis for 2-6 cycle/s rhythm in sympathetic nerve discharge. *Am. J. Physiol.* 239 (*Regulatory Integrative Comp. Physiol.* 8): R48-R56, 1980.
31. GEBBER, G. L., AND S. M. BARMAN. Origin of the cardiac-related rhythm in sympathetic nerve discharge. In: *Arterial Baroreceptors and Hypertension,* edited by P. Sleight, New York: Oxford Univ. Press, 1980, p. 141-148.
32. GEBBER, G. L., AND S. M. BARMAN. Rhythmogenesis in the sympathetic nervous system. *Federation Proc.* 39: 2526-2530, 1980.
33. GEBBER, G. L., AND S. M. BARMAN. Sympathetic-related activity of brain stem neurons in baroreceptor-denervated cats. *Am. J. Physiol.* 240 (*Regulatory Integrative Comp. Physiol.* 9): R348-R355, 1981.
34. GEBBER, G. L., D. G. TAYLOR, AND R. B. McCALL. Organization of central vasomotor system. *Proc. 6th Int. Congr. Pharmacol.* 4: 49-58, 1975.
35. GEBBER, G. L., D. G. TAYLOR, AND L. C. WEAVER. Electrophysiological studies on organization of central vasopressor pathways. *Am. J. Physiol.* 224: 470-481, 1973.
36. GOOTMAN, P. M., AND M. I. COHEN. Periodic modulation (cardiac and respiratory) of spontaneous and evoked sympathetic discharge. *Acta Physiol. Pol.* 24: 97-109, 1973.
37. GOOTMAN, P. M., AND M. I. COHEN. The interrelationships between sympathetic discharge and central respiratory drive. In: *Central Rhythmic and Regulation,* edited by W. Umbach, and H. P. Koepchen. Stuttgart, W. Germany: Hippokrates, 1974, p. 195-209.
38. GOOTMAN, P. M., AND M. I. COHEN. Origin of rhythms common to sympathetic outflows at different spinal levels. In: *Arterial Baroreceptors and Hypertension,* edited by P. Sleight. New York: Oxford Univ. Press, 1980, p. 154-160.
39. GOOTMAN, P. M., M. I. COHEN, M. P. PIERCEY, AND P. WOLOTSKY. A search for medullary neurons with activity patterns similar to those in sympathetic nerves. *Brain Res.* 87: 395-406, 1975.
40. GREEN, J. H., AND P. F. HEFFRON. Observations on the origin and genesis of a rapid sympathetic rhythm. *Arch. Int. Pharmacodyn.* 169: 403-411, 1967.
41. GREEN, J. H., AND P. F. HEFFRON. Studies upon the relationship between baroreceptor and sympathetic activity. *Q.J. Exp. Physiol.* 53: 23-32, 1968.
42. GUYTON, A. C., AND J. W. HARRIS. Pressoreceptor-autonomic oscillation: A probable cause of vasomotor waves. *Am. J. Physiol.* 165: 158-166, 1951.
43. HEYMANS, C., AND E. NEIL. *Reflexogenic Areas of the Cardiovascular System.* Boston, MA: Little, Brown, 1958.
44. HUGELIN, A. Anatomical organization of bulbopontine respiratory oscillators. *Federation Proc.* 36: 2390-2394, 1977.
45. HUKUHARA, T. Neuronal organization of the central respiratory mechanisms in the brain stem of the cat. *Acta Neurobiol. Exp.* 33: 219-244, 1973.
46. HUKUHARA, T., AND R. TAKEDA. Neuronal organization of central vasomotor control mechanisms in the brain stem of the cat. *Brain Res.* 87: 419-429, 1975.
47. HUMPHREY, D. R. Neuronal activity in the medulla oblongata of cat evoked by stimulation of the carotid sinus nerve. In: *Baroreceptors and Hypertension,* edited by P. Kezdi. New York: Pergamon, 1967, p. 131-167.
48. KAMINSKI, R. J., G. A. MEYER, AND D. L. WINTER. Sympathetic unit activity associated with Mayer waves in the spinal dog. *Am. J. Physiol.* 219: 1768-1771, 1970.
49. KOEPCHEN, H. P. *Die Blutdruckrhythmik.* Darmstadt, W. Germany: Steinkopff, 1962.
50. KOEPCHEN, H. P., P. LANGHORST, AND H. SELLER. The problem of identification of autonomic neurons in the lower brain stem. *Brain Res.* 87: 375-393, 1975.
51. KOIZUMI, K., H. SELLER, A. KAUFMAN, AND C. McC. BROOKS. Pattern of sympathetic discharges and their relation to baroreceptor and respiratory activities. *Brain Res.* 27: 281-294, 1971.
52. LIPSKI, J. M., R. M. McALLEN, AND K. M. SPYER. The sinus nerve and baroreceptor input to the medulla of the cat. *J. Physiol. London* 251: 61-78, 1975.
53. MAGNES, J., G. MORUZZI, AND O. POMPEIANO. Synchronization of the EEG produced by low frequency electrical stimulation of the solitary tract. *Arch. Ital. Biol.* 99: 33-67, 1961.
54. McCALL, R. B., AND G. L. GEBBER. Brain stem and spinal synchronization of sympathetic nervous discharge. *Brain Res.* 89: 139-143, 1975.

55. MIDDLETON, S., C. N. WOOLSEY, H. BURTON, AND J. E. ROSE. Neural activity with cardiac periodicity in medulla oblongata of cat. *Brain Res.* 50: 297–314, 1973.

56. MIURA, M. Postsynaptic potentials recorded from nucleus of the solitary tract and its subjacent reticular formation elicited by stimulation of the carotid sinus nerve. *Brain Res.* 100: 437–440, 1975.

57. MIURA, M., AND D. J. REIS. Termination and secondary projections of carotid sinus nerve in the cat brain stem. *Am. J. Physiol.* 217: 142–153, 1969.

58. MOORE, G. P., J. P. SEGUNDO, D. H. PERKEL, AND H. LEVITAN. Statistical signs of synaptic interaction in neurons. *Biophys. J.* 10: 876–900, 1970.

59. OWSJANNIKOW, P. Die tonischen und reflectorischen Centren der Gefassnerven. *Ber. Verh. Saechs. Wiss. Leipzig Math. Phys. K.* 23: 135–147, 1871.

60. PREISS, G., F. KIRSCHNER, AND C. POLOSA. Patterning of sympathetic preganglionic neuron firing by the central respiratory drive. *Brain Res.* 87: 363–374, 1975.

61. PREISS, G., AND C. POLOSA. Patterns of sympathetic neuron activity associated with Mayer waves. *Am. J. Physiol.* 226: 724–730, 1974.

62. PRZYBYLA, A. C., AND S. C. WANG. Neurophysiological characteristics of cardiovascular neurons in the medulla oblongata of the cat. *J. Neurophysiol.* 30: 645–660, 1967.

63. RANSON, S. W., AND P. R. BILLINGSLEY. Vasomotor reactions from stimulation of the floor of the fourth ventricle. *Am. J. Physiol.* 41: 85–90, 1916.

64. RICHTER, D. W., AND H. SELLER. Baroreceptor effects on medullary respiratory neurones of the cat. *Brain Res.* 86: 168–171, 1975.

65. SALMOIRAGHI, G. C. 'Cardiovascular' neurons in brain stem of cat. *J. Neurophysiol.* 25: 182–197, 1962.

66. SCHEIBEL, M. E., AND A. B. SCHEIBEL. Structural substrates for integrative patterns in the brainstem reticular core. In: *Reticular Formation of the Brain,* edited by H. H. Jasper, L. D. Proctor, R. S. Knighton, W. C. Noshay, and R. T. Costello. Boston, MA: Little, Brown, 1958, p. 31–55.

67. SCHEIBEL, M. E., AND A. B. SCHEIBEL. Anatomical basis of attention mechanisms in vertebrate brains. In: *The Neurosciences: A Study Program,* edited by G. C. Quarton, T. Melnechuk, and F. O. Schmitt. New York: Rockefeller Univ. Press, 1967, p. 577–602.

68. SCHWABER, J., AND N. SCHNEIDERMAN. Aortic nerve-activated cardioinhibitory neurons and interneurons. *Am. J. Physiol.* 229: 783–789, 1975.

69. STROH-WERZ, M., P. LANGHORST, AND H. CAMERER. Neuronal activity with relation to cardiac rhythm in lower brainstem of the dog. *Brain Res.* 106: 293–305, 1976.

70. STROH-WERZ, M., P. LANGHORST, AND H. CAMERER. Neuronal activity with cardiac rhythm in the nucleus of the solitary tract in cats and dogs. I. Different discharge patterns related to the cardiac cycle. *Brain Res.* 133: 65–80, 1977.

71. STROH-WERZ, M., P. LANGHORST, AND H. CAMERER. Neuronal activity with cardiac rhythm in the nucleus of the solitary tract in cats and dogs. II. Activity modulation in relation to the respiratory cycle. *Brain Res.* 133: 81–93, 1977.

72. TAYLOR, D. G., AND G. L. GEBBER. Baroreceptor mechanisms controlling sympathetic nervous rhythms of central origin. *Am. J. Physiol.* 228: 1002–1013, 1975.

73. WALLER, H. J., AND S. M. FELDMAN. Correlations between somatosensory thalamic activity and cortical rhythms. *Brain Res.* 57: 417–441, 1973.

74. WANG, S. C., AND S. W. RANSON. Autonomic responses to electrical stimulation of the lower brain stem. *J. Comp. Neurol.* 71: 437–455, 1939.

75. WEISS, G. K., AND K. G. KASTELLA. Medullary single unit activity: response to periodic pressure changes in the carotid sinus. *Proc. Soc. Exp. Biol. Med.* 141: 314–317, 1972.

CHAPTER 10

Peripheral Muscarinic Control of Norepinephrine Release in the Cardiovascular System

E. MUSCHOLL
Department of Pharmacology, University of Mainz,
Mainz, Federal Republic of Germany

The purpose of this chapter is to integrate information from pharmacological, physiological, biochemical, and morphological studies carried out mainly on the cardiovascular system and to present a concept that may be valid for other organ systems as well. The subject of the review is the inhibitory mechanism mediated via muscarine receptors that controls the release of norepinephrine (NE) from peripheral adrenergic nerve fibers. These receptors are not only activated by drugs but also by the endogenous transmitter released by cholinergic nerve stimulation.

Although presynaptic inhibitory receptors in the autonomic nervous system were unknown until 1968, a great amount of work has since been carried out that is reflected by several detailed review articles (46, 62, 64). Thus, not only activation of muscarinic acetylcholine (ACh) receptors but also activation of α-adrenoceptors or receptors for dopamine, prostaglandin E, adenosine, histamine, and opiates inhibits adrenergic neurotransmission. The muscarinic inhibition of NE release, however, was the first peripheral receptor-controlled mechanism to be elaborated, and this was achieved only by measuring transmitter release as the genuine neuronal response rather than its consequence, the response of the end organ. There have been early (25, 40, 41) and recent review articles (11, 42, 58) that deal specifically with the muscarinic inhibitory mechanism.

According to current nomenclature, presynaptic inhibitory receptors are defined by the function that they control, and presynaptic does not imply that evidence as to their location has been obtained (47). The terms liberation and release have been used in a general sense, and it is accepted that the actual quantity of the transmitter detected in perfusates (output from an organ) or incubation media (overflow from an isolated tissue) is only a certain fraction of the amount that has been emitted from all the nerve terminals on the secretory stimulus. Muscarinic effects of ACh include negative chrono-, ino-, and dromotropic actions on the heart, the vasodilatation, the stimulation of gastrointestinal or bronchiolar smooth muscle, and the stimulation of exocrine glands. These effects are blocked by atropine. Nicotinic actions of ACh refer to stimulation of autonomic ganglion cells and the neuromuscular junction, effects that can be blocked by hexamethonium and d-tubocurarine, respectively.

Initial Pharmacological Findings

The studies that ultimately led to the concept of muscarinic inhibition of NE release were initiated in 1967 with the purpose of investigating the effects of nicotinic drugs on the peripheral adrenergic nerve fiber. The perfused rabbit heart preparation had been used for this work because the adrenergic nerve fibers are exclusively postganglionic; the endogenous transmitter released into the perfusion medium can be reliably detected by spectrofluorometry, and the mechanical responses of the organ can be related to the quantity of NE determined. The nicotinic actions of ACh were compared with those of dimethylphenylpiperazinium (DMPP). Whereas ACh has both muscarinic and nicotinic effects, DMPP is a purely nicotinic agent. As a pharmacological tool it is preferred to nicotine because its action is readily reversed by perfusion with a drug-free medium. According to the then prevailing practice, the experiments with ACh were done in the presence of a muscarinic antagonist such as atropine to prevent cardiac arrest. The consensus maintained that muscarine receptor blockade affected end organ responses only. However, after the accidental omission of atropine from the perfusion fluid, it was noted that the transmitter output into the perfusion fluid was greatly reduced if ACh but not if DMPP was administered (ref. 32; see Fig. 1). Furthermore, the property of ACh to cause a large release of NE only in the presence of atropine could be mimicked by a combination of DMPP with the muscarinic agonists methacholine or pilocarpine. Hence pharmacological analysis revealed that the nicotinic NE release was unrestricted only if the concomitant muscarinic activity (present in the ACh molecule) was blocked.

It was soon established that muscarine receptor activation or inhibition did not interfere with the neuronal uptake of NE and therefore its output from the heart, that atropine did not enhance the amine-liberating effect of ACh by delaying its metabolism, and that the postsynaptic effects of cholinomimetic drugs on heart rate, myocardial tension development, and coronary flow could not in any way explain the facilitatory action of atropine on NE overflow evoked by ACh (32).

Hypothesis of Muscarinic Presynaptic Inhibition

The above findings were explained by the hypothesis "that the peripheral adrenergic nerve fibre contains inhibitory muscarine receptors in addition to

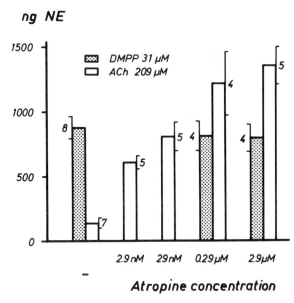

Fig. 1. Effect of atropine on norepinephrine (NE) output evoked by nicotinic effects of dimethylphenylpiperazinium(DMPP) or acetylcholine (ACh) on the isolated rabbit heart. Height of columns, NE output in nanograms after 2-min perfusion of DMPP or ACh in absence (−) or presence of atropine (2.9 nM–2.9 μM perfused 15–60 min before and during DMPP or ACh). Number of experiments from which means + SE were obtained indicated at tops of columns. NE release evoked by ACh (muscarinic and nicotinic activity) is increasingly enhanced as its muscarinic action is progressively antagonized by atropine. Releasing effect of DMPP (only nicotinic activity) is unaltered after atropine. [Data from Lindmar, Löffelholz, and Muscholl (32).]

the excitatory nicotine receptors mediating NE release" (32). Subsequently it was shown that the NE output from the perfused rabbit heart evoked by electrical stimulation of the postganglionic sympathetic nerves also was inhibited by ACh 55 nM–5.5 μM in a concentration-dependent manner (Fig. 2; refs. 35, 39). The inhibitory effect of ACh was blocked by atropine (Fig. 2) but not by hexamethonium, and it was not shared by DMPP (35, 40), thus confirming the clear distinction between muscarinic inhibitory and nicotinic excitatory effects on transmitter release suggested earlier (32, 39). Furthermore, in a study using several muscarinic agonists with greatly differing potencies on cardiac mechanical parameters, a correponding inhibition of NE release evoked by DMPP (12) or electrical stimulation of sympathetic nerves was found (Fig. 3). Again, the effects were sensitive to atropine. Because the NE release caused by the indirectly acting amine, tyramine, was not blocked either by methacholine or by omission of calcium ions from the perfusion medium, it was suggested that muscarine receptor activation inhibits those mechanisms of release that are linked to electrical events on the membrane and to the entry of calcium (35). The original findings and the hypothesis derived afforded a lead to search for muscarinic inhibition of NE release in organs other than heart and to explore its possible physiological significance.

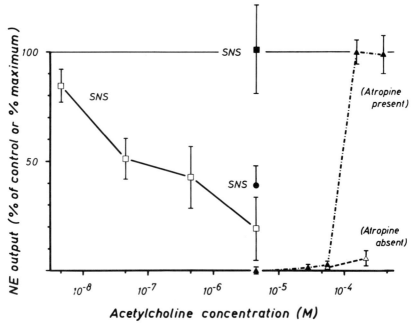

Fig. 2. Muscarinic and nicotinic effects of ACh on release of NE from perfused rabbit heart. In sympathetic nerve stimulation (SNS, 600 impulses at 10 Hz, □) ordinate gives NE output as percent of control output obtained in absence of ACh. ACh was added 1 min before SNS. Hexamethonium 6.6 μM (●) or atropine 2.9 μM (■) was perfused before ACh and SNS. In nicotinic NE release evoked by ACh above 10 μM [*broken lines*, presence (▲) or absence (△) of atropine 1.4 μM] ordinate is expressed as percent of maximum output. Abscissa, molar concentration of ACh, log scale. Values are means ± SE of 3–8 observations. Note that muscarinic inhibition of NE release evoked by SNS occurs at concentrations of ACh that are below threshold for nicotinic release of NE. Inhibitory effect of ACh 5.5 μM is reversed by atropine but not by hexamethonium. [Data from Löffelholz and Muscholl (35) and Muscholl (40).]

Occurrence of Muscarinic Inhibition of NE Release in Various Organ Systems

The facilitatory effect of atropine on the NE release evoked by ACh as observed originally in rabbit and guinea pig isolated hearts (32) was soon confirmed for the cat heart (19). Likewise inhibition by muscarinic drugs of transmitter release evoked by electrical stimulation of adrenergic nerves has been demonstrated in various organs and species. In a few instances the release of endogenous NE has been measured, e.g., on the perfused cat spleen (23, 24), the heart of the chicken (10) or guinea pig (26), and the dog saphenous vein (57), but in most of the studies on smooth muscle preparations release has been monitored after labeling the transmitter stores with radioactive NE (for review of work using guinea pig vas deferens, rabbit pulmonary and ear artery, dog retractor penis, and various vessel preparations see refs. 11, 42, 46, 64). In those studies in which the radioactive compounds were separated chemically it was ascertained that ACh caused a decrease in overflow of [^{3}H]NE (9, 59, 60). Interestingly, ACh inhibited the stimulation-evoked [^{3}H]NE overflow from

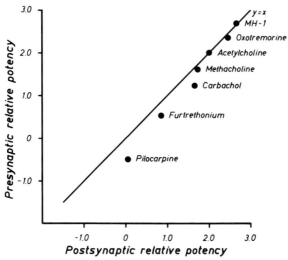

Fig. 3. Relative potencies of seven muscarinic agonists on pre- and postsynaptic responses of isolated perfused rabbit heart. Inhibition by drugs of NE output evoked by sympathetic nerve stimulation at 10 Hz, 600 impulses, taken as presynaptic response. Decrease of atrial systolic tension development by drugs taken as postsynaptic response. Potencies (log scale) of agonists were calculated from negative log molar concentrations causing 50% inhibition (NE) or decrease (tension) and expressed relative to that of ACh (at as log 100 = 2.0). MH-1, N-methyl-1, 2, 5, 6-tetrahydronicotinic acid prop-2-yne ester. Line y = x represents identical potencies of agonists at pre- and postsynaptic muscarine receptors. [Values from Fozard and Muscholl (12).]

rabbit pulmonary artery to the same degree as the overflow of [³H]dihydroxyphenylglycol (derived from released NE taken up and deaminated by intraneuronal monoamine oxidase) and [³H]normetanephrine (derived from NE metabolized by catechol-O-methyltransferase extraneuronally) (9). If ACh decreased NE overflow by enhancing uptake and/or metabolism, then these metabolite fractions should have been increased. In agreement with this, the inhibitory effect of ACh on ³H overflow was unaltered in the presence of cocaine or corticosterone that block neuronal and extraneuronal amine uptake, respectively (9). However, pharmacological interventions with the uptake and degradation of the transmitter raise its concentration in the biophase and thereby activate the α-adrenoceptor-mediated negative feedback control of NE release (46, 62, 64). On the other hand, inhibition by ACh of evoked NE release from the dog saphenous vein was still obtained after blockade of pre- and postsynaptic α-adrenoceptors with phentolamine or phenoxybenzamine (60), ruling out the possibility that the muscarinic inhibition acts through facilitation of the presynaptic effect of NE. The above-mentioned observations that the presynaptic muscarinic inhibition also operates after an elevation of NE levels agree with results on the rabbit heart according to which the inhibitory effect of ACh on NE output persisted in the presence of amphetamine, which increased the stimulation-evoked transmitter overflow sixfold (41). Finally, muscarinic inhibition does not require the functional integrity of the prostaglandin-mediated feedback system (14), which in various organs restricts the NE liberation from adrenergic nerves (46, 50, 64).

Presynaptic Muscarinic Effects Evoked
by Stimulation of Cholinergic Nerves

The possibility of cholinergic-adrenergic interactions has been the matter of numerous experiments and several review articles (3, 4, 11, 25, 29). Since the discovery by Hillarp (21) of the autonomic ground plexus in which terminal cholinergic and adrenergic axons run side by side, such juxtaposed fibers, without intervention of insulating Schwann cell processes, have been found in various organs, e.g., iris (8, 54), vas deferens (53, 54), pancreatic arterioles (17), myenteric plexus of the gut (37) and heart, especially atrium (8, 53). Widely held views suggested that the liberation of ACh causes (3, 8, 28) or facilitates (17) NE release from nearby adrenergic fibers. However, in the light of the evidence that the muscarinic inhibitory receptors have a much higher affinity for ACh than the nicotinic excitatory ones, it was pointed out that in the absence of atropine the most likely consequence would be inhibition of NE release rather than its facilitation (32, 35, 40).

This open question was put to the test by experiments on perfused rabbit atria that were isolated with the vagus and sympathetic nerves intact (36). Supramaximal stimulation at 20 Hz of both vagus nerves decreased by 48% the NE output evoked by simultaneous stimulation at 10 Hz of the right postganglionic sympathetic nerves. The perfusion medium contained amphetamine in order to enhance the NE output to a level that could be determined by spectrofluorometry. The ventricles were discarded because preliminary results on the whole rabbit heart had shown that vagus stimulation did not significantly decrease the evoked output of NE (K. Löffelholz and E. Muscholl, unpublished observations), which may be related to the fact that the ACh concentration in the ventricles is only 19% that of the atria (22).

Recently two additional isolated organ studies have strengthened the view that cholinergic nervous activity inhibits adrenergic transmission presynaptically. First, on a perfused rabbit lung preparation with the sympathetic and parasympathetic innervation intact the evoked output of endogenous NE (10 Hz) was inhibited by concurrent stimulation of the vagi at 10 Hz or by methacholine (38). Second, a combined morphological and functional study was carried out on the rabbit jejunum preparation incubated with either $[^3H]NE$ or $[^3H]$choline (37). The overflow of radioactivity from the gut during perivascular stimulation of sympathetic nerves at 4 Hz was found to be decreased by exogenous ACh, and this effect was reversed by atropine. Furthermore, electrical field stimulation of the jejunum, which activates all nerves indiscriminately and released $[^3H]ACh$, evoked an overflow of tritium derived from $[^3H]NE$ that was larger in the presence than in the absence of atropine. In the myenteric plexus there was ultrastructural evidence of axoaxonic synapses between adrenergic and nonadrenergic, probably cholinergic, terminal fibers. Finally, perivascular nerve stimulation (in the presence of physostigmine) decreased the overflow of $[^3H]ACh$ evoked by field stimulation of the gut, thus supporting the idea of a reciprocal adrenergic-cholinergic inhibitory interaction (37). The impressive case for an α-adrenoceptor-mediated inhibition of ACh release has recently been reviewed (62), but for anatomical reasons it has not been settled whether inhibition occurs through an action on the perikaryon or on the terminal fiber of the cholinergic neuron.

The strongest evidence for a physiological role of the muscarinic inhibition of NE release comes from investigations on dog hearts in vivo. In the first study (31) electrical stimulation of the left cardiac sympathetic nerves at 2 and 4 Hz increased ventricular force of contraction by 40%–50% and evoked NE overflow into the coronary sinus blood. Concurrent stimulation of both vagi at 15 Hz decreased ventricular force by 22% and 26%, and NE output by 33% and 29%, respectively. At 2 Hz sympathetic stimulation the vagally induced changes in ventricular force and NE overflow were blocked by atropine. The modulation by vagus stimulation of both ventricular force of contraction and NE output, which is predominantly derived from terminal fibers in the ventricles, shows that in the dog the muscarinic inhibition must have affected the bulk of cardiac adrenergic nervous supply. In the second study (27) stimulation of the right cardioaccelerator nerves at 10 Hz caused an increase in coronary sinus blood concentration of catecholamines that was progressively attenuated by simultaneous stimulation of the right cervical vagus at frequencies increasing from 1–16 Hz. The reductions of catecholamine outputs (up to 66%) were associated with decreased responses of heart rate and left ventricular dP/dt. Furthermore, the effect of vagus stimulation on amine output was mimicked by methacholine and antagonized by atropine. A functional role for a presynaptic control of NE release is suggested by the fact that significant inhibitory effects were obtained at vagus stimulation frequencies of 1–4 Hz.

Cholinergic-Adrenergic Interactions Not Due to Presynaptic Muscarinic Effects

In the past, various examples of cholinergic-adrenergic interactions have been described that can only in part be attributed to a presynaptic muscarinic mechanism (4, 6, 20, 29, 30). Briefly, the antagonistic interactions have been classified (30) into two main categories: 1) interneuronal mechanisms comprising the muscarinic inhibition, the nicotinic (or vagally induced) excitation, particularly in the presence of atropine, and the excitation by stimulation of sympathetic fibers running in the vagus nerve, that has been called a false interaction (30); and 2) intracellular mechanisms that are not dependent on a presynaptic effect of ACh and that are observed also when adrenergic stimulation is brought about by exogenous NE. These interactions are currently explained by the hypothesis that NE evokes a positive inotropic effect by elevating the intracellular level of cyclic adenosine monophosphate, and that this effect is antagonized by muscarinic stimulation either directly or through increasing the concentration of cyclic guanosine monophosphate (for review see refs. 20, 29, 30).

It has often been observed that addition of ACh to preparations of mammalian ventricular myocardium has little effect on contractility, whereas strong negative inotropic actions are seen in the presence of a background level of adrenergic activity, evoked either by nerve stimulation or by exogenous NE (20). Reinterpretation of these findings should take into account both pre- and postsynaptic sites for reciprocal actions, as outlined above.

Because of the reliable functioning of electrical vagus stimulation, the majority of interaction studies have been performed on the dog, which provides a further phenomenon, the postvagal tachycardia (5). In this case there is firm evidence that vagus stimulation releases catecholamines from extraneuronal stores, possibly chromaffin cells, by a nicotinic action that outlasts the postsynaptic vagus effect after cessation of the stimulation (1, 5, 33). Release of NE from adrenergic nerve fibers can be excluded as the cause of the postvagal tachycardia because the tachycardia was neither prevented by pretreatment of the dogs with 6-hydroxydopamine (1, 33) nor enhanced after administration of desipramine, a blocker of neuronal uptake (33).

Caution must be used when results obtained with electrical field stimulation of incubated organs are interpreted. This mode of excitation simultaneously and indiscriminately activates all kinds of neurons present. Moreover, the transmitters overflowing into the incubation bath rather than escaping into the circulatory system of a perfused organ build up to larger concentrations because of longer diffusion distances and may reach targets not accessible under physiological conditions. Thus, anticholinergic drugs have been found to enhance, and anticholinesterases to decrease, NE overflow evoked by field stimulation of isolated organs known to be innervated by cholinergic and adrenergic nerves (see refs. 11, 42, 64). It remains to be established, however, whether extraneous generation of cholinergic activity decreases NE release in a similar manner to field stimulation.

Mechanisms Underlying Muscarinic Inhibition

Type of Receptor Involved. The pharmacological characterization of a receptor is achieved by 1) its sensitivity to reversible blockade by low concentrations of selective antagonists and 2) by its sensitivity, in relative terms, to various selective agonists that for a given receptor should exhibit a typical order of potency. As to the latter procedure, the potency ratio, relative to ACh, of six muscarinic agonists that was obtained for inhibition of atrial tension development on the rabbit heart was similar to that obtained for inhibition of NE output evoked by sympathetic nerve stimulation (Fig. 3). In a complementary way, the potencies of six muscarinic antagonists against methacholine at post- and presynaptic sites were determined and expressed as pA_2 values (pA_2 negative log molar concentration of the antagonist that requires a twofold concentration of the agonist, i.e., methacholine, to achieve control response). Despite a range of affinities of 5 log units among the antagonists, their pA_2 values determined on atrial tension and NE output did not differ (Fig. 4). These results suggest that the pre- and postsynaptic muscarine receptors occurring in the rabbit heart are similar. This situation is different from that revealed by corresponding studies of the pre- and postsynaptic α-adrenoceptors of various organs including the rabbit heart where a clear dissimilarity exists (46, 62, 64). In contrast to the presynaptic nicotine receptor (19, 34, 40) the inhibitory muscarine receptor is not desensitized by the continuous presence of an agonist. Thus methacholine was just as effective as an inhibitor of NE release from the rabbit heart if administered 0.5 or 15 min before a high potassium solution (43).

Site of Control. In a topographical sense the evidence for the location of inhibitory muscarine receptors on the terminal adrenergic nerve fiber is only indirect. Originally, this location had been suggested (32) because inhibitory muscarine receptors were already known to be present on the postsynaptic

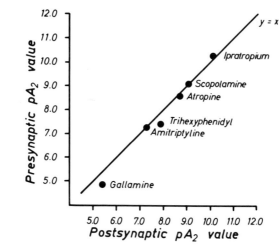

Fig. 4. Pre- and postsynaptic pA_2 values of six muscarinic antagonists against metha-choline on isolated rabbit heart. Inhibition of NE output evoked by sympathetic nerve stimulation at 3 Hz, 540 impulses (presynaptic response) and decrease of atrial tension development (postsynaptic response). *Line* $y = x$ represents identical affinities of antag-onists to pre- and postsynaptic muscarine receptors. [Values from Fuder, Muscholl, et al. (13) and determined according to Arunlakshana and Schild (2).]

membrane of the sympathetic ganglion cell (52) and because this assumption provided the most reasonable explanation of the results obtained. This still holds true after taking into account the subsequent investigations (see pages 228–233) and the lack of evidence of any other mechanism that could serve as an alternative, e.g., transmission of the inhibitory signal from extraneuronal tissue to the neuron by a prostaglandin-based link or involvement of mechanical, vascular, or metabolic factors. Strong evidence of a functional coupling and therefore topographically close association of membranal events and the mus-carinic inhibition is provided by the following studies.

Calcium Availability for Release. It was noted at an early stage that calcium-dependent processes of NE release such as those evoked by electrical nerve stimulation or by nicotinic drugs are subject to muscarinic inhibition, whereas NE release by tyramine, which occurs independently of calcium ions, is not modified by muscarinic activity (35). These findings have been amply confirmed and extended to various conditions and tissues (Table 1). Furthermore, the release of α-methyladrenaline that had been incorporated into the adrenergic nerves of the rabbit heart as a false transmitter was inhibited by methacholine if evoked by the depolarizing stimuli listed in Table 1 but was unaffected if elicited by tyramine (15). Three observations are particularly suggestive of muscarinic inhibition affecting the process of calcium-dependent electrosecre-tory coupling. First, a high potassium concentration causes NE release by a local depolarization of the adrenergic terminal fiber (19), and any modulation must be exerted at the very site of secretion (41). Second, a decrease in external calcium concentration enhanced the inhibitory effect of methacholine on po-tassium-evoked NE release (7). Third, muscarinic inhibition was inversely related to the frequency of stimulation of the splenic nerves (24). If the rate of

Table 1. Relation Between Occurrence of Muscarinic Inhibition and Mechanisms
Underlying NE Release From Peripheral Adrenergic Neurons

Mechanism of NE Release*	Stimulus Used	Muscarinic Inhibition of NE Release	Ref. No.
Membrane depolarization, Ca^{2+} requirement; $D\beta H$ release, exocytosis	Electrical nerve stimulation	Established	12, 23, 35, 36, 49, 56, 59
	Nicotinic drugs	Established	12, 19, 32, 40, 41, 65
	K^+	Established	7, 16, 41, 44, 60, 61
Carrier-mediated release in absence of depolarization, Ca^{2+} not required, no $D\beta H$ release	Tyramine	Not found	15, 35, 57, 59
	Low Na^+ solutions	Not found	7, 16

NE, norepinephrine; $D\beta H$, dopamine β-hydroxylase.   * For reviews see refs. 42, 46, 62, and 64.   [Adapted from Muscholl (41, 42).]

influx of calcium ions is low (either as a result of a reduced external concentration or at a low stimulation frequency) a deceleration by muscarinic agonists would lead to a proportionally greater effect than at a maximum rate of calcium influx.

The calcium availability hypothesis has been criticized on the grounds that the interaction between presynaptic inhibition and the rate-limiting role of calcium may be due to a functional antagonism (46) that does not permit any conclusions about the mode of action to be drawn. Recently, Stjärne (50) proposed an alternative hypothesis centered on the modulation of recruitment of varicosities for release (see below). Stjärne presented a kinetic analysis of data obtained on the isolated guinea pig vas deferens, which allows discrimination between the effect of a modulator of NE release on calcium availability on the one hand and on control of recruitment on the other. If the latter process predominates, modulation of the transmitter release evoked by potassium is absent.

Recruitment of Terminal Fibers. It has been suggested that the endogenous α-adrenoceptor-mediated control is primarily directed toward invasion of impulses into the more distal parts of the terminal fibers, thus depressing the recruitment of varicosities (50). Because calcium dependency of release, as an additional factor, is compatible with the latter proposal, the possibility has been considered that muscarinic presynaptic inhibition also may be attributed to a decrease of the number of secretory units invaded by the impulse (50). The relative contribution of the latter mechanism to in vivo modulation of release has not been determined for any presynaptic control system. However, in vitro a nearly complete inhibition by methacholine of the NE output from the rabbit heart evoked by 135 mM potassium (7) contrasts with the inefficiency of oxymetazoline, a potent α-adrenoceptor agonist at presynaptic sites, to inhibit NE output evoked by 80 mM potassium (48). Likewise, the [³H]NE effux from the rat vas deferens evoked by 117 mM potassium was inhibited by carbachol but not altered by clonidine (44). On the other hand, reintroduction of calcium during depolarization of nerve fibers of the rabbit heart by a calcium-deficient medium containing 140 mM potassium evoked NE release that was found to be decreased by both methacholine and oxymetazoline (16). Because in the latter experiments the NE output was much smaller than in the experiments that failed to show an α-adrenoceptor-mediated inhibition (48), it is possible that maximal activation of the presynaptic α-adrenoceptor by a high concentration of endogenously liberated NE precluded an inhibitory effect of oxymetazoline. Thus a greater response from the muscarinic compared to the α-adrenoceptor-mediated inhibitory mechanism may be created by the conditions of the experiment, and such results cannot be used as an argument for a preferential action of muscarinic inhibition on electrosecretory coupling as opposed to recruitment. Therefore, the recruitment hypothesis remains an interesting possibility that should be tested in future experiments.

Hyperpolarization of Terminal Fiber. If muscarinic drugs inhibit the invasion of the terminal fiber by the action potential, they might do so by causing hyperpolarization of the membrane (50) in a manner similar to the block of invasion of the soma by an antidromic spike (18). Hyperpolarization has also been suggested to be the cause of the inhibitory effect of ACh on the NE overflow evoked by a high potassium solution on the dog saphenous vein (61). It is, however, unlikely that a complete depolarization by potassium in excess of 100 mM could be antagonized by muscarine receptor activation despite clear evidence that the NE releasing action is inhibited by methacholine or carbachol (7, 16, 43, 44). Indirect evidence of a hyperpolarizing effect of muscarinic drugs on the adrenergic terminal neuron is provided by two observations. First, on cardiac sympathetic nerves of the cat the amplitude of retrograde asynchronous discharges elicited by infusion of ACh was decreased by pilocarpine (19). Second, the muscarinic compounds found to inhibit the NE output of the rabbit heart in response to nerve stimulation (see Fig. 3) or DMPP are known to cause hyperpolarization of the sympathetic ganglion cell (12).

Unfortunately, from these findings rather different conclusions could be drawn, and it has even been pointed out that hyperpolarization might increase the amplitude of the action potential of the varicosity membrane and thereby the quantity of NE released (18). It is felt that the state of knowledge of the electrical events in the adrenergic terminal fiber is unsatisfactory, and that this will hardly change until electrophysiological investigations on terminal axons are feasible.

Prospects of Further Research

Since the neuronal location of presynaptic muscarine receptors is only suggestive, efforts will be made to obtain direct evidence. One experimental approach is to label muscarine receptors with radioligands in normal tissue and after sympathectomy. Thus it has been reported (45) that the high-affinity binding of [^{3}H]quinuclidinyl benzilate to muscarine receptors of a membrane fraction of rat heart was decreased after treatment of the animals with 6-hydroxydopamine, which destroys adrenergic nerve terminals. However, in another study (51) binding of the ligand to the muscarine receptors of rat ventricle membranes was found to be unaltered by 6-hydroxydopamine although the binding of [^{3}H]dihydroergocryptine was decreased, indicating the loss of neuronal and presumably presynaptic α-receptors. It seems important not only to resolve this apparent discrepancy but also to obtain functional evidence of presynaptic muscarine receptors in the rat heart. The latter question has recently been studied, and presynaptic inhibitory muscarine receptors have been found that do not differ from the postsynaptic receptors of the rat heart in their susceptibility to blockade by N-methylatropine or pirenzepine (15a).

Useful models for the study of cholinergic-adrenergic interactions may be developed in future. Favored targets will be organs in which an autonomic ground plexus has been detected. For example, an innervated perfused rabbit atria preparation has been described recently in which the NE and ACh stores are labeled with different isotopes (42a). If related to the fractional rate of [^{14}C]ACh release, the [^{3}H]NE overflow due to 3 Hz sympathetic stimulation was optimally inhibited at 3 Hz vagus nerve stimulation; 10 or 20 Hz vagus stimulation caused a larger [^{14}C]ACh release but failed to further enhance muscarinic presynaptic inhibition suggesting that the vagally induced control is exerted at discrete sites rather than ubiquitously in the tissue.

On the other hand, positive functional evidence of cholinergic control of the release of NE like that mentioned in section **Presynaptic Muscarinic Effects Evoked by Stimulation of Cholinergic Nerves** should encourage anatomists to search for the morphological basis of such observations. One stimulating hint is given by a recent report that on the blood-perfused stomach of the dog vagal stimulation at 10 Hz, through an atropine-sensitive mechanism, inhibits the vasoconstrictor response to sympathetic stimulation at 5 Hz more than that produced by exogenous NE (55). Such an interaction, if present in humans, might well explain the beneficial effect of vagotomy on diffuse gastric bleeding. Another challenge to further research, on a morphological and biochemical level, comes from the observation that vagosympathetic interactions are less pronounced in the dog atrioventricular junction than in the sinoatrial node (63). The variables presumed to be involved—axoaxonic distances, presence of cholinesterase, density of autonomic receptors—are now liable to experimental manipulations. This field of physiological interest has a clear bearing on clinical problems, such as dysrhythmias, notably those resulting from myocardial infarction (6).

Another direction for research will be to probe further the link between the inhibitory muscarine receptor and the releasing process. This will lead to a more rigorous trial of the calcium availability and recruitment of terminals hypotheses. Finally, because muscarine receptors modulate the exocytotic transmitter release, cholinergic drugs may be used as pharmacological tools for neurochemical studies.

Acknowledgments

The experiments reported from this laboratory have been supported by Deutsche Forschungsgemeinschaft Grants Mu 118/10, 118/11, and 118/13.

References

1. ALTER, W. A., III, G. K. WEISS, D. V. PRIOLA, AND H. A. SPURGEON. The effects of 6-hydroxydopamine on vagal cardioaccelerator systems. *J. Pharmacol. Exp. Ther.* 187: 99–104, 1973.
2. ARUNLAKSHANA, O., AND H. O. SCHILD. Some quantitative uses of drug antagonists. *Br. J. Pharmacol.* 14: 48–58, 1959.

3. BURN, J. H., AND M. J. RAND. A new interpretation of the adrenergic nerve fibre. *Adv. Pharmacol.* 1: 1–30, 1962.

4. CAMPBELL, G. Autonomic nervous supply to effector tissues. In: *Smooth Muscle*, edited by E. Bülbring, A. F. Brady, A. W. Jones, and T. Tomita. London: Arnold, 1970, p. 451–495.

5. COPEN, D. L., D. P. CIRILLO, AND M. VASSALLE. Tachycardia following vagal stimulation. *Am. J. Physiol.* 215: 696–703, 1968.

6. CORR, P. B., AND R. A. GILLIS. Autonomic neural influences on the dysrhythmias resulting from myocardial infarction. *Circ. Res.* 43: 1–9, 1978.

7. DUBEY, M. P., E. MUSCHOLL, AND A. PFEIFFER. Muscarinic inhibition of potassium-induced noradrenaline release and its dependence on the calcium concentration. *Naunyn-Schmiedelbergs Arch. Pharmacol.* 291: 1–15, 1975.

8. EHINGER, B., B. FALCK, AND B. SPORRONG. Possible axo-axonal synapses between peripheral adrenergic and cholinergic nerve terminals. *Z. Zellforsch.* 107: 508–521, 1970.

9. ENDO, T., K. STARKE, A. BANGERTER, AND H. D. TAUBE. Presynaptic receptor systems on the noradrenergic neurones of the rabbit pulmonary artery. *Naunyn-Schmiedebergs Arch. Pharmacol.* 296: 229–247, 1977.

10. ENGEL, U., AND K. LÖFFELHOLZ. Presence of muscarinic inhibitory and absence of nicotinic excitatory receptors at the terminal sympathetic nerves of chicken hearts. *Naunyn-Schmiedebergs Arch. Pharmacol.* 295: 225–230, 1976.

11. FOZARD, J. R. Cholinergic mechanisms in adrenergic function. In: *Trends in Autonomic Pharmacology*, edited by S. Kalsner. Baltimore, MD: Urban & Schwarzenberg, 1979, p. 145–194.

12. FOZARD, J. R., AND E. MUSCHOLL. Effects of several muscarinic agonists on cardiac performance and the release of noradrenaline from sympathetic nerves of the perfused rabbit heart. *Br. J. Pharmacol.* 45: 616–629, 1972.

13. FUDER, H., C. MEISER, H. WORMSTALL, AND E. MUSCHOLL. The effects of several muscarinic antagonists on pre- and postsynaptic receptors in the isolated rabbit heart. *Naunyn-Schmiedebergs Arch. Pharmacol.* 316: 31–37, 1981.

14. FUDER, H., AND E. MUSCHOLL. The effect of methacholine on noradrenaline release from the rabbit heart perfused with indomethacin. *Naunyn-Schmiedebergs Arch. Pharmacol.* 285: 127–132, 1974.

15. FUDER, H., E. MUSCHOLL, AND R. WEGWART. The effects of methacholine and calcium deprivation on the release of the false transmitter, α-methyladrenaline, from the isolated rabbit heart. *Naunyn-Schmiedebergs Arch. Pharmacol.* 293: 225–234, 1976.

15a.FUDER, H., D. RINK, AND B. ALT. Pirenzepine and functional pre- and postsynaptic muscarine receptors in rat and rabbit heart. *Naunyn-Schmiedebergs Arch. Pharmacol.* 316: (Suppl.) R52, 1981.

16. GÖTHERT, M. Effects of presynaptic modulators on Ca^{2+}-induced noradrenaline release from cardiac sympathetic nerves. *Naunyn-Schmiedelbergs Arch. Pharmacol.* 300: 267–272, 1977.

17. GRAHAM, J. D. P., J. D. LEVER, AND T. L. B. SPRIGGS. An examination of adrenergic axons around pancreatic arterioles of the cat for the presence of acetylcholinesterase by high resolution autoradiographic and histochemical methods. *Br. J. Pharmacol.* 33: 15–20, 1968.

18. HAEFELY, W. Electrophysiology of the adrenergic neuron: In: *Catecholamines. Handbook of Experimental Pharmacology*, edited by H. Blaschko and E. Muscholl. Berlin: Springer-Verlag, 1972, vol. 33, p. 661–725.

19. HAEUSLER, G., H. THOENEN, W. HAEFELY, AND A. HUERLIMANN. Electrical events in cardiac adrenergic nerves and noradrenaline release from the heart induced by acetylcholine and KCl. *Naunyn-Schmiedebergs Arch. Pharmacol.* 261: 389–411, 1968.

20. HIGGINS, C. B., S. F. VATNER, AND E. BRAUNWALD. Parasympathetic control of the heart. *Pharmacol. Rev.* 25: 119–155, 1973.

21. HILLARP, N.-A. The construction and functional organization of the autonomic innervation apparatus. *Acta Physiol. Scand.* 46, Suppl. 157: 1–68, 1959.

22. KILBINGER, H. Gas chromatographic estimation of acetylcholine in the rabbit heart using a nitrogen selective detector. *J. Neurochem.* 21: 421–429, 1973.

23. KIRPEKAR, S. M., J. C. PRAT, M. PUIG, AND A. R. WAKADE. Modification of the evoked release of noradrenaline from the perfused cat spleen by various ions and agents. *J. Physiol. London.* 221: 601–615, 1972.

24. KIRPEKAR, S. M., J. C. PRAT, AND A. R. WAKADE. Effect of calcium on the relationship between frequency of stimulation and release of noradrenaline from the perfused spleen of the cat. *Naunyn-Schmiedebergs Arch. Pharmacol.* 287: 205–212, 1975.

25. KOSTERLITZ, H. W., AND G. M. LEES. Interrelationships between adrenergic and cholinergic mechanisms. In: *Catecholamines. Handbook of Experimental Pharmacology*, edited by H. Blaschko and E. Muscholl. Berlin: Springer-Verlag, 1972, vol. 33, p. 762–812.

26. LANGLEY, A. E., AND R. W. GARDIER. Effect of atropine and acetylcholine on nerve stimulated output of noradrenaline and dopamine-beta-hydroxylase from isolated rabbit and guinea pig hearts. *Naunyn-Schmiedebergs Arch. Pharmacol.* 297: 251–256, 1977.

27. LAVALLÉE, M., J. DE CHAMPLAIN, R. A. NADEAU, AND N. YAMAGUCHI. Muscarinic inhibition of endogenous myocardial catecholamine liberation in the dog. *Can. J. Physiol. Pharmacol.* 56: 642–649, 1978.

28. LEADERS, F. E. Local cholinergic-adrenergic interaction: mechanism for the biphasic chronotropic response to nerve stimulation. *J. Pharmacol. Exp. Ther.* 42: 31–38, 1963.

29. LEVY, M. N. Sympathetic-parasympathetic interactions in the heart. *Circ. Res.* 19: 437–445, 1971.

30. LEVY, M. N. Parasympathetic control of the heart. In: *Neural Regulation of the Heart*, edited by W. C. Randall. New York: Oxford Univ. Press, 1977, p. 95–129.

31. LEVY, M. N., AND B. BLATTBERG. Effect of vagal stimulation on the overflow of norepinephrine into the coronary sinus during cardiac sympathetic nerve stimulation in the dog. *Circ. Res.* 38: 81–85, 1976.

32. LINDMAR, R., K. LÖFFELHOLZ, AND E. MUSCHOLL. A muscarinic mechanism inhibiting the release of noradrenaline from peripheral adrenergic nerve fibres by nicotinic agents. *Br. J. Pharmacol.* 32: 280–294, 1968.

33. LOEB, J. M., AND M. VASSALLE. Adrenergic mechanisms in postvagal tachycardia. *J. Pharmacol. Exp. Ther.* 210: 56–63, 1979.

34. LÖFFELHOLZ, K. Autoinhibition of nicotinic release of noradrenaline from postganglionic sympathetic nerves. *Naunyn-Schmiedebergs Arch. Pharmacol.* 267: 49–63, 1970.

35. LÖFFELHOLZ, K., AND E. MUSCHOLL. A muscarinic inhibition of the noradrenaline release evoked by postganglionic sympathetic nerve stimulation. *Naunyn-Schmiedebergs Arch. Pharmacol.* 265: 1–15, 1969.

36. LÖFFELHOLZ, K., AND E. MUSCHOLL. Inhibition by parasympathetic nerve stimulation of the release of the adrenergic transmitter. *Naunyn-Schmiedebergs Arch. Pharmacol.* 267: 181–184, 1970.

37. MANBER, L., AND M. D. GERSHON. A reciprocal adrenergic-cholinergic axoaxonic synapse in the mammalian gut. *Am. J. Physiol.* 236 (*Endocrinol. Metab. Gastrointest. Physiol.* 5): E738–E745, 1979.

38. MATHÉ, A. A., E. Y. TONG, AND P. W. TISHER. Norepinephrine release from the lung by sympathetic nerve stimulation inhibition by vagus and methacholine. *Life Sci.* 20: 1425–1430, 1977.

39. MUSCHOLL, E. Discussion of J. H. Burn. The mechanism of the release of noradrenaline. In: *Adrenergic Neurotransmission*, edited by G. E. W. Wolstenholme and M. O'Connor. London: Churchill, 1968, p. 16–25.

40. MUSCHOLL, E. Cholinomimetic drugs and release of the adrenergic transmitter. In: *New Aspects of Storage & Release Mechanisms of Catecholamines*, edited by H. J. Schümann and G. Kroneberg. Berlin: Springer-Verlag,, 1970, p. 168–186.

41. MUSCHOLL, E. Muscarinic inhibition of the norepinephrine release from peripheral sympathetic fibres. In: *Proc. 5th Int. Congr. Pharmacol.* Basel: Karger, 1973, vol. 4, p. 440–457.

42. MUSCHOLL, E. Presynaptic muscarine receptors and inhibition of release. In: *The Release of Catecholamines from Adrenergic Neurons*, edited by D. M. Paton. Oxford: Pergamon, 1979, p. 87–110.

42a. MUSCHOLL, E., AND A. MUTH. The output of labelled transmitters evoked by sympathetic and vagal nerve stimulation from the isolated perfused rabbit atria. *Naunyn-Schmiedebergs Arch. Pharmacol.* 316: (Suppl.) R52, 1981.

43. MUSCHOLL, E., H. RITZEL, AND K. RÖSSLER. Presynaptic muscarinic control of neuronal noradrenaline release. In: *Presynaptic Receptors*, edited by S. Z. Langer, K. Starke, and M. L. Dubocovich. Oxford: Pergamon, 1979, p. 287–291.

44. ROSS, S. B., AND D. KELDER. Release of ^{3}H-noradrenaline from the rat vas deferens under various in vitro conditions. *Acta Physiol. Scand.* 105: 338–349, 1979.

45. SHARMA, V. K., AND S. P. BANERJEE. Presynaptic muscarinic cholinergic receptors. *Nature London.* 272: 276–278, 1978.

46. STARKE, K. Regulation of noradrenaline release by presynaptic receptor systems. *Rev. Physiol. Biochem. Pharmacol.* 77: 1–124, 1977.

47. STARKE, K., AND S. Z. LANGER. A note in terminology for presynaptic receptors. In: *Presynpatic Receptors*, edited by S. Z. Langer, K. Starke, and M. L. Dubocovich. Oxford: Pergamon, 1979, p. 1-3.

48. STARKE, K., AND H. MONTEL. Influence of drugs with affinity for α-adrenoceptors on noradrenaline release by potassium, tyramine and dimethylphenylpiperazinium. *Eur. J. Pharmacol.* 27: 273-280, 1974.

49. STEINSLAND, O. S., R. F. FURCHGOTT, AND S. M. KIRPEKAR. Inhibition of adrenergic neurotransmission by parasympathomimetics in the rabbit ear artery. *J. Pharmacol. Exp. Ther.* 184: 346-356, 1973.

50. STJÄRNE, L. Facilitation and receptor-mediated regulation of noradrenaline secretion by control of recruitment of varicosities as well as by control of electro-secretory coupling. *Neuroscience* 3: 1147-1155, 1978.

51. STORY, D. D., M. S. BRILEY, AND S. Z. LANGER. The effects of chemical sympathectomy with 6-hydroxydopamine on α-adrenoceptor and muscarinic cholinoceptor binding in rat heart ventricle. *Eur. J. Pharmacol.* 57: 423-426, 1979.

52. TAKESHIGE, C., A. J. PAPPANO, W. C. DE GROAT, AND R. L. VOLLE. Ganglionic blockade produced in sympathetic ganglia by cholinomimetic drugs. *J. Pharmacol. Exp. Ther.* 141: 333-342, 1963.

53. THOENEN, H., AND J. P. TRANZER. Chemical sympathectomy by selective destruction of adrenergic nerve endings with 6-hydroxydopamine. *Naunyn-Schmiedebergs Arch. Pharmacol.* 261: 271-288, 1968.

54. THOENEN, H., J. P. TRANZER, A. HUERLIMANN, AND W. HAEFELY. Untersuchungen zur Frage eines cholinergischen Gliedes in der postganglionären sympathischen Transmission. *Helv. Physiol. Pharmacol. Acta* 24: 229-246, 1966.

55. VAN HEE, R. H., AND P. M. VANHOUTTE. Cholinergic inhibition of adrenergic neurotransmission in the canine gastric artery. *Gastroenterology* 74: 1266-1270, 1978.

56. VANHOUTTE, P. M. Inhibition by acetylcholine of adrenergic neurotransmission in vascular smooth muscle. *Circ. Res.* 34: 317-326, 1974.

57. VANHOUTTE, P. M., E. P. COEN, W. J. DE RIDDER, AND T. J. VERBEUREN. Evoked release of endogenous norepinephrine in the canine saphenous vein. Inhibition by acetylcholine. *Circ. Res.* 45: 608-614, 1979.

58. VANHOUTTE, P. M., AND M. N. LEVY. Prejunctional cholinergic modulation of adrenergic neurotransmission in the cardiovascular system. *Am. J. Physiol.* 238 (*Heart Circ. Physiol.* 7): H275-H281, 1980.

59. VANHOUTTE, P. M., R. R. LORENZ, AND G. M. TYCE. Inhibition of norepinephrine-^{3}H release from sympathetic nerve endings in veins by acetylcholine. *J. Pharmacol. Exp. Ther.* 185: 386-394, 1973.

60. VANHOUTTE, P. M., AND T. J. VERBEUREN. Inhibition by acetylcholine of ^{3}H-norepinephrine release in cutaneous veins after alpha-adrenergic blockade. *Arch. Int. Pharmacodyn.* 221: 344-346, 1976.

61. VANHOUTTE, P. M., AND T. J. VERBEUREN. Inhibition by acetylcholine of the norepinephrine release evoked by potassium in canine saphenous veins. *Circ. Res.* 39: 263-269, 1976.

62. VIZI, E. S. Presynaptic modulation of neurochemical transmission. *Progr. Neurobiol.* 12: 181-290, 1979.

63. WALLICK, D. W., D. FELDER, AND M. N. LEVY. Autonomic control of pacemaker activity in the atrioventricular junction of the dog. *Am. J. Physiol.* 235 (*Heart Circ. Physiol.* 4): H308-H313, 1978.

64. WESTFALL, T. C. Local regulation of adrenergic neurotransmission. *Physiol. Rev.* 57: 659-728, 1977.

65. WESTFALL, T. C., AND P. E. HUNTER. Effect of muscarinic agonists on the release of [^{3}H]-noradrenaline from the guinea-pig perfused heart. *J. Pharm. Pharmacol.* 26: 458-460, 1974.

CHAPTER 11

Prejunctional Cholinergic Modulation of Adrenergic Neurotransmission and the Cardiovascular System

PAUL M. VANHOUTTE AND MATTHEW N. LEVY

Department of Physiology and Biophysics, Mayo Clinic and Foundation,
Rochester, Minnesota, and Department of Investigative
Medicine, Mount Sinai Hospital, and Departments of Medicine,
Physiology, and Biomedical Engineering, Case Western
Reserve University, Cleveland, Ohio

The heart is regulated by both divisions of the autonomic nervous system. Acetylcholine released by the vagal nerves inhibits and norepinephrine liberated by the sympathetic nerves facilitates the function of the myocardium. The effects of these two autonomic transmitters on the cardiac effector cells do not summate linearly, but pronounced interactions prevail between them; part of these interactions occur at the level of the myocardiac cells themselves (3, 16, 55, 60). Certain blood vessels, in particular those supplying cardiac and skeletal muscle, are also innervated by both cholinergic and adrenergic neurons (4, 5, 44). The direct effects of the two main autonomic transmitters on the peripheral effector cells oppose each other, with norepinephrine exciting and acetylcholine depressing vascular tone (42, 44).

In the intact organism the balance between the two types of autonomic nerves is achieved in part by central modulations, which usually result in oppositely directed adjustments of the outputs of the vagal nuclei and of the vasomotor centers. In addition evidence has accumulated strongly suggesting that acetylcholine, released in the vicinity of adrenergic nerve endings in the heart or the blood vessel walls, exerts a profound inhibitory effect on the release of norepinephrine and thus reduces the amount of adrenergic neurotransmitter present in the junctional cleft between the adrenergic nerve terminal and the cardiovascular effector cells (14, 15, 16–18, 23, 25–27, 35, 42, 48, 54, 61). This chapter focuses on this prejunctional ("presynaptic") inhibitory effect of the cholinergic transmitter and the role it plays in cardiovascular control.

Prejunctional Inhibitory Effect of Exogenous Acetylcholine

In the dog, intracoronary infusions of acetylcholine have only a minimal depressant effect on ventricular contractility in the absence of a background of sympathetic activity. During tonic sympathetic stimulation, however, the same infusions of acetylcholine have substantial negative inotropic effects (12). Experiments performed on isolated cardiac tissues demonstrate that the depressant effect of the cholinergic transmitter during sympathetic nerve stimulation is due, at least in part, to an inhibitory effect on the release of norepinephrine from the adrenergic neurons in the heart. Thus in the isolated rabbit atrium, exogenous acetylcholine reduces the quantity of norepinephrine overflowing during sympathetic nerve activation (22).

In the intact organism, acetylcholine is a potent vasodilator. In contrast in most isolated blood vessels acetylcholine either has no effect or even causes contraction (42, 44); if relaxations are observed they are due to an inhibitory signal generated by the endothelial cells rather than to a direct inhibitory effect of the cholinergic transmitter on the vascular smooth muscle cells (7, 10, 11, 45). Acetylcholine markedly depresses the constrictor responses to sympathetic nerve stimulation in various isolated blood vessels, however, including the ear artery of the rabbit (1, 2, 6, 13, 29, 36), the femoral and gastric arteries of the dog (39, 41), the mesenteric arteries of the dog (41) and rat (24), the pulmonary artery of the dog (41), and the saphenous veins of the dog, rabbit, and human (8, 32, 41, 51). The available evidence demonstrates that the prejunctional inhibitory effect of acetylcholine plays a major role in the depression of the response of vascular smooth muscle cells to sympathetic nerve stimulation. This conclusion is based on the following observations (1, 2, 32, 36, 41, 46, 50): 1) the inhibitory effect of exogenous acetylcholine on the responses to neurally released norepinephrine occurs at concentrations that have variable effects on the response to exogenous norepinephrine (Fig. 1); 2) in rabbit ear arteries, in dog pulmonary arteries, and in canine and human saphenous veins previously incubated with [^{3}H]norepinephrine, acetylcholine decreases the quantity of tritiated transmitter released by electrical stimulation of the adrenergic nerve endings (Fig. 2); and 3) in the canine saphenous vein acetylcholine depresses the evoked release of endogenous norepinephrine, measured by a radioenzymatic assay (Fig. 3). In blood vessels such as the dog mesenteric and pulmonary veins, acetylcholine augments the contractile response to sympathetic nerve stimulation because the vascular smooth muscle cells in these vessels are exquisitely sensitive to the direct stimulatory effect of the cholinergic transmitter (41). However, the cholinergic transmitter still causes depression of the evoked release of [^{3}H]norepinephrine during sympathetic nerve stimulation (44). From these different observations on isolated blood vessels it can be concluded that acetylcholine inhibits adrenergic neurotransmission throughout the vascular tree.

The prejunctional effect of acetylcholine on adrenergic neurotransmission in the blood vessel wall has also been demonstrated in the intact animal. Thus in the anesthetized dog, acetylcholine does not depress the constrictor response of

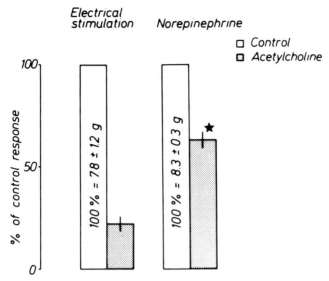

Fig. 1. In the isolated human saphenous vein the same concentration of acetylcholine (10^{-7} M) depresses the contractile responses to electrical stimulation of the adrenergic nerve endings significantly (★) more than contractions of comparable amplitude evoked by exogenous norepinephrine. This difference indicates a prejunctional effect of the cholinergic transmitter. [From Rorie et al. (32) by permission of the American Heart Association, Inc.]

the lateral saphenous vein to exogenous norepinephrine but markedly reduces those to lumbar sympathetic chain stimulation (51). In the perfused hindleg of the same species, vasodilators such as isoproterenol depress the vasoconstrictor response to exogenous norepinephrine more than that to sympathetic nerve stimulation; acetylcholine has the opposite effect (45, 48). Likewise in the kidney, acetylcholine exerts a prejunctional inhibitory effect (31).

Both in the heart and in the blood vessel wall the inhibition by acetylcholine of the release of norepinephrine evoked by nerve impulses is due to activation of muscarinic receptors on the adrenergic nerve endings. This conclusion is based on the following evidence: 1) inhibitors of α-adrenergic and β-adrenergic receptors, or of prostaglandin synthesis, do not interfere with the depressing effect of acetylcholine on adrenergic neurotransmission (9, 52); 2) the action of the cholinergic transmitter is inhibited by muscarinic antagonists (e.g., atropine; see Fig. 2) but not by nicotinic antagonists [e.g., hexamethonium (8, 13, 24, 29, 36, 39, 41, 51, 57); and 3) other muscarinic agonists also specifically depress the response to sympathetic nerve stimulation (1, 29, 30, 36). Higher concentrations of acetylcholine can activate nicotinic receptors on adrenergic nerve endings, which results in the release of norepinephrine (42, 44). Because the muscarinic effect occurs at lower concentrations of acetylcholine and causes inhibition of the release evoked by nicotine (62), nicotinic prejunctional effects of acetylcholine are only to be expected in preparations treated with muscarinic antagonists.

The binding of acetylcholine to the muscarinic receptors of the adrenergic nerve endings results in inhibition of the exocytotic process, because the cholinergic transmitter depresses the release of [³H]norepinephrine evoked by

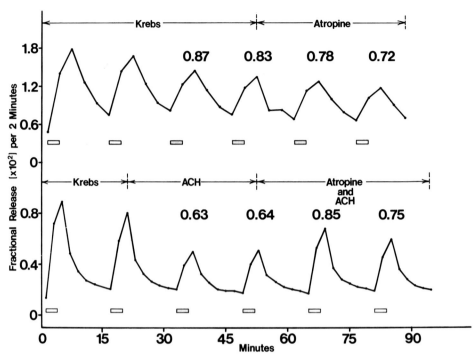

Fig. 2. Demonstration of the prejunctional inhibitory effect of acetylcholine in isolated human saphenous veins. *Upper:* fractional release of [³H]norepinephrine in veins previously incubated with labeled transmitter. Atropine had no effect on the fractional release. *Lower:* acetylcholine causes a depression of the fractional release which is depressed by atropine. [From Rorie et al. (32) by permission of the American Heart Association, Inc.]

electrical impulses, nicotine, or moderate increases in K^+ concentration but does not inhibit its pharmacologic displacement caused by indirect sympathomimetic amines (41, 42, 44, 50–52, 56). The inhibitory effect of acetylcholine on adrenergic neurotransmission is not blocked by tetrodotoxin, which suggests that it is not due to a direct interference with Na^+ exchanges or with spike electrogenesis (52). Altering the extracellular Ca^{2+} concentration does not affect the inhibitory response to acetylcholine during sympathetic nerve stimulation (54); the cholinergic transmitter augments the overflow of [³H]norepinephrine evoked by the ionophore A-23187, which demonstrates that its prejunctional effect is not due to direct interference with the availability of Ca^{2+} for the electrosecretory process in the adrenergic nerve terminals (54). Acetylcholine inhibits the release evoked by partial but not by full depolarization of the adrenergic nerve endings (52). Thus the most likely explanation for the prejunctional muscarinic inhibitory effect of the cholinergic transmitter is that it causes hyperpolarization of the sympathetic nerve terminals that would prevent or delay the occurrence of action potentials at the adrenergic varicosities.

Prejunctional Autonomic Interactions in the Heart

It has been known for more than 40 years that the decrease in heart rate evoked by vagal stimulation is considerably greater when the background level

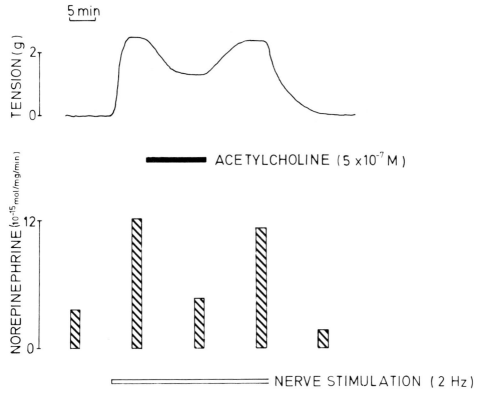

Fig. 3. Effect of acetylcholine on tension development (*top*) and endogenous norepinephrine overflow (*bottom*) evoked by nerve stimulation in a superfused canine saphenous vein. [From Vanhoutte et al. (46) by permission of the American Heart Association, Inc.]

of sympathetic tone is increased, thereby demonstrating the preponderance of vagal over sympathetic influences on the sinus node (33, 34). In the dog, the frequencies of sympathetic and vagal stimulation can be adjusted so that, when these neural elements are stimulated separately, they each produce chronotropic responses of equal magnitude but of opposite direction (Fig. 4A, C). When such vagal and sympathetic stimuli are applied simultaneously (Fig. 4B) the heart rate decreases almost as if there were no sympathetic stimulation at all (21). This type of response has been termed "accentuated antagonism" (16, 19).

In the dog, vagal stimulation has only a minimal depressant effect on ventricular contractility in the absence of tonic sympathetic stimulation but markedly reduces the depressant effect if imposed during activation of the adrenergic nerves (20). In the experiment illustrated in Figure 5, vagal stimulation produced a 20% reduction in peak systolic pressure in an isovolumetric left ventricle preparation when there was no background level of sympathetic activity (Fig. 5A, B, C). The same vagal stimulus produced a 34% reduction in peak pressure when the cardiac sympathetic nerves were stimulated concomitantly (Fig. 5D). More recently it has been shown that this type of accentuated antagonism applies to the atrial as well as to the ventricular myocardium (37).

The inhibition of adrenergic neurotransmission by acetylcholine released from the vagal nerves modulates both the chronotropic and inotropic behavior

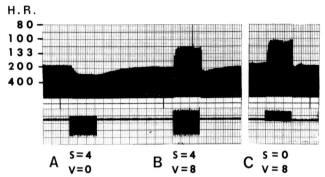

Fig. 4. Changes in heart rate, H.R., in beats/min, evoked by autonomic neural stimulation in an anesthetized dog. Right stellate (S) ganglion and left vagus (V) nerve were each stimulated with 10-V pulses, 2 ms wide, at the following frequencies. A: stellate 4 Hz, vagus 0 Hz; B: stellate 4 Hz, vagus 8 Hz; C: stellate 0 Hz, vagus 8 Hz. A, with sympathetic stimulation alone heart rate increased by 78 beats/min. C, with vagal stimulation alone heart rate decreased by 70 beats/min. B, during combined stimulation heart rate also decreased by 70 beats/min. [From Vanhoutte and Levy (49).]

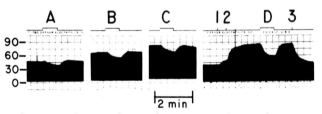

Fig. 5. Effects of autonomic neural stimulation on peak systolic pressure in a canine isovolumetric left ventricle preparation. During events marked A, B, C, and D, the right vagus nerve was stimulated at 20 Hz. Left ventricular volumes were increased between A and B and between B and C, and then decreased again between C and D. In D, the left stellate ganglion was stimulated at 3 V, 2 Hz, beginning at mark 1; the voltage was increased to 3.5 at mark 2, and maintained at that level until mark 3. Note greater depressant effect of vagal stimulation during concomitant sympathetic stimulation, D, than in absence of any background of sympathetic activity, A, B, C. [From Levy et al. (20) by permission of the American Heart Association, Inc.]

of the heart (15, 18, 19, 48). Thus in the isolated rabbit atrium vagal stimulation reduces the quantity of norepinephrine overflowing during sympathetic nerve stimulation (23, 27). Further, in the anesthetized dog stimulation of the left stellate ganglion markedly augments the overflow of norepinephrine into the coronary sinus blood. When the vagus nerves are activated simultaneously with the sympathetic nerves, the rate of norepinephrine overflow is markedly reduced (18). The inhibitory influence of vagal stimulation varies with the frequency of neural stimulation (15). Vagal stimulation at a frequency of only 1 Hz produces a significant decline in the quantity of catecholamines overflowing into the coronary sinus in response to stimulation of the cardiac sympathetic nerves (Fig. 6). When the vagal stimulation frequency is increased to 16 Hz, the catecholamine output is reduced to 34% of the output obtained with sympathetic stimulation alone. The curtailment of norepinephrine overflow by vagal stimu-

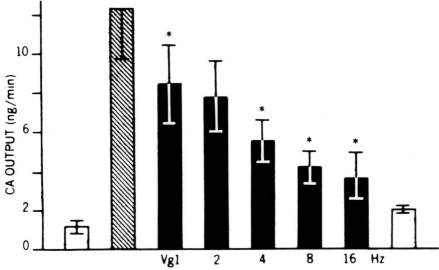

Fig. 6. Coronary sinus blood catecholamine output (ng/min) in absence of neural stimulation (*open bars*), during sympathetic stimulation with 10 Hz alone (*hatched bars*), and during combined sympathetic (10 Hz) and vagal stimulation (*solid bars*). Frequencies of vagal (Vg) stimulation (in Hz) are indicated at bottoms of solid bars. Values are means ± SE of data obtained from 7 dogs. [From Lavallée et al. (15), reproduced by permission of the National Research Council of Canada from the *Canadian Journal of Physiology and Pharmacology*, vol. 56, p. 642–649, 1978.]

lation can be prevented by atropine. In the intact dog heart most of the norepinephrine appearing in the coronary sinus blood in response to left stellate ganglion stimulation is derived from nerve terminals in the ventricular walls, and hence the depression of norepinephrine release during vagal stimulation must occur in the ventricles. Thus inhibition of norepinephrine release by the cholinergic transmitter is part of the normal regulatory mechanisms in both atria and ventricles (15, 18, 19, 27, 48). This prejunctional interaction between the vagal and sympathetic nerves together with the interaction of the cholinergic and adrenergic neurotransmitters at the level of the myocardial effector cells (3, 38, 60) explains the accentuated antagonism between the two divisions of the autonomic nervous system in controlling cardiac function.

Prejunctional Autonomic Interactions in Blood Vessel Wall

Arteries, and in particular arterioles of several vascular beds, are innervated by both adrenergic and cholinergic nerve terminals that are in close apposition (see refs. 4, 5, 42, 44, 48, 54). In these vessels the direct effect of acetylcholine on the vascular smooth muscle cells is inhibitory, but prejunctional inhibition of adrenergic neurotransmission could contribute to the vasodilator response during cholinergic nerve activation. In the isolated gastric artery of the dog, transmural electrical field stimulation presumably activates both adrenergic and cholinergic nerve endings. The contractile response to such stimulation is potentiated by the muscarinic antagonist atropine but depressed by the inhibitor

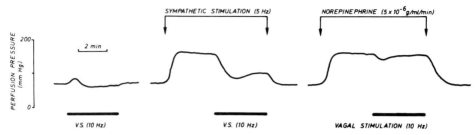

Fig. 7. In blood-perfused stomach of dog, vagal stimulation inhibits markedly the vaso-constrictor response to sympathetic nerve activation (*middle*), but has little effect in basal conditions (*left*) or during vasoconstrictions induced with exogenous norepineph-rine (*right*). This demonstrates that acetylcholine released from vagal nerve endings inhibits adrenergic neurotransmission in blood vessel wall. [From Van Hee and Van-houtte (39).]

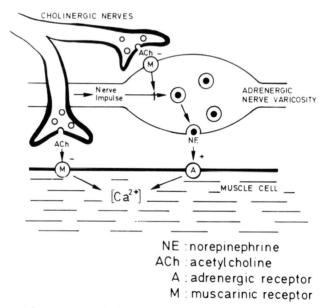

NE : norepinephrine
ACh : acetylcholine
A : adrenergic receptor
M : muscarinic receptor

Fig. 8. Summary of prejunctional effects of acetylcholine on adrenergic nerve terminals in heart and blood vessel wall. Acetylcholine, whether liberated from cholinergic nerve endings or exogenously added, inhibits release of norepinephrine, presumably by causing hyperpolarization of the terminal. This inhibitory effect combines with direct action on effector cells (cardiac or vascular smooth muscle); both in heart and in most vessels innervated by cholinergic nerves the latter is inhibitory; +, activation; −, inhibition. [From Vanhoutte and Levy (48).]

of acetylcholinesterase, physostigmine (39). These observations imply that dur-ing field stimulation acetylcholine is liberated, which causes partial inhibition of adrenergic neurotransmission. In the blood-perfused stomach of the dog studied in situ, vagal stimulation reduces the constrictor response to exogenous norepinephrine significantly less than that to sympathetic nerve stimulation (Fig. 7). The most logical explanation for these findings is that the cholinergic transmitter released by the vagal nerve endings has a prejunctional inhibitory

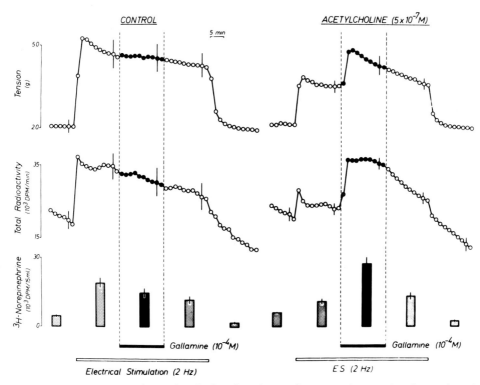

Fig. 9. Experiments performed on isolated canine saphenous veins previously incubated with [³H]norepinephrine. The muscle relaxant gallamine does not affect tension (*top*), ³H efflux (*middle*), or overflow of [³H]norepinephrine (*bottom*) during sympathetic nerve stimulation applied in control conditions (*left*). However, if the response to nerve stimulation is first depressed by the presence of acetylcholine in the superfusion fluid (*right*), gallamine markedly augments the contractile response and overflow of adrenergic transmitter, illustrating how muscarinic antagonists can have an "indirect sympathomimetic" activity. [From Vercruysse et al. (57).]

effect on the adrenergic nerve endings in the gastric blood vessel wall. Thus cholinergic neurogenic vasodilatation involves decreased release of norepinephrine in those vascular beds where the sympathetic nerves are activated, and prejunctional inhibition of adrenergic neurotransmission must be part of the normal regulatory mechanisms also in the blood vessel wall (25, 35, 42, 44, 48, 54).

Conclusion

In cardiac tissue and in the blood vessel wall, acetylcholine markedly and reversibly inhibits the response to sympathetic nerve stimulation. This is due in part to activation of muscarinic receptors on the adrenergic neuronal terminals, presumably by hyperpolarizing the membranes of the nerve endings. The subsequent depression of the exocytotic process decreases the release of norepinephrine evoked by the sympathetic impulses. If exogenous acetylcholine is given to the heart or to a vascular bed when the sympathetic traffic is increased, the prejunctional inhibitory effect of acetylcholine greatly reinforces the direct depressant effect it has on the myocardial cells and the vasodilatation that it induces, which presumably is mediated by the endothelial cells. Hence the action of acetylcholine on the adrenergic nerve terminals explains part of its pharmacological effects.

The only physiological source of acetylcholine is the transmitter released by cholinergic nerve endings, which affects only cells located in the immediate vicinity of these neurones, since the transmitter is degraded almost instantaneously by acetylcholinesterase. In the heart the pattern of cholinergic innervation allows direct control of the myocardial cells by the cholinergic transmitter. In the blood vessel wall the direct effect of acetylcholine on the vascular smooth muscle cells usually is to cause their contraction, albeit in high concentrations; since acetylcholine does not circulate in the blood, the absence of evidence of cholinergic innervation of endothelial cells makes it unlikely that endothelium-mediated vasodilatations evoked by the cholinergic transmitter when given exogenously play a physiological role (45). Because both in the heart and in the blood vessel wall close associations exist between the cholinergic terminals and the adrenergic nerve endings, however, acetylcholine released in the vicinity of the latter can curtail the existing release of norepinephrine (Fig. 8). Conversely when muscarinic antagonists are given under conditions in which the vagal and the sympathetic cholinergic vasodilator tone is high, they cause cardiovascular activation in part by withdrawal of the prejunctional inhibition due to the released acetylcholine; this is examplified best by the increased release of norepinephrine caused by muscle relaxants such as gallamine and pancuronium bromide [Fig. 9; (57)].

The prejunctional inhibitory effect of acetylcholine greatly increases the efficacy of the cholinergic messenger to counteract the excitatory effects of the adrenergic transmitter. A similar pattern of inhibition has been described for tissues in which acetylcholine is the excitatory transmitter and in which exogenously added or endogenously released norepinephrine inhibits its liberation by the cholinergic nerve endings (e.g., refs. 28, 58, 59). It thus appears that autonomic alteration of the activity of peripheral effector cells does not depend only on the summation of the opposing direct effects of the liberated amounts of acetylcholine and norepinephrine, but that on release of the inhibitory agonist the liberation of the excitatory transmitter is automatically suppressed.

References

1. ALLEN, G. S., M. J. RAND, AND D. F. STORY. Effect of the muscarinic agonist McN-A-343 on the release by sympathetic nerve stimulation of (^{3}H)-noradrenaline from rabbit isolated ear arteries and guinea-pig atria. Br. J. Pharmacol. 51: 29–34, 1974.
2. ALLEN, G. S., A. B. GLOVER, M. W. MCCULLOCH, M. J. RAND, AND D. F. STORY. Modulation by acetylcholine of adrenergic transmission in the rabbit ear artery. Br. J. Pharmacol. 54: 49–53, 1975.
3. BAILEY, J. C., A. M. WATANABE, H. R. BESCH, AND D. A. LATHROP. Acetylcholine antagonism of the electrophysiological effects of isoproterenol on canine cardiac Purkinje fibers. Circ. Res. 44: 378–383, 1979.
4. BRODY, M. J. Histaminergic and cholinergic vasodilator systems. In: Mechanisms of Vasodilatation, edited by P. M. Vanhoutte and I. Leusen. Basel: Karger, 1978, p. 266–277.
5. BURNSTOCK, G. Cholinergic and purinergic regulation of blood vessels. In: Handbook of Physiology. The Cardiovascular System: Vascular Smooth Muscle, edited by D. F. Bohr, A. P. Somlyo, and H. V. Sparks, Jr. Bethesda, MD: Am. Physiol. Soc., 1980, sect. 1, vol. II, chapt. 19, p. 567–612.

6. DE LA LANDE, I. S., AND M. J. RAND. A simple isolated nerve-blood vessel preparation. *Aust. J. Exp. Biol. Med. Sci.* 43: 639–656, 1965.

7. DE MEY, J. G., AND P. M. VANHOUTTE. Comparison of the responsiveness of cutaneous veins of dog and rabbit to adrenergic and cholinergic stimulation. *Blood Vessels* 17: 27–43, 1980.

8. DE MEY, J. G., AND P. M. VANHOUTTE. Role of the intima in cholinergic and purinergic relaxation of isolated canine femoral arteries. *J. Physiol. London* 316: 347–355, 1981.

9. FUDER, H., AND E. MUSCHOLL. The effect of metacholine on noradrenaline release from the rabbit heart perfused with indometacin. *Naunyn-Schmiedebergs Arch. Pharmacol.* 285: 127–132, 1974.

10. FURCHGOTT, R. F., AND J. V. ZAWADSKI. The obligatory role of endothelial cells in the relaxation of arterial smooth muscle by acetylcholine. *Nature* 288: 373–376, 1980.

11. FURCHGOTT, R. F., J. V. ZAWADSKI, AND P. D. CHERRY. Role of endothelium in the vasodilator response to acetylcholine. In: *Vasodilatation*, edited by P. M. Vanhoutte and I. Leusen. New York: Raven, p. 49–66, 1981.

12. HOLLENBERG, M., S. CARRIERE, AND A. C. BARGER. Biphasic action of acetylcholine on ventricular myocardium. *Circ. Res.* 16: 527–536, 1965.

13. HUME, W. R., I. S. DE LA LANDE, AND J. G. WATERSON. Effect of acetylcholine on the response of the isolated rabbit ear artery to stimulation of the perivascular sympathetic nerves. *Eur. J. Pharmacol.* 17: 227–233, 1972.

14. LANGER, S. Z. Presynaptic regulation of the release of catecholamines. *Pharmacol. Rev.* 32: 337–362, 1981.

15. LAVALLÉE, M., J. DE CHAMPLAIN, R. A. NADEAU, AND N. YAMAGUCHI. Muscarinic inhibition of endogenous myocardial catecholamine liberation in the dog. *Can. J. Physiol. Pharmacol.* 56: 642–649, 1978.

16. LEVY, M. N. Sympathetic-parasympathetic interactions in the heart. *Circ. Res.* 29: 437–445, 1971.

17. LEVY, M. N. Neural control of the heart: sympathetic-vagal interactions. In: *Cardiovascular System Dynamics*, edited by J. Baan, A. Noordergraaf, and J. Raines. Cambridge, MA: MIT Press, 1978, p. 365–370.

18. LEVY, M. N., AND B. BLATTBERG. Effect of vagal stimulation on the overflow of norepinephrine into the coronary sinus during cardiac sympathetic nerve stimulation in the dog. *Circ. Res.* 38: 81–85, 1976.

19. LEVY, M. N., AND P. J. MARTIN. Neural control of the heart. In: *Handbook of Physiology. Cardiovascular System: The Heart*, edited by R. M. Berne and N. Sperelakis. Bethesda, MD: Am. Physiol. Soc., 1979, sect. 1, vol. I, chapt. 16, p. 581–620.

20. LEVY, M. N., M. NG, P. MARTIN, AND H. ZIESKE. Sympathetic and parasympathetic interactions upon the left ventricle of the dog. *Circ. Res.* 19: 5–10, 1966.

21. LEVY, M. N., AND H. ZIESKE. Autonomic control of cardiac pacemaker activity and atrioventricular transmission. *J. Appl. Physiol.* 27: 465–470, 1969.

22. LÖFFELHOLZ, K., AND E. MUSCHOLL. A muscarinic inhibition of the noradrenaline release evoked by postganglionic sympathetic nerve stimulation. *Naunyn-Schmiedebergs Arch. Pharmacol.* 265: 1–15, 1969.

23. LÖFFELHOLZ, K., AND E. MUSCHOLL. Inhibition by parasympathetic nerve stimulation of the release of the adrenergic transmitter. *Naunyn-Schmiedebergs Arch. Pharmacol.* 267: 181–184, 1970.

24. MALIK, K. U., AND G. M. LING. Modification by acetylcholine of the response of rat mesenteric arteries to sympathetic stimulation. *Circ. Res.* 25: 1–9, 1969.

25. MCGRATH, M. A., AND P. M. VANHOUTTE. Vasodilatation caused by peripheral inhibition of adrenergic neurotransmission. In: *Mechanisms of Vasodilatation*, edited by P. M. Vanhoutte and I. Leusen. Basel: Karger, 1978, p. 248–257.

26. MUSCHOLL, E. Muscarinic inhibition of the norepinephrine release from peripheral sympathetic fibers. In: *Pharmacology and the Future of Man. Proc. Int. Congr. Pharmacol., 5th, San Francisco.* 4: 440–457, 1973.

27. MUSCHOLL, E. Neural control of the heart. Peripheral muscarinic control of the heart. *Am. J. Physiol.* 239(*Heart Circ. Physiol.* 8): H713–H720, 1980.

28. POWELL, D. W., AND E. J. TAPPER. Intestinal ion transport: cholinergic-adrenergic interactions. In: *Mechanisms of Intestinal Secretion*, edited by H. J. Binder, Kroc Foundation. New York: Alan Hiss, 1979, p. 175–192.

29. RAND, M. J., AND B. VARMA. The effects of cholinomimetic drugs on responses to sympathetic nerve stimulation and noradrenaline in the rabbit ear artery. *Br. J. Pharmacol.* 38: 758–770, 1970.

30. RAND, M. J., AND B. VARMA. Effects of the muscarinic agonist McN-A-343 on responses to sympathetic nerve stimulation in the rabbit ear artery. Br. J. Pharmacol. 43: 536–542, 1971.

31. ROBIE, N. W. Presynaptic inhibition of canine renal adrenergic nerves by acetylcholine in vivo. Am. J. Physiol. 237(Heart Circ. Physiol. 6): H326–H331, 1979.

32. RORIE, D. K., N. J. RUSCH, J. T. SHEPHERD, P. M. VANHOUTTE, AND G. M. TYCE. Prejunctional inhibition of norepinephrine release caused by acetylcholine in the human saphenous vein. Circ. Res. 49: 337–341, 1981.

33. ROSENBLUETH, A., AND F. A. SIMEONE. The interrelations of vagal and accelerator effects on the cardiac rate. Am. J. Physiol. 110: 42–55, 1934.

34. SAMAAN, A. The antagonistic cardiac nerves and heart rate. J. Physiol. London 83: 332–340, 1935.

35. SHEPHERD, J. T., R. R. LORENZ, G. M. TYCE, AND P. M. VANHOUTTE. Acetylcholine inhibition of transmitter release from adrenergic nerve terminals mediated by muscarinic receptors. Federation Proc. 37: 191–194, 1978.

36. STEINSLAND, O. S., R. F. FURCHGOTT, AND S. M. KIRPEKAR. Inhibition of adrenergic neurotransmission by parasympathomimetics in the rabbit ear artery. J. Pharmacol. Exp. Ther. 184: 346–356, 1973.

37. STUESSE, S. L., D. W. WALLICK, AND M. N. LEVY. Autonomic control of right atrial contractile strength in the dog. Am. J. Physiol. 236 (Heart Circ. Physiol. 5): H860–H865, 1979.

38. STULL, J. T., AND S. E. MAYER. Biochemical mechanisms of adrenergic and cholinergic regulation of myocardial contractility. In: Handbook of Physiology. Cardiovascular System: The Heart, edited by R. M. Berne and N. Sperelakis. Bethesda, MD: Am. Physiol. Soc., 1979, sect. 1, vol. I, chapt. 21, p. 741–774.

39. VAN HEE, R., AND P. M. VANHOUTTE. Cholinergic inhibition of adrenergic neurotransmission in the canine gastric artery. Gastroenterology 74: 1266–1270, 1978.

40. VANHOUTTE, P. M. Cholinergische Mechanismen in Vasculair Glad Spierweefsel. Wilrijk, Belgium: Univ. of Antwerp, 1973, p. 1–181.

41. VANHOUTTE, P. M. Inhibition by acetylcholine of adrenergic neurotransmission in vascular smooth muscle. Circ. Res. 34: 317–326, 1974.

42. VANHOUTTE, P. M. Cholinergic inhibition of adrenergic transmission. Federation Proc. 36: 2444–2449, 1977.

43. VANHOUTTE, P. M. Adrenergic neuroeffector interaction in the blood vessel wall. Federation Proc. 37: 181–186, 1978.

44. VANHOUTTE, P. M. Heterogeneity in vascular smooth muscle. In: Microcirculation, edited by G. Kaley and B. M. Altura. Baltimore, MD: University Park, 1978, vol. II, p. 181–309.

45. VANHOUTTE, P. M. Why is acetylcholine a vasodilator? In: Vasodilatation, edited by P. M. Vanhoutte and I. Leusen. New York: Raven, p. 67–72, 1981.

46. VANHOUTTE, P. M., E. P. COEN, W. J. DE RIDDER, AND T. J. VERBEUREN. Evoked release of engdenous norepinephrine in the canine saphenous vein. Inhibition by acetylcholine. Circ. Res. 45: 608–614, 1979.

47. VANHOUTTE, P. M., M. G. COLLIS, W. J. JANSSENS, AND T. J. VERBEUREN. Calcium dependency of prejunctional inhibitory effects of adenosine and acetylcholine on adrenergic neurotransmission in canine saphenous veins. Eur. J. Pharmacol. 72: 189–198, 1981.

48. VANHOUTTE, P. M., AND M. N. LEVY. Cholinergic inhibition of adrenergic neurotransmission in the cardiovascular system. In: Integrative Functions of the Autonomic Nervous System, edited by C. McBrooks, K. Koizumi, and A. Sato. Tokyo: Tokyo Univ. Press, 1979, p. 159–167.

49. VANHOUTTE, P. M., AND M. N. LEVY. Prejunctional cholinergic modulation of adrenergic neurotransmission in the cardiovascular system. Am. J. Physiol. 238 (Heart, Circ. Physiol. 7) H275–H281, 1980.

50. VANHOUTTE, P. M., R. R. LORENZ, AND G. M. TYCE. Inhibition of norepinephrine-^{3}H release from sympathetic nerve endings in veins by acetylcholine. J. Pharmacol. Exp. Ther. 135: 386–394, 1973.

51. VANHOUTTE, P. M., AND J. T. SHEPHERD. Venous relaxation caused by acetylcholine acting on the sympathetic nerves. Circ. Res. 32: 259–267, 1973.

52. VANHOUTTE, P. M., AND T. J. VERBEUREN. Inhibition by acetylcholine of the norepinephrine release evoked by potassium in canine saphenous veins. Circ. Res. 39: 263–269, 1976.

53. VANHOUTTE, P. M., AND T. J. VERBEUREN. Inhibition by acetylcholine of ^{3}H-norepinephrine release in cutaneous veins after alpha-adrenergic blockade. Arch. Int. Pharmacodyn. 221: 344–346, 1976.

54. VANHOUTTE, P. M., T. J. VERBEUREN, AND R. C. WEBB. Local modulation of adrenergic neuroeffector interaction in the blood vessel wall. *Physiol. Rev.* 61: 151–247, 1981.

55. VATNER, S. F., J. D. RUTHERFORD, AND H. R. OCHS. Baroreflex and vagal mechanisms modulating left ventricular contractile responses to sympathomimetic amines in conscious dogs. *Circ. Res.* 44: 195–207, 1979.

56. VERBEUREN, T. J., AND P. M. VANHOUTTE. Acetylcholine inhibits potassium evoked release of ^{3}H-norepinephrine in different blood vessels of the dog. *Arch. Int. Pharmacodyn.* 221: 347–350, 1976.

57. VERCRUYSSE, P., P. BOSSUYT, G. HANEGREEFS, T. J. VERBEUREN, AND P. M. VANHOUTTE. Gallamine and pancuronium inhibit pre- and post-junctional muscarinic receptors in canine saphenous veins. *J. Pharmacol. Exp. Ther.* 209: 225–230, 1979.

58. VERMEIRE, P. A., AND P. M. VANHOUTTE. Inhibitory effects of catecholamines in isolated canine bronchial smooth muscle. *J. Appl. Physiol.: Respirat. Environ. Exercise Physiol.* 46: 787–791, 1979.

59. VIZI, E. S., AND J. KNOLL. The effects of sympathetic nerve stimulation and guanethidine on parasympathetic neuroeffector transmission; the inhibition of acetylcholine release. *J. Pharm. Pharmacol.* 23: 918–925, 1971.

60. WATANABE, A. M., AND H. R. BESCH. Interaction between cyclic adenosine monophosphate and cyclic guanosine monophosphate in guinea pig ventricular myocardium. *Circ. Res.* 37: 309–317, 1975.

61. WESTFALL, T. C. Local regulation of adrenergic neurotransmission. *Physiol. Rev.* 57: 659–728, 1977.

62. WESTFALL, T. C., AND P. E. HUNTER. Effect of muscarinic agonists on the release of ^{3}H-noradrenaline from the guinea-pig perfused heart. *J. Pharm. Pharmacol.* 26: 458–460, 1974.

CHAPTER 12

Release of Acetylcholine in the Isolated Heart

KONRAD LÖFFELHOLZ

*Department of Pharmacology, University of Mainz,
Mainz, Federal Republic of Germany*

**Anatomy of Parasympathetic
  Innervation**
**Methods Used to Evoke Release
  of ACh**
Vagal Stimulation
Field Stimulation
Chemical Stimulation
**Methods Used to Quantify Release
  of ACh**

Sources of ACh Release
Mechanism of ACh Release
Synthesis of ACh
Inactivation of Released ACh
Enzymatic Cleavage
Extracellular Washout t½ of ACh
Cellular Uptake of ACh
*Time Course of ACh-Induced
  Cardiac Effects*

Although the isolated heart with its parasympathetic neuroeffector junction played a crucial role in the discovery of the chemical nature of synaptic transmission (43, 59, 60, 62), it never found a widespread use as a tool for studying cholinergic mechanisms or as a model for cholinergic synapse (62). This neglect is in sharp contrast to the importance of the parasympathetic control of cardiac functions (38, 52).

The present review summarizes the scattered data on the release of acetylcholine (ACh) in the isolated heart. This summary explains why the isolated heart preparation gained so little interest as a tool for studying cholinergic mechanisms, and it also explains that the preparation is nevertheless excellent and even unique for research on certain synaptic mechanisms.

The results of a simple experiment shown in Figure 1 illustrate some of the most interesting features of ACh release in an isolated heart preparation. They raise several questions concerning organization of release, synthesis, and inactivation of ACh in the heart. Figure 1 shows that large amounts of ACh released by vagal (preganglionic) stimulation for only 5 s escape unhydrolyzed into the perfusate, although the slopes of the monoexponential washout curves exhibit a t½ of about 10–15 s. This phenomenon of a high "overflow of unhydrolyzed ACh" into the heart perfusate is unique, as seen from the biochemical studies on peripheral cholinergic mechanisms in other organs. The reasons for the apparently slow rate of hydrolysis and its physiological significance for the neuroeffector transmission in the heart, among other problems, are analyzed and discussed in this present chapter.

257

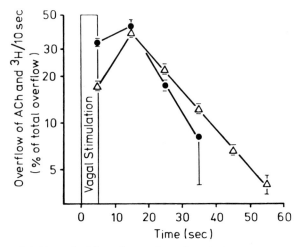

Fig. 1. Overflow of ACh and of ^{3}H ([^{3}H]ACh + [^{3}H]choline) into perfusate of chicken heart evoked by a 5-s period of vagal stimulation. Cervical vagus nerves of heart perfused with Tyrode solution (20 ml/min) were stimulated for 5 s (as indicated) at 40 Hz (1 ms, 40 V). Perfusate was collected in 10-s periods beginning at zero time (start of stimulation). Ordinate, overflow of ACh determined by bioassay on guinea pig ileum (*circles*) and overflow of ^{3}H-labeled material (*triangles*) ([^{3}H]ACh and [^{3}H]choline; chromatographic identification; for details on rate of hydrolysis see Fig. 3 and section **Inactivation of Released ACh**). Overflows were expressed as total overflow evoked by vagal stimulation (100% = sum of overflows determined in 4 or 6 samples as indicated in diagram). Total overflows were 33.3 ± 1.7 pmol/g ACh and 12.3 ± 1.7 × 10^3 disintegrations/min ^{3}H. ACh store was labeled by infusing tracer dose of [^{3}H]choline (100 μmol) into heart 20 min before stimulation (for details of labeling technique, see ref. 56). Note occurrence of unhydrolyzed ACh even 30 s after vagal stimulation in amounts that can be determined by conventional methods. Washout $t\frac{1}{2}$, which follows first-order kinetics, is about 10–15 s for both ACh and ^{3}H material. Moreover, note differences between two curves. Ratio between efflux of unhydrolyzed ACh (*circles*) and ^{3}H efflux (*triangles*) decreases with time, indicating that hydrolysis of released ACh increased gradually with prolonged duration of washout (see section **Inactivation of Released ACh**). Given are means ± SE of 5 experiments.

Anatomy of Parasympathetic Innervation

This section is not an overview on the anatomy of the parasympathetic cardiac innervation but recapitulates only what is necessary for the understanding of this chapter.

The parasympathetic innervation of the vertebrate heart involves two cholinergic synapses arranged in series. The preganglionic nerves originate from various regions of the brain stem (e.g., nucleus ambiguus and dorsal motor nucleus of the vagus) (29) and synapse on cardiac ganglionic neurons that are mostly associated with the posterior aspect of atria at a subepicardial level (84). Postganglionic cholinergic nerves, in common with other autonomic nerves, differ from motor nerves supplying skeletal muscle in that they have extensive varicose terminal fibers in effector organs from which ACh is released "en passage" (12, 69), i.e., ACh is released by action potentials invading the terminal cholinergic fiber and passing from one varicosity (assumed site of release) to the next. Cholinergic fibers running in parallel with adrenergic nerves form the "autonomic groundplexus" (39).

A pattern of nonuniform distribution for cholinergic innervation of the mammalian heart has been suggested by the greater histochemical staining for cholinesterase (see section **Inactivation of Released ACh**) in the atria compared to the ventricles in cat (40), dog, and human heart (47). The possession of ventricular ganglionic cells is a constant feature of avian cardiac innervation (84), which supports the view that the ventricular innervation of the avian heart is more developed than that of the mammalian heart (9).

Differences in the regional distribution of ACh and of choline acetyltransferase (CAT) (see section **Sources of ACh Release**) supplement the histological findings on the density of cholinergic innervation. A detailed review on the regional innervation of pacemaker areas, ventricular specialized and nonspecialized tissues, and coronaries goes beyond the scope of this chapter.

Methods Used to Evoke Release of ACh

Vagal Stimulation. Most of the experiments on the release of ACh in the heart were carried out by stimulating electrically the cervical vagus nerves (Fig. 1), which are easy to isolate (68). Even after isolation of the heart with the vagus nerves attached, vagal transmission remains functional without essential decay of the end-organ response for hours (71). When the cervical trunks of the vagus nerves are stimulated, afferent and cardioaccelerator fibers are excited in addition to the preganglionic parasympathetic nerves. Afferent discharges are not expected to alter ACh release or cardioinhibition after interruption of the reflex arc. Randall and Armour (77) localized the levels of entry and exit of afferent fibers to and from the vagus trunk. Finally, after administration of atropine, cardioaccelerator nerves have been detected in the vagus trunk (70, 78, 83), which presumably are adrenergic according to fluorescence histochemical studies of cat and dog vagus nerves (72). These fibers may modulate cardioinhibition by vagal stimulation even under in vitro conditions.

Field Stimulation. Field stimulation has been applied to incubated heart preparations in the last century (76), because electrical discharges of the nerve endings had been considered as the natural stimulus of the heart cell (43, 62). After discovery of the chemical nature of neuroeffector transmission, field stimulation has been used to stimulate the nerves within the end-organ tissue that was exposed to the electrical field. Electrode size, shape, and positioning have been "tailored" for each experimental condition. A systematic study on the spread of current through the tissue is still lacking.

Because suprathreshold stimulation of the myocardium masks the neurally mediated effects, such as heart arrest, various methods have been developed to surmount this difficulty. First, suprathreshold stimulation at high frequencies was applied for only 5 s, which was followed by atropine-sensitive inhibition and propranolol-sensitive acceleration of the heart rate (1, 74). Second, the nerves of isolated heart preparations have been stimulated with pulses that were subthreshold with respect to the myocardial cell, because the latter have a higher threshold excitability than the nerves (6, 53, 93). Third, application of a train of suprathreshold stimuli during the absolute refractory period of electrically paced cardiac preparations caused a direct twitch depression that was followed by indirect effects that were mediated by the release of neuro-transmitters (10).

There are, however, disadvantages of field stimulation. Field stimulation is not selective. It may release numerous endogenous substances in addition to ACh and may alter K^+ and Ca^{2+} in the extracellular environment (28), effects that may interact with synthesis and release of ACh and with receptor stimulation. Moreover, the deep cardiac nerve plexus may be stimulated below threshold levels for excitation. Because part of these deep fibers originate from subepicardial cell bodies, they may release ACh by action potentials generated at a distance from the nerve terminals. Finally, current strength should not be increased beyond a certain level, because transmitters might be released by a Ca^{2+}-independent mechanism (46, 81).

ACh release or cardioinhibition evoked by field stimulation was found to be resistant to blockade of ganglionic transmission by hexamethonium (53) and by d-tubocurarine (56). It follows that ganglionic transmission is not linked into transmitter release evoked by field stimulation. This is important when the total cardiac release of ACh is studied, which originates mainly from the postganglionic neuron (section **Sources of ACh Release**).

Chemical Stimulation. The early observation that nicotine caused atropine-sensitive cardioinhibition has been interpreted in 1920 as stimulation of the cardiac ganglion (23), i.e., the correct interpretation was achieved 1 yr before ACh was identified as transmitter (59). In autonomic ganglia, preganglionic stimulation and electrophoretic application of ACh give rise to a graded relatively long-lasting synaptic potential that is sensitive to blockade by hexamethonium and related drugs (97). In experiments on parasympathetic ganglia of the frog heart, the (postsynaptic) permeability changes initiated by the synaptic transmitter and by the applied ACh were identical, because the ionic fluxes for both responses had the same equilibrium potential (19). These results demonstrate the identity of the ganglionic transmitter with ACh.

The kind of presynaptic receptors, such as cholinoceptors, adrenoceptors, and other receptors modulating ACh release from pre- as well as postganglionic nerve terminals, is still unknown for many organs including the heart. Haefely (35) has reviewed the literature on nonnicotinic agents that induce generation of action potentials in autonomic ganglion cells and/or facilitate the generation in response to orthodromic or direct electrical stimulation or to chemical stimulation. Nothing is known yet about the release of ACh in the heart evoked by nonnicotinic agents such as muscarinic drugs, 5-hydroxytryptamine, histamine, polypeptides, cardiac glycosides, and veratrum alkaloids. Only recently the release of ACh during infusion of a high-K^+, low-Na^+ solution (108 mM K^+, 44 mM Na^+) into the perfusate of the isolated chicken heart has been studied in the absence and presence of physostigmine (55). Experiments of this kind require direct determination of the released ACh, because its cardiac effects are masked by a K^+-induced arrest and contracture.

Methods Used to Quantify Release of ACh

Release of ACh in the heart can be studied indirectly by recording postsynaptic effects or directly by measuring ACh and/or its metabolite choline in the perfusate.

Although cardiac postsynaptic effects of released ACh usually are directly related to the amount of transmitter released, they are inappropriate parameters for a quantitative assessment of the release. Any experiment designed to observe changes in the release of ACh may change factors affecting electrical and/or contractile properties of the heart cell. Although autonomic ganglia are better suited for electrophysiological experiments on quantal aspects of transmitter release than the heart, a comparative review on neurochemical and electrophysiological data yielded discrepancies between these direct and indirect measures of ACh release even in the former preparation (66). Moreover, because the concentration-response curves for hyperpolarization (31) and cardioinhibition evoked by exogenous ACh are flat and extend over a range of 4–5 log units, changes in postsynaptic effects may be small and insignificant even at large changes of the release of ACh. For example, pentobarbital in a concentration that approximately halved the amount of ACh released by vagal stimulation caused a reduction of the concomitant cardioinhibition by only 10%–20% (56). Because miniature potentials as an effect of quantal release have not been observed in the heart, Glitsch and Pott (32), in a quantitative analysis, compared atropine-sensitive hyperpolarization of the cell membrane of quiescent atrial cells evoked by subthreshold field stimulation (see section **Methods Used to Evoke Release of ACh**) at 50 Hz for 1 s with the hyperpolarization caused by exogenous ACh. An advantage of this method may be that regional differences of transmitter release and inactivation can be investigated.

The most reliable method to study mechanisms of release, synthesis, and inactivation of ACh in the heart is to determine directly the transmitter and its metabolite choline. Even in the absence of cholinesterase inhibition, ACh was liberated on vagal stimulation into the medium filling the ventricle of the frog heart (59) or into the perfusion medium of hearts of birds and mammals (21). According to the latter study, the chicken heart shows by far the highest overflow of ACh into the perfusate. Overflow represents the fraction of ACh released within the heart that escapes into the perfusate (Figs. 1 and 3). The hydrolyzed fraction of ACh released by vagal stimulation at 20 Hz for 1 min appeared as choline in the perfusate and amounted to 71% of the total release of ACh (section **Inactivation of Released ACh**). Measurements of the overflow of ACh into the perfusate of the chicken heart by gas chromatography (20, 49), by the radioenzymatic method described by Goldberg and McCaman (33), and by bioassay using the guinea pig ileum (the two latter determinations were used comparatively, ref. 56) revealed similar results.

Exposure of peripheral organs, such as the guinea pig ileum (85), the superior cervical ganglion of the cat (16), and the chicken heart (57), to radioactive choline labeled the ACh that is released subsequently by electrical or chemical stimulation. This technique has been found useful, for example, in the study of ACh release from brain slices in the absence of physostigmine (86).

Sources of ACh Release

The fine structure of the autonomic cholinergic innervation of peripheral organs such as that of the heart indicates that a comparatively small amount of ACh released from the preganglionic nerve is sufficient to stimulate the cardiac ganglionic cell, which, in turn, releases a larger amount of ACh en passage from the numerous varicosities lined up on a single postganglionic fiber (see section

Anatomy of Parasympathetic Innervation). Indeed, block of the ganglionic transmission by d-tubocurarine reduced the overflow of ACh evoked by vagal stimulation of the perfused chicken heart to $31 \pm 5\%$ ($n = 7$) (56) and, using a lower frequency of stimulation (10 Hz), to $12 \pm 6\%$ ($n = 7$) (R. Lindmar, K. Löffelholz, and W. Weide, unpublished observations). It is not clear yet whether the residual release originated exclusively from the preganglionic nerve, because nonnicotinic ganglionic transmission cannot be excluded to occur in the presence of d-tubocurarine.

Because choline acetyltransferase (CAT) activity can be used, with some reservations (90), as a marker for the parasympathetic innervation of the heart (79), it had been hoped that changes in the enzyme activity after preganglionic denervation reflect the preganglionic contribution to the overall CAT activity in a quantitative manner. Transplantation of the rat heart caused a decrease of the CAT activity in all regions by at least 50% (63); the most pronounced decrease was found in the right atrium (-98%). The data indicated a nonuniform distribution of the preganglionic innervation with a maximum in both atria, i.e., in the regions showing the most frequent occurrence of ganglionic cells (see section **Anatomy of Parasympathetic Innervation**). The authors of the latter study had reason to consider the possibility that the changes overestimate the preganglionic CAT activity, because a postganglionic reduction in the enzyme activity is likely to occur after the inductive influence of nervous activity is withdrawn (27, 73). The latter argument is important for interpreting previously observed changes in the ACh content after vagotomy. In the right atrium of the rat heart, Tucek and Vlk (91) observed a 13% (sinoatrial node area) and a 25% (remaining atrial tissue) loss of ACh after left vagotomy and a 15% and 24% loss, respectively, after right vagotomy. When the values obtained after left and right vagotomy are added to each other, a theoretical loss of 28% and 49%, respectively, would be expected for the bilaterial vagotomy.

The above results are complicated by a second factor. In addition to the presumed secondary loss of postganglionic CAT (and ACh) after denervation, Lund et al. (64) demonstrated also a compensatory increase in CAT activity after unilateral vagotomy, which developed during the first wk after vagotomy and may be related to collateral sprouting (18).

In summary, changes in total CAT activity and in ACh content after vagotomy are inadequate or at least complicated parameters for studying the pre- and postganglionic fractions of the total CAT activity or of the total ACh content of the heart.

The regional variations in ACh and in CAT activity in the heart are composed of differences in the preganglionic as well as in the postganglionic innervation. The above data indicate a nonuniform preganglionic innervation (see section **Anatomy of Parasympathetic Innervation**). Because the bulk of the released ACh originates from postganglionic neurons, regional differences in CAT activity and ACh content are probably dominated by the postganglionic distribution. In the guinea pig heart the rank order of cardiac regions with decreasing CAT activity was right atrium > right ventricle > left atrium > left ventricle (80). The mean CAT activity of dog ventricles was only 18% of that of the atria (89). The ACh contents of the cat heart regions decreased in the order right atrium > left atrium > right ventricle > left ventricle (11). Similar data were described for the rabbit heart (48). These observations confirm the histologist's view (section **Anatomy of Parasympathetic Innervation**) that the ventricular region of the mammalian heart exhibits a low density of (postganglionic) cholinergic innervation. In contrast, the cholinergic innervation of the chicken ventricle seems to be relatively dense (section **Anatomy of Parasympathetic Innervation;** ref. 49) and may have functional significance even for the nonspecialized ventricular tissue (8, 9).

Mechanism of ACh Release

Despite the interruption of parasympathetic discharges by removing the heart from the body, synthesis and release of ACh do not cease even under resting conditions. At rest, ACh was continuously formed in isolated rat atria (90, 96). Likewise, uptake of choline (57) and release of ACh (34, 55, 88) were still present.

Although the opinions on the subcellular events leading to resting and evoked release of ACh differ considerably (13, 101), the release evoked by K^+ or propagated action potentials follows certain features of stimulus-secretion coupling (24, 58). Dependency on extracellular Ca^{2+} was shown for the release of ACh in the chicken heart evoked by high K^+ (55) and by vagal stimulation (21). Studies on postsynaptic responses to vagal stimulation such as reduction of sinoatrial node activity in rabbit atria (87, 94) and hyperpolarization in quiescent guinea pig atria (32) also indicated Ca^{2+} dependence of ACh release. Glitsch and Pott (32) described also a weak inhibitory effect of Mg^{2+}. 4-Aminopyridine, a drug known to decrease K^+ conductance and to increase transmitter release in motor nerves (65) and in parasympathetic nerves of the Auerbach plexus (95) by promoting influx of Ca^{2+}, was also effective in experiments on the release of ACh in the chicken heart (99). In the same preparation the depressant effect of pentobarbital on the ACh release was most likely due to a noncompetitive inhibition of the Ca^{2+} transport (56, 99). This conclusion agrees well with the direct proof of a noncompetitive antagonism between pentobarbital and $^{45}Ca^{2+}$ uptake into brain synaptosomes (4).

In conclusion the mechanism of ACh release in the heart fits well with the general feature of stimulus-secretion coupling. Peculiarities of the heart preparation that markedly influence the release and overflow of ACh in response to vagal (preganglionic) stimulation are given by the intramural ganglionic transmission modulating the input-output ratio (sections **Anatomy of Parasympathetic Innervation** and **Sources of ACh Release**), by synthesis (section **Synthesis of ACh**) and inactivation (section **Inactivation of Released ACh**) of ACh rather than by the mechanism of transmitter release itself.

Synthesis of ACh

In general, ACh is synthesized mostly at the site of its release, i.e., in the terminal parts of the nerve. CAT, however, which catalyzes the synthesis, is transported by proximodistal flow from the cell body to the site of synthesis and release of ACh. Because CAT activity can be used as a marker for cholinergic nerves, it has been determined in various regions of the heart to study the distribution of cholinergic innervation (section **Sources of ACh Release**). As long as further data on the cardiac CAT are lacking, it is merely concluded that the enzyme is present in pre- and postganglionic nerves of the heart. In consideration of the data obtained from brain synaptosomes, however, it may be assumed that the enzyme activity is not rate limiting for the synthesis of ACh under in vitro conditions (41, 42).

The immediate precursors of ACh synthesis are acetyl-CoA and choline. Experiments to identify the source of the acetyl groups have been done almost exclusively on brain tissue (90), but also on rat diaphragm (25). The most likely candidates are pyruvate and acetate under physiological conditions. The considerable variation between tissues and species does not allow any prediction for the heart.

That the second precursor choline is taken up from the extracellular space is generally accepted (Fig. 2; ref. 66). Thus, cardioinhibition observed in rabbit atria during vagal stimulation and reduced by hemicholinium-3 was almost completely restored after treating the atria with choline chloride (14). Haubrich et al. (36) found that intravenously administered [³H]choline was rapidly incorporated into ACh of the guinea pig heart. The plasma level of free choline in animals of various species (usually 5–10 μmol/liter) (5, 21) is well above the apparent K_T of the high-affinity choline uptake, which is generally in the range of 0.5–2.0 μmol/liter (42). Thus, under physiological conditions the supply of choline by the blood is sufficient to maintain ACh synthesis in peripheral organs.

Even after isolation of the heart had removed this source, the release of ACh in hearts of various species was maintained during a 20-min period of field stimulation (20 Hz) at a steady-state level that ranged between 29% (cat and rabbit heart) and 58% (guinea pig heart) of the initial release (21). This observation is remarkable, because the turnover of ACh during sustained high-frequency stimulation is rather high. In the chicken heart, nerve stimulation (>15 Hz) caused a steady-state release of about 3%/min of the ACh tissue content in the absence of exogenous choline and in the presence of physostigmine (62); the tissue content of ACh was unchanged after a 20-min stimulation period. Thus, the formation of an "extracellular choline pool" (Fig. 2) in the

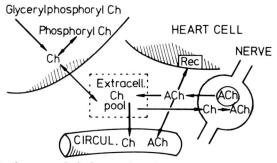

Fig. 2. Schematic diagram of choline and ACh distribution in heart. Note that choline (Ch) is released from bound form in presumably nonneuronal store, which represents major source for the "extracellular choline pool" under conditions of isolation and perfusion. Nerve stimulation for 1 min causes biphasic change of extracellular choline pool (Fig. 3). Increase due to hydrolysis of released ACh is followed by decrease due to neuronal uptake of choline, which, after acetylation, substitutes released ACh. Changes in amount of choline appearing in the circulation (Circul.) and subsequently in perfusate reflect formation or removal of extracellular choline.

isolated heart guarantees a steady-state release and resynthesis of ACh during high-frequency stimulation.

In general, extracellular choline is formed in an isolated organ: first, from released and subsequently hydrolyzed ACh and, second, from breakdown of phospholipids, glycerophosphocholine, and phosphorylated choline. A schematic diagram of the distribution and movements of choline and ACh in the heart is shown in Figure 2. Because inhibition of the cholinesterase activity did not alter the level of the steady state release of ACh in the heart (62), it is concluded that the second source of extracellular choline, i.e., its formation from phospholipids, is sufficient for maintaining resynthesis of ACh during constant release of the transmitter from the postganglionic neuron of the heart. This conclusion is corroborated by the high rate of choline efflux from isolated hearts of various species (21); the mean rates ranged from 0.4 to 2.1 $nmol \cdot g^{-1} \cdot min^{-1}$. Because this efflux was markedly increased by hypoxia (61), it may be responsible also for the postmortem accumulation of tissue choline (26). Further studies have to elucidate whether this tissue source of choline plays an essential role in vivo for the synthesis of ACh besides the circulating choline.

As to the sources of extracellular choline, a difference between the isolated heart and the isolated sympathetic ganglion is striking. Approximately 50% of the choline formed from ACh released during preganglionic nerve stimulation was found to be recaptured for ACh synthesis in the superior cervical ganglion of the cat (15, 75). Moreover, the isolated ganglion was dependent on ACh hydrolysis during repetitive nerve stimulation to maintain a steady-state release of ACh (17, 67). Bennett and McLachlan (2, 3), who used intracellularly recorded postsynaptic potentials as a measure of transmitter release in guinea pig ganglia, showed that the effect of physostigmine was to limit the amount of choline available for neuronal uptake and for ACh synthesis, an effect that finally must lead to transmission failure (98).

Partial transmission failure during infusion with choline-free solutions has been reported also for the parasympathetic ganglion of isolated mammalian hearts in contrast to that of the chicken heart (21). This was evident from two observations using both vagal (preganglionic) and field stimulation, which bypasses ganglionic transmission (section **Methods Used to Evoke Release of ACh**). First, it was found that both kinds of stimulation released approximately the same amount of ACh in the chicken heart, whereas in mammalian hearts only field stimulation evoked an overflow of ACh that was above the limit of the assay. Second, infusion of 10^{-5} M choline into the cat heart increased the overflow of ACh evoked by vagal stimulation from below the assay limit (≤ 3 $pmol \cdot g^{-1} \cdot min^{-1}$) to a value that was 23 times the assay limit. It was concluded that transmission in the cardiac ganglion of isolated perfused mammalian hearts was depressed by a choline deficiency during perfusion with choline-free media, just as transmission was depressed in isolated sympathetic ganglia.

With pyrolysis gas chromatography, Green et al. (34) were able to measure even the resting overflow of ACh into the perfusate of the rabbit heart after inhibition of the cholinesterase activity (16 ± 3 ng/heart in 20 min, $n = 4$). However, vagal stimulation at 20 Hz caused only a 1.7-fold increase in the resting overflow. Likewise, the release of ACh in the perfused cat heart evoked by vagal stimulation was very low (≤ 3 $pmol \cdot g^{-1} \cdot min^{-1}$, which was the limit of the assay); the release was ≤0.05%/min of the tissue content of ACh (6.3 nmol/g) (21). In contrast, vagal stimulation of the chicken heart caused a 15.1-fold increase in the resting overflow of ACh (62), which was equal to a release of 8%/min of the tissue content of ACh.

Ganglionic transmission failure after isolation of the rabbit heart explains also certain unexpected effects of hemicholinium-3, an inhibitor of the neuronal choline uptake and resynthesis of ACh in both pre- and postganglionic cholinergic nerves of the heart. Vagal stimulation in the presence of hemicholinium-3 did not reduce the ACh content of atria in vitro when choline-free incubation media were used, whereas in vivo the ACh content was reduced by 67% (54). The negative in vitro results had been found difficult to explain by the authors of the latter study, because the end-organ response to vagal stimulation vanished rapidly in the presence of hemicholinium-3. Both apparently contradictory effects are understandable today as ganglionic transmission failure in mammalian hearts. Again, the results obtained from the isolated chicken heart were different: the ACh content (7.2 nmol/g) was reduced significantly by 70% (mean of 4 experiments) after 20 min of vagal stimulation in the presence of hemicholinium-3 (62). The reason for the species difference with respect to transmission through the cardiac ganglion under isolation conditions is not clear yet.

It may be assumed that the extracellular choline concentration is not uniform within the heart and that the local concentration depends on numerous factors, such as the topographical source of choline formation, rate, and direction of the interstitial flow, interstitial pressure, cardiac contractions, and oxygenation. The superficial localization of many ganglia, single or in clusters, plus diffusion barriers are hypothetical reasons for the observation that the supply of choline is critical for the preganglionic synthesis of ACh and presumably is not rate limiting for the synthesis in the varicosities of the autonomic groundplexus.

It can be argued that fatigue of vagal stimulation in causing cardioinhibition has not been observed even after several hours in, for example, cat atria incubated in choline-free solution (71). It has been pointed out in section **Methods Used to Quantify Release of ACh** that large changes in the release of ACh may have little consequences for the postsynaptic effects of vagal stimulation. Nevertheless, it seems justified to recommend addition of choline (e.g., 10^{-5} M) to the perfusion or incubation medium whenever the preganglionic parasympathetic nerves of isolated mammalian heart preparations are stimulated. Likewise, choline is routinely added to the medium when the release of ACh is studied in perfused ganglia (e.g., ref. 15).

An advantage of measuring overflow of ACh and of choline into the heart perfusate lies in the fact that changes in the overall extracellular concentration of ACh and choline can be measured with high temporal resolution and little delay due to the extracellular washout ($t_{1/2}$ <1 min). As shown in Figure 3, the time courses of ACh overflow and of the subsequent removal of extracellular choline due to activation of the neuronal uptake could be monitored (57). The release occurred during stimulation without noticeable delay, whereas the neuronal uptake of choline reached a maximum activation 1–2 min after the 1-min period of nerve stimulation and gradually returned to the prestimulation level within several minutes. These results were obtained in the presence of physostigmine, but even in the absence of any anticholinesterase agent the two

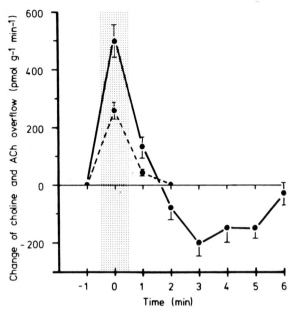

Fig. 3. Effects of vagal stimulation on choline and ACh overflows into the perfusate of the chicken heart in absence of cholinesterase inhibition. Cervical vagus nerves were stimulated at 20 Hz for 1 min (*shaded area*). Ordinate, change in overflow of choline (*solid line*) and of ACh (*broken line*); mean prestimulation values were 2.4 ± 0.2 nmol$\cdot$g$^{-1}\cdot$min^{-1} and 24 ± 3 pmol$\cdot$g$^{-1}\cdot$min^{-1}. Given are means $\pm$ SE of 20 experiments. For further explanations see Fig. 2. [From Lindmar et al. (56).]

events, i.e., formation of choline from released ACh and subsequent neuronal removal of choline, could be monitored as biphasic change in the choline overflow (Fig. 3). The uptake was highly sensitive to hemicholinium-3 and low Na$^+$ and therefore was attributed to the high-affinity uptake (see above). Its delayed activation in the heart is not compatible with the view of a direct coupling of nervous activity and choline uptake (82), but may be governed by the restoration of the equilibrium between ACh and choline compartments affected by the release. General aspects of the high-affinity choline transport and its driving forces were summarized recently by, among others, Jope (42) and Vaca and Pilar (92).

The question as to whether the release of ACh can be increased by exogenous choline has received physiological (41) and clinical interest (44, 100). Undoubtedly, administration of choline in vitro and in vivo has effects on cholinergic neurons, but the mechanisms involved are often not understood. This situation is reflected by the conflicting data obtained on the heart. Injection of choline chloride, 1.2 mmol/kg subcutaneously into rats (51) and 0.4 mmol/kg intravenously into guinea pigs (37), increased the cardiac content of ACh. Moreover, infusion of 10^{-5} M choline chloride into isolated chicken and cat hearts caused a twofold increase in the release of ACh evoked by field stimulation (21). On the other hand, resynthesis of ACh in the chicken heart following release of large amounts of ACh met exactly the synaptic demands to maintain a certain level of the tissue content (62). Moreover, the amount of choline taken up in response to a 1-min period of nerve stimulation in the same preparation was equimolar to the preceding release of ACh in the presence of physostigmine (57). Despite large variations in the resting rate of choline overflow (1.3–4.0 nmol·g^{-1}·min^{-1}) from the chicken heart, a significant correlation to the release was absent (57). In conclusion, these data suggest that the extracellular concentration of endogenous choline in the isolated chicken heart is not an essential determinant for the activated neuronal uptake of choline and for the release of ACh and its resynthesis. In other words, in isolated heart preparations the basic neuronal uptake of choline works adequately to meet the basic requirements, including even large synaptic demands. Only in cardiac ganglia of mammalian species the supply of choline may be critical and rate limiting (see above).

It is difficult to explain how administration of exogenous choline affected the ACh content and release in the heart under conditions where intraneuronal mechanisms seem to control the rate of the high-affinity uptake (K_T for choline ~0.5–2.0 μM). Among several alternative explanations, only two are mentioned. First, high concentrations of choline may increase the cytoplasmic concentration of choline via the low-affinity uptake ($K_T > 30\ \mu$M) and thereby facilitate influx of choline (by countertransport). Second, the high-affinity uptake system may be under regulatory restraint by endogenous promoting or inhibiting metabolic factors; their absence or presence, respectively, may decrease the affinity of the choline carrier. The various alternatives have been discussed in detail recently (41).

Inactivation of Released ACh

ACh released into the neuroeffector cleft evokes cardioinhibition that is terminated by the removal of the transmitter (Fig. 2). In general, ACh can be removed from the biophase by enzymatic cleavage, by washout of the extracellular space, and by cellular uptake.

Enzymatic Cleavage. Although acetylcholinesterase activity can be demonstrated in some adrenergic nerves and in nonnervous tissue, cholinergic nerves usually contain higher levels of this enzyme and therefore can be localized histochemically by enzymatic staining (12). This technique has been used to map cholinergic neurons within the heart (50; section **Anatomy of Parasympathetic Innervation**), for example, in its specialized tissue (7). The overall cholinesterase activity of hearts of various species, such as rabbit (49), rat (30), frog (30), and chicken (20), were found to be in the same range. Both acetylcholinesterase and butyrylcholine esterase activities are higher in the atria than in the ventricles (50).

These data are, however, not very meaningful with respect to the rate of extracellular hydrolysis of released ACh in the heart as compared to other tissues. The fact that a fraction of released ACh could escape unhydrolyzed into the solution filling the frog ventricle (59) or into the perfusate (Figs. 1 and 3; section **Methods Used to Quantify Release of ACh**) is clear evidence that the apparent rate of hydrolysis of released ACh is relatively slow in the heart as compared to other organs. The recovery of ACh from the perfusion stream of sympathetic ganglia (98) or from the incubation medium of skeletal muscle (45) does not seem to be possible using conventional assay methods.

McMahan and Kuffler (69) observed in the frog heart that many of the postganglionic cholinergic varicosities were located at a distance of even "many micrometers" from any obvious target cell. They were puzzled by this observation considering the general view "that ACh must be released quite close to the target organ if it is to escape inactivated by cholinesterase." The overflow of ACh into the cardiac perfusate is a phenomenon that is incompatible with the general validity of this view, at least under in vitro conditions. This phenomenon is not a species-dependent curiosity, because physostigmine increased the overflow of ACh to a similar extent in hearts of chickens, guinea pigs, cats, and rabbits (21).

Finally it is mentioned that when the rate of hydrolysis is estimated on the basis of cholinesterase inhibition one has to consider alterations in the release of ACh. Physostigmine (10^{-6} M) reduced the evoked release of ACh by 32% (57) and increased the cardiac content of ACh by 30% (20).

Extracellular Washout $t_{1/2}$ of ACh. The $t_{1/2}$ of extracellular washout of ACh released by vagal stimulation of the perfused chicken heart was found to be about 10–15 s (Fig. 1). The appearance of unhydrolyzed ACh in the perfusate 30 s after nerve stimulation favors the view that washout of released ACh is an important mechanism for the inactivation of this transmitter in the heart.

Cellular Uptake of ACh. When 8 nmol of ACh were infused into the chicken heart ($n = 4$) during 1 min, most of the ACh was recovered from the perfusate unmetabolized; of the remaining fraction, 71% appeared as choline (57). The total recovery (ACh + choline = 95%) was not significantly different from the infused amount of ACh. Thus there was no loss of ACh other than by hydrolysis. In other tissues, cellular uptake of released ACh does not seem to play a significant role except after certain pretreatments of the tissue (66).

Time Course of ACh-Induced Cardiac Effects. The effects of released ACh on atrial membrane potential and on heart rate are augmented by inhibition of the cholinesterase activity (1, 31, 60). The above considerations show, however, that cardioinhibition evoked by vagal stimulation is terminated by both hydrolysis and washout of ACh. Both mechanisms of inactivation exhibit a close interaction. First, increasing the *rate of washout* (rate of cardiac perfusion was increased from 10 to 30 ml/min) caused a reduction in the rate of hydrolysis (22). Second, as shown in Figure 1, hydrolysis increased with increasing *duration of washout*. Thus, the relative contribution of either of the two mechanisms is expected to vary according to the experimental design. Incubation favors the hydrolysis (31) and perfusion increases the washout factor (22). Finally it is presumed that isolation and perfusion (or incubation) with a saline solution lacking oncotic pressure and cholinesterase activity markedly alter the modes of removal. Conclusions on the inactivation of released ACh in vivo should be drawn with caution when based on experiments carried out under unphysiological conditions.

Acknowledgments

This paper was completed during the author's tenure in 1980 at the Istituto di Farmacologia, Florence. The author takes this occasion to thank Prof. Giancarlo Pepeu and his colleagues for their hospitality and help in preparing the manuscript and to thank Dr. Stanislav Tuček, Prague, for providing stimulating discussions and helpful comments.

References

1. AMORY, D. W., AND T. C. WEST. Chronotropic response following direct electrical stimulation of the sinoatrial node: a pharmacological evaluation. *J. Pharmacol. Exp. Ther.* 137: 14–23, 1962.
2. BENNETT, M. R., AND E. M. McLACHLAN. An electrophysiological analysis of the storage of acetylcholine in preganglionic nerve terminals. *J. Physiol. London* 221: 657–668, 1972.
3. BENNETT, M. R., AND E. M. McLACHLAN. An electrophysiological analysis of the synthesis of acetylcholine in preganglionic nerve terminals. *J. Physiol. London* 221: 669–682, 1972.
4. BLAUSTEIN, M. P., AND A. C. ECTOR. Barbiturate inhibition of calcium uptake by depolarized nerve terminals in vitro. *Mol. Pharmacol.* 11: 369–378, 1975.
5. BLIGH, J. The level of free choline in plasma. *J. Physiol. London* 117: 234–240, 1952.
6. BLINKS, J. R. Field stimulation as a means of effecting the graded release of autonomic transmitters in isolated heart muscle. *J. Pharmacol. Exp. Ther.* 151: 221–235, 1966.
7. BOJSEN-MØLLER, F., AND J. TRANUM-JENSEN. Rabbit heart nodal tissue, sinuatrial ring bundle and atrioventricular connexions identified as a neuromuscular system. *J. Anat.* 112: 367–382, 1972.
8. BOLTON, T. B. Intramural nerves in the ventricular myocardium of the domestic fowl and other animals. *Br. J. Phamacol. Chemother.* 31: 253–268, 1967.
9. BOLTON, T. B. Nervous system. In: *Avian Physiology* (3rd ed.), edited by P. D. Sturkie. New York: Springer-Verlag, 1976, p. 1–28.
10. BRADY, A. J., B. C. ABBOTT, AND W. F. MOMMAERTS. Inotropic effects of trains of impulses applied during the contraction of cardiac muscle. *J. Gen. Physiol.* 44: 415–432, 1960.
11. BROWN, O. M. Cat heart acetylcholine: structural proof and distribution. *Am. J. Physiol.* 231: 781–785, 1976.
12. BURNSTOCK, G. The ultrastructure of autonomic cholinergic nerves and junctions. *Prog. Brain Res.* 49: 3–21, 1979.
13. CECCARELLI, B., AND W. P. HURLBUT. Vesicle hypothesis of the release of quanta of acetylcholine. *Physiol. Rev.* 60: 396–441, 1980.
14. CHANG, V., AND M. J. RAND. Transmission failure in sympathetic nerves produced by hemicholinium. *Br. J. Pharmacol.* 15: 588–600, 1960.
15. COLLIER, B., AND H. S. KATZ. Acetylcholine synthesis from recaptured choline by a sympathetic ganglion. *J. Physiol. London* 238: 639–655, 1974.
16. COLLIER, B., AND C. LANG. The metabolism of choline by a sympathetic ganglion. *Can. J. Physiol. Pharmacol.* 47: 119–126, 1969.
17. COLLIER, B., AND F. C. MACINTOSH. The source of choline for acetylcholine synthesis in a sympathetic ganglion. *Can. J. Physiol. Pharmacol.* 47: 127–135, 1969.
18. COURTNEY, K., AND S. ROPER. Sprouting of synapses after partial denervation of frog cardiac ganglion. *Nature London* 259: 317–319, 1976.
19. DENNIS, M. J., A. J. HARRIS, AND S. W. KUFFLER. Synaptic transmission and its duplication by focally applied acetylcholine in parasympthetic neurons in the heart of frog. *Proc. R. Soc. London Ser. B.* 177: 509–539, 1971.
20. DIETERICH, H. A., H. KAFFEI, H. KILBINGER, AND K. LÖFFELHOLZ. The effects of physostigmine on cholinesterase activity, storage and release of acetylcholine in the isolated chicken heart. *J. Pharmacol. Exp. Ther.* 199: 236–246, 1976.
21. DIETERICH, H. A., R. LINDMAR, AND K. LÖFFELHOLZ. The role of choline in the release of acetylcholine in isolated hearts. *Naunyn-Schmiedebergs Arch. Pharmacol.* 301: 207–215, 1978.
22. DIETERICH, H. A., AND K. LÖFFELHOLZ. Effect of coronary perfusion rate on the hydrolysis of exogenous and endogenous acetylcholine in the isolated heart. *Naunyn-Schmiedebergs Arch. Pharmacol.* 296: 143–148, 1977.
23. DIXON, W. E. Nicotin, Coniin, Piperidin, Lupetidin, Cytisin, Lobelin, Spartein, Gelsemin. Mittel, welche auf bestimmte Nervenzellen wirken. In: *Handbuch der Experimentellen Pharmakologie*, edited by A. Heffter. Berlin: Springer, 1920, vol. 2, p. 656–736.
24. DOUGLAS, W. W. Stimulus-secretion coupling: the concept and clues from chromaffin and other cells. *Br. J. Pharmacol.* 34: 451–474, 1968.
25. DREYFUS, M. P. Identification de l'acétate comme précurseur de radical acétyle de l'acétylcholine des jonctions neuromusculaires du rat. *C. R. Acad. Sci. Paris Ser. D* 280: 1893–1894, 1975.
26. DROSS, K., AND H. KEWITZ. Concentration and origin of choline in the rat brain. *Naunyn-Schmiedebergs Arch. Pharmacol.* 274: 91–106, 1972.

27. EKSTROM, J. Fall in choline acetyltransferase activity in ventricles of rat heart after treatment with ganglion blocking drug. *Acta Physiol. Scand.* 102: 116–119, 1978.

28. ERULKAR, S. D., AND F. F. WEIGHT. Ionic environment and the modulation of transmitter release. *Trends Neurosci.* 2: 298–301, 1979.

29. GEIS, G. S., J. W. KOZELKA, AND R. D. WURSTER. Organization and reflex control of vagal cardiomotor neurons. *J. Autonomic Nerv. Syst.* 3: 437–450, 1981.

30. GIRARDIER, L., F. BAUMANN, AND J. M. POSTERNAK. Recherches sur les cholinesterases cardiaques. *Helv. Physiol. Pharmacol. Acta* 18: 467–481, 1960.

31. GLITSCH, H. G., AND L. POTT. Effects of acetylcholine and parasympathetic nerve stimulation on membrane potential in quiescent guinea-pig atria. *J. Physiol. London* 279: 655–668, 1978.

32. GLITSCH, H. G., AND L. POTT. Effect of divalent cations on acetylcholine release from cardiac parasympathetic nerve endings. *Pfluegers Arch.* 377: 57–63, 1978.

33. GOLDBERG, A. M., AND R. E. McCAMAN. An enzymatic method for the determination of picomole amounts of choline and acetylcholine. In: *Choline and Acetylcholine. Handbook of Chemical Assay Methods*, edited by I. Hanin. New York: Raven, 1974, p. 47–61.

34. GREEN, J. P., D. L. ALKON, D. E. SCHMIDT, AND P. I. A. SZILAGYI. Confirmation of acetylcholine in perfusates of the stimulated vagus nerve by pyrolysis gas chromatography. *Life Sci.* 9: 741–745, 1970.

35. HAEFELY, W. E. Non-nicotinic chemical stimulation of autonomic ganglia. In: *Pharmacology of Ganglionic Transmission. Handbook of Experimental Pharmacology*, edited by D. A. Kharkevich. New York: Springer-Verlag, 1980, vol. 53, p. 313–357.

36. HAUBRICH, D. R., P. F. L. WANG, AND P. W. WEDEKING. Distribution and metabolism of intravenously administered choline (methyl-^{3}H) and synthesis in vivo of acetylcholine in various tissues of guinea pigs. *J. Pharmacol. Exp. Ther.* 193: 246–255, 1975.

37. HAUBRICH, D. R., P. W. WEDEKING, AND P. F. L. WANG. Increase in tissue concentration of acetylcholine in guinea pigs in vivo induced by administration of choline. *Life Sci.* 14: 921–927, 1974.

38. HIGGINS, C. B., S. F. VATNER, AND E. BRAUNWALD. Parasympathetic control of the heart. *Pharmacol. Rev.* 25: 119–155, 1973.

39. HILLARP, N.-Å. The construction and functional organization of the autonomic innervation apparatus. *Acta Physiol. Scand.* 46, Suppl. 157: 1–38, 1959.

40. JACOBOWITZ, D., T. COOPER, AND H. B. BARBER. Histochemical and chemical studies of the localization of adrenergic and cholinergic nerves in normal and denervated cat hearts. *Circ. Res.* 20: 289–298, 1967.

41. JENDEN, D. J. An overview of choline and acetylcholine metabolism in relation to the therapeutic uses of choline. In: *Choline and Lecithin in Brain Disorders. Nutrition and the Brain*, edited by A. Barbeau, J. H. Growdon, and R. J. Wurtman. New York: Raven, 1979, vol. 5, p. 13–24.

42. JOPE, R. S. High affinity choline transport and acetyl CoA production in brain and their roles in the regulation of acetylcholine synthesis. *Brain Res. Rev.* 1: 313–344, 1979.

43. KARCZMAR, A. G. Introduction: history of the research with anticholinesterase agents. In: *Anticholinesterase Agents*, edited by A. G. Karczmar. New York: Pergamon, 1970, vol. 1, p. 1–44.

44. KARCZMAR, A. G. Overview: cholinergic drugs and behavior—what effects may be expected from a "cholinergic diet"? In: *Choline and Lecithin in Brain Disorders. Nutrition and the Brain*, edited by A. Barbeau, J. H. Growdon, and R. J. Wurtman. New York: Raven, 1979, vol. 5, p. 141–175.

45. KATZ, B., AND R. MILEDI. The binding of acetylcholine to receptors and its removal from the synaptic cleft. *J. Physiol. London* 231: 549–574, 1973.

46. KATZ, R. I., AND I. J. KOPIN. Electrical field-stimulated release of norepinephrine-H^3 from rat atrium: effects of ions and drugs. *J. Pharmacol. Exp. Ther.* 169: 229–236, 1969.

47. KENT, K. M., S. E. EPSTEIN, T. COOPER, AND D. M. JACOBOWITZ. Cholinergic innervation of the canine and human ventricular conducting system. *Circulation* 50: 948–955, 1974.

48. KILBINGER, H. Gas chromatographic estimation of acetylcholine in the rabbit heart using a nitrogen selective detector. *J. Neurochem.* 21: 421–429, 1973.

49. KILBINGER, H., AND K. LÖFFELHOLZ. The isolated perfused chicken heart as a tool for studying acetylcholine output in the absence of cholinesterase inhibition. *J. Neural Transm.* 38: 9–14, 1976.

50. KOELLE, G. B. Cytological distribution and physiological functions of cholinesterases. In:

Cholinesterases and Anticholinesterase Agents. Handbuch der Experimentellen Pharmakologie, edited by G. B. Koelle. Berlin: Springer, 1963, vol. 15, p. 187–298.

51. KUNTSCHEROVÁ, J. Effect of short-term starvation and choline on the acetylcholine content of organs of albino rats. *Physiol. Bohemoslov.* 21: 655–660, 1972.

52. LEVY, M. N. Parasympathetic control of the heart. In: *Neural Regulation of the Heart*, edited by W. C. Randall. New York: Oxford Univ. Press, 1977, p. 95–129.

53. LEWARTOWSKI, B. Selective stimulation of intra-cardiac post-ganglionic fibres. *Nature London* 199: 76–77, 1963.

54. LEWARTOWSKI, B., AND K. BIELECKI. The influence of hemicholinium no. 3 and vagal stimulation on acetylcholine content of rabbit atria. *J. Pharmacol. Exp. Ther.* 142, 24–30, 1963.

55. LINDMAR, R., K. LÖFFELHOLZ, AND H. POMPETZKI. Acetylcholine overflow during infusion of a high potassium-low sodium solution in the perfused chicken heart in the absence and presence of physostigmine. *Naunyn-Schmiedebergs Arch. Pharmacol.* 299: 17–21, 1977.

56. LINDMAR, R., K. LÖFFELHOLZ, AND W. WEIDE. Inhibition by phentobarbital of the acetylcholine release from the postganglionic parasympathetic neuron of the heart. *J. Pharmacol. Exp. Ther.* 210: 166–173, 1979.

57. LINDMAR, R., K. LÖFFELHOLZ, W. WEIDE, AND J. WITZKE. Neuronal uptake of choline following release of acetylcholine in the perfused heart. *J. Pharmacol. Exp. Ther.* 215: 710–715, 1980.

58. LLINÁS, R. R., AND J. E. HEUSER. Depolarization-release coupling systems in neurons. *Neurosci. Res. Prog. Bull.* 15: 557–687, 1977.

59. LOEWI, O. Über humorale Übertragbarkeit der Herznervenwirkung. I. Mitteilung. *Pfluegers Arch.* 189: 239–242, 1921.

60. LOEWI, O., AND E. NAVRATIL. Über humorale Übertragbarkeit der Herznervenwirkung. XI. Mitteilung. Über den Mechanismus der Vaguswirkung von Physostigmin und Ergotamin. *Pfluegers Arch.* 214: 689–696, 1926.

61. LÖFFELHOLZ, K., R. LINDMAR, AND W. WEIDE. The relationship between choline and acetylcholine release in the autonomic nervous system. In: *Choline and Lecithin in Brain Disorders. Nutrition and the Brain*, edited by A. Barbeau, J. H. Growdon, and R. J. Wurtman. New York: Raven, 1979, vol. 5, p. 233–241.

62. LÖFFELHOLZ, K., R. LINDMAR, AND W. WEIDE. The isolated perfused heart and its parasympathetic neuro-effector junction. In: *Model Cholinergic Synapses. Progress in Cholinergic Biology*, edited by I. Hanin and A. M. Goldberg. New York: Raven. In press.

63. LUND, D. D., P. G. SCHMID, S. E. KELLEY, R. J. CORRY, AND R. ROSKOSKI, JR. Choline acetyltransferase activity in rat heart after transplantation. *Am. J. Physiol.* 235 (*Heart Circ. Physiol.* 4): H367–H371, 1978.

64. LUND, D. D., P. G. SCHMID, AND R. ROSKOSKI, JR. Choline acetyltransferase activity in rat and guinea pig heart following vagotomy. *Am. J. Physiol.* 236 (*Heart Circ. Physiol.* 5): H620–H623, 1979.

65. LUNDH, H., AND S. THESLEFF. The mode of action of 4-aminopyridine and guanidine on transmitter release from motor nerve terminals. *Eur. J. Pharmacol.* 42: 411–412, 1977.

66. MACINTOSH, F. C., AND B. COLLIER. Neurochemistry of cholinergic terminals. In: *Neuromuscular Junction. Handbook of Experimental Pharmacology*, edited by E. Zaimis. New York: Springer-Verlag, 1976, vol. 42, p. 99–228.

67. MATTHEWS, E. K. The effects of choline and other factors on the release of acetylcholine from the stimulated perfused superior cervical ganglion of the cat. *Br. J. Pharmacol.* 21: 244–249, 1963.

68. McEWENS, L. M. The effect on the isolated rabbit heart of vagal stimulation and its modification by cocaine, hexamethonium and ouabain. *J. Physiol. London* 131: 678–689, 1956.

69. McMAHAN, U. J., AND S. W. KUFFLER. Visual identification of synaptic boutons on living ganglion cells and of varicosities in postganglionic axons in the heart of the frog. *Proc. R. Soc. London Ser. B.* 177: 485–508, 1971.

70. MIDDLETON, S., H. H. MIDDLETON, AND J. TODA. Adrenergic mechanisms of vagal cardiostimulation. *Am. J. Physiol.* 158: 31–37, 1949.

71. MISU, Y., AND S. M. KIRPEKAR. Effects of vagal and sympathetic nerve stimulation on the isolated atria of the cat. *J. Pharmacol. Exp. Ther.* 163, 330–342, 1968.

72. MURYOBAYASHI, T., J. MORI, M. FUJIWARA, AND K. SHIMAMOTO. Fluorescence histochemical demonstration of adrenergic nerve fibers in the vagus nerve of cats and dogs. *Jpn. J. Pharmacol.* 18: 285–293, 1968.

73. OESCH, F. Trans-synaptic induction of choline acetyltransferase in the preganglionic neuron of the peripheral sympathetic nervous system. *J. Pharmacol. Exp. Ther.* 188: 439–446, 1974.

74. PAPPANO, A. J., AND K. LÖFFELHOLZ. Ontogenesis of adrenergic and cholinergic neuroeffector transmission in chick embryo heart. *J. Pharmacol. Exp. Ther.* 191: 468–478, 1974.

75. PERRY, W. L. M. Acetylcholine release in the cat's superior cervical ganglion. *J. Physiol. London* 119: 439–454, 1953.

76. PICKERING, J. W. Experiments on the hearts of mammalian and chick-embryos, with special reference to action of electric currents. *J. Physiol. London* 20: 165–222, 1896.

77. RANDALL, W. C., AND J. A. ARMOUR. Regional vagosympathetic control of the heart. *Am. J. Physiol.* 227: 444–452, 1974.

78. RANDALL, W. C., J. S. WECHSLER, J. B. PACE, AND M. SZENTIVANYI. Alterations in myocardial contractility during stimulation of the cardiac nerves. *Am. J. Physiol.* 214: 1205–1212, 1968.

79. ROSKOSKI, R., JR., R. J. McDONALD, L. M. ROSKOSKI, W. J. MARVIN, AND K. HERMSMEYER. Choline acetyltransferase activity in heart: evidence for neuronal and myocardial origin. *Am. J. Physiol.* 233 (*Heart Circ. Physiol.* 2): H642–H646, 1977.

80. SCHMID, P. G., G. B. GREIF, D. D. LUND, AND R. ROSKOSKI, JR. Regional choline acetyltransferase activity in the guinea pig heart. *Circ. Res.* 42: 657–660, 1978.

81. SCHROLD, J., AND O. A. NEDERGAARD. ³H-noradrenaline outflow induced from isolated adventitia and intima-media of rabbit aorta by electrical field stimulation. *Eur. J. Pharmacol.* 39: 423–428, 1976.

82. SIMON, J. R., AND M. J. KUHAR. Impulse-flow regulation of high affinity choline uptake in brain cholinergic nerve terminals. *Nature London* 255: 162–163, 1975.

83. SMITH, D. C. Synaptic sites in sympathetic and vagal cardioaccelerator nerves of the dog. *Am. J. Physiol.* 218: 1618–1623, 1970.

84. SMITH, R. B. Intrinsic innervation of the avian heart. *Acta Anat.* 79: 112–119, 1971.

85. SZERB, J. C. Storage and release of labeled acetylcholine in the myenteric plexus of the guinea-pig ileum. *Can. J. Physiol. Pharmacol.* 54: 12–22, 1976.

86. SZERB, J. C., AND G. T. SOMOGYI. Depression of acetylcholine release from cerebral cortical slices by cholinesterase inhibition and by oxotremorine. *Nature New Biol. London* 241: 121–122, 1973.

87. TODA, N., AND T. C. WEST. Interaction between Na, Ca, Mg, and vagal stimulation in the S-A node of the rabbit. *Am. J. Physiol.* 212: 424–430, 1967.

88. TRAUTWEIN, W., W. J. WHALEN, AND E. GROSSE-SCHULTE. Elektrophysiologischer Nachweis spontaner Freisetzung von Acetylcholin im Vorhof des Herzens. *Pfluegers Arch.* 270: 560–570, 1960.

89. TUČEK, S. Changes in choline acetyltransferase activity in the cardiac auricles of dogs during postnatal development. *Physiol. Bohemoslov.* 14: 530–535, 1965.

90. TUČEK, S. *Acetylcholine Synthesis in Neurons.* London: Chapman and Hall, 1978.

91. TUČEK, S., AND J. VLK. The effect of vagotomy on the acetylcholine content and cholinesterase activity in various regions of the rat heart atria. *Physiol. Bohemoslov.* 11: 319–328, 1962.

92. VACA, K., AND G. PILAR. Mechanisms controlling choline transport and acetylcholine synthesis in motor nerve terminals during electrical stimulation. *J. Gen. Physiol.* 73: 605–628, 1979.

93. VINCENZI, F. F., AND T. C. WEST. Release of autonomic mediators in cardiac tissue by direct subthreshold electrical stimulation. *J. Pharmacol. Exp. Ther.* 141: 185–194, 1963.

94. VINCENZI, F. F., AND T. C. WEST. Modification by calcium of the release of autonomic mediators in the isolated sinoatrial node. *J. Pharmacol. Exp. Ther.* 150: 349–360, 1965.

95. VIZI, E. S., J. VAN DIJK, AND F. F. FOLDES. The effect of 4-aminopyridine on acetylcholine release. *J. Neural Transm.* 41: 265–274, 1977.

96. VLK, J., AND S. TUČEK. The formation of acetylcholine in isolated heart auricles of white rats and guinea-pigs. *Physiol. Bohemoslov.* 13: 310–314, 1964.

97. VOLLE, R. L. Nicotinic ganglion-stimulating agents. In: *Pharmacology of Ganglionic Transmission. Handbook of Experimental Pharmacology,* edited by D. A. Kharkevich. New York: Springer-Verlag, 1980, vol. 53, p. 281–312.

98. VOLLE, R. L. Ganglionic actions of anticholinesterase agents, catecholamines, neuro-muscular blocking agents, and local anaesthetics. In: *Pharmacology of Ganglionic Transmission. Handbook of Experimental Pharmacology,* edited by D. A. Kharkevich. New York: Springer-Verlag, 1980, vol. 53, p. 385–410.

99. WEIDE, W., AND K. LÖFFELHOLZ. 4-Aminopyridine antagonizes the inhibitory effect of pento-

barbital on acetylcholine release in the heart. *Naunyn-Schmiedebergs Arch. Pharmacol.* 312: 7–13, 1980.

100. WURTMAN, R. J. Precursor control of transmitter synthesis. In: *Choline and Lecithin in Brain Disorders. Nutrition and the Brain,* edited by A. Barbeau, J. H. Growdon, and R. J. Wurtman. New York: Raven, 1979, vol. 5, p. 1–12.
101. ZIMMERMANN, H. Vesicle recycling and transmitter release. *Neurosci.* 4: 1773–1804, 1979.

CHAPTER 13

Parasympathetic Cardiovascular Control in Humans

DWAIN L. ECKBERG

Departments of Medicine and Biophysics, Veterans Administration
Medical Center and Medical College of Virginia,
Richmond, Virginia

Parasympathetic cardiovascular responses play an important but poorly understood role in human physiology and pathophysiology. Changes of vagal outflow produced experimentally may be subnormal (42) or supranormal (21, 107) in patients with cardiovascular diseases, and changes of efferent parasympathetic activity triggered by disease processes, such as acute myocardial infarction, may be life threatening (4). Efferent sympathetic nervous traffic can be measured directly in conscious human volunteers (58); efferent vagal traffic has not been measured directly, but has been inferred from changes of reflex responses produced by atropine administration. This approach is justifiable in the case of sinus node responses, which are related nearly linearly to fluxes of efferent cardiac vagal activity in experimental animals (67). (In this chapter, the terms vagal and parasympathetic are used interchangeably to refer to cardiovascular changes such as bradycardia and arterial hypotension that can be reversed or prevented by administration of clinical doses of atropine sulfate.)

This chapter focuses on human parasympathetic cardiovascular mechanisms. It deals with response patterns of normal subjects and patients with cardiovascular diseases to representative tests used commonly to alter vagal outflow. It stresses three recurrent themes. First, although tests of human autonomic function may be simple to perform, mechanisms underlying cardiovascular responses to these interventions may be exceedingly complex and difficult to unravel in the human. Second, although abnormal response patterns of patients to tests of autonomic function have been documented exhaustively, very little is known about the mechanisms responsible for these abnormalities. Third, available evidence, however fragmentary, suggests that human parasympathetic responses may have great pathophysiological and public health importance.

A surprisingly diverse and powerful array of techniques is available for research into human autonomic cardiovascular control mechanisms: afferent sensory activity can be modified by numerous physical and pharmacological maneuvers; central autonomic modulation can be altered (wittingly or unwittingly); changes of efferent autonomic activity can be measured directly or inferred from changes of effector organ function; and efferent mediation of responses can be studied with autonomic blocking drugs.

A Complex Intervention: The Valsalva Maneuver

Probably the most widely used test of human autonomic function is the Valsalva maneuver; responses to the Valsalva maneuver have been studied extensively in normal subjects and patients with cardiovascular diseases. The Valsalva maneuver holds great attraction because it is safe (76, 104), can be performed without sophisticated equipment, and yields reproducible, quantitative results (76).

The Valsalva maneuver comprises an abrupt transient voluntary elevation of intrathoracic and intra-abdominal pressures provoked by straining. A representative Valsalva maneuver is illustrated in Figure 1. This widely used test, as described in published reports, is heterogeneous. Straining may be initiated after a maximal inspiration. (104), a full inspiration (65), or at the end of a normal inspiration (70). Mouth pressure during the period of straining [used to estimate intrapleural pressure (47)] is commonly raised to 40 mmHg (19, 90, 100), but may be raised to a constant level (113), to 40–50 mmHg (83), to 40–60 mmHg (65), or to a level in excess of 35 mmHg (75). The duration of straining is commonly 10 s (35, 65, 114), but it may be 12–15 s (19), 15 s (46, 55), 20 s (90), 15–30 s (114), 22–73 s (83), or as long as the patient is able (53).

Other aspects of Valsalva testing, including body position and the position of the glottis during the period of straining, may not be uniform from study to study. Also, very few investigators appear to control respiration after release of straining. It is highly probable that the breathing pattern during this terminal phase of the Valsalva maneuver influences important late reflex responses significantly (43, 44).

Notwithstanding the methodological differences that exist among investigators who use the Valsalva maneuver, there appears to be nearly universal acceptance of the proposal of Hamilton, Woodbury, and Harper (59) that responses of normal subjects to the maneuver be divided into four phases. These sequential phases are as follows: 1) an evanescent rise of arterial pressure and reduction of heart rate immediately after the onset of straining; 2) a fall, and later partial recovery of arterial pressure, and a speeding of heart rate during the period of straining; 3) a sudden brief further reduction of arterial pressure and elevation of heart rate immediately following the release of straining; and 4) a terminal elevation of arterial pressure above control levels and slowing of heart rate.

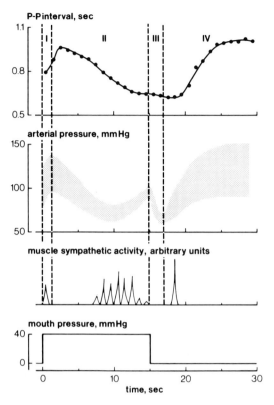

Fig. 1.   Schematic representation of normal responses to the Valsalva maneuver. Four phases are indicated by Roman numerals, according to Hamilton, Woodbury, and Harper (59). Fluctuations of P-P intervals comprise computer averaged responses of 3 healthy adult volunteers to multiple Valsalva maneuvers. [Changes of integrated efferent muscle sympathetic activity were redrawn from Fig. 3 of Wallin, Delius, and Hagbarth (121).]

In phase 1 it is likely that several mechanisms contribute to abrupt elevation of arterial pressure during the first seconds of the Valsalva maneuver. Increased intrathoracic and intra-abdominal pressures compress the aorta and probably propel relatively incompressible blood into peripheral arteries. In addition, left ventricular stroke volume increases (100). Increased stroke volume does not appear to result importantly from squeezing of blood out of the pulmonary vascular bed (101), because the increase of left ventricular dimensions during phase 1 is small and inconstant (97). An alternative explanation is that left ventricular function is improved transiently by reduced afterload (22). Elevation of arterial pressure during phase 1 does not appear to be sympathetically mediated, because it occurs in patients with high cervical spinal cord transections (123) and in normal subjects after administration of α-adrenergic blocking drugs (70). Elevation of arterial pressure is associated with inhibition of efferent sympathetic outflow (32) and slowing of heart rate [mediated by increased efferent parasympathetic activity (75).]

Continued straining in phase 2 impedes return of venous blood to the heart (24, 114) and leads to displacement of large quantities of blood from the thorax and abdomen to the limbs (105). Vein-to-artery circulation time is increased (114). Left atrial (98) and left ventricular (19, 90, 98) dimensions, left ventricular stroke volume (19, 90, 100), aortic cross-sectional area (56), and arterial pressure decline. Cardiac output falls (70) despite tachycardia [caused by withdrawal of parasympathetic activity (75)]. Efferent sympathetic nervous outflow to limb muscles (32), plasma epinephrine concentration (97), and total peripheral resistance (32, 70, 84) increase. Within 4 s after the augmentation of efferent sympathetic activity, the decline of arterial pressure is arrested (32) and pressure begins to return toward control levels, even though left ventricular stroke volume continues to fall (90). Arterial pressure falls more if tachycardia is prevented by cardiac autonomic blockade with propranolol and atropine (70) or if vasoconstriction is prevented by tetraethylammonium chloride (55) or phentolamine and guanethedine (70).

In phase 3, immediately after expiratory pressure is released, arterial pressure falls transiently. This reduction occurs after α-adrenergic blockade (70) and, therefore, results from mechanical factors, including perhaps sudden augmentation of left ventricular afterload (22) and sudden expansion of intrathoracic vessels. There is an additional burst of efferent sympathetic activity during this phase (121) and a further minor elevation of heart rate.

During phase 4 venous inflow to the heart (126), left ventricular stroke volume (19, 90), and cardiac output (19, 56, 100) return toward normal. Arterial pressure rises above control levels, because the normal left ventricular stroke volume is ejected into a constricted arterial bed; the rise of pressure can be prevented or reduced substantially, acutely by pharmacologic adrenergic blockade (70), and chronically by surgical sympathectomy (128). Arterial pressure elevation leads to protracted reduction of efferent sympathetic activity (32) and to bradycardia [mediated by increased parasympathetic activity (48, 75)].

Hemodynamic Responses of Patients with Cardiovascular Diseases. Several patterns of responses to the Valsalva maneuver are illustrated in Figure 2.

In phase 1, patients with advanced cardiovascular diseases (the most striking abnormalities are found in patients with frank congestive heart failure) experience normal elevation of arterial pressure at the outset of straining (46, 53, 65) but subnormal cardiac slowing (46).

As straining continues, in phase 2 arterial pressure may remain elevated [square-wave response, *left panel*, Fig. 2 (104)] or it may fall slightly (53, 104); heart rate may remain constant (104) or it may accelerate slightly (46, 53, 65, 90); left atrial (98) and left ventricular (90) dimensions decline, but insignificantly; stroke volume remains virtually fixed (90); and circulation time is only slightly prolonged (114). Arterial pressure does not rise [during the latter portion of phase 2 (113)].

In phase 4, after release of straining, arterial pressure usually returns to but does not rise above control levels (53, 113); in some patients, however, it may rise above control levels, but the rate of rise [in patients with mitral stenosis (83)] and the extent of pressure elevation (53, 83) are less than those which occur in normal subjects. It appears that in some patients terminal arterial pressure elevation may occur without associated cardiac slowing (83); however, this response pattern has not been sought out and studied systematically.

Valsalva Maneuver as Test of Autonomic Function. In normal subjects there is compelling evidence that the Valsalva maneuver provokes major alterations

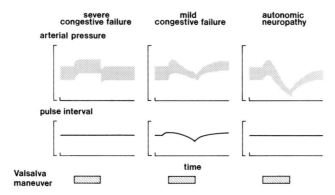

Fig. 2. Three abnormal response patterns to Valsalva maneuver. The square wave pressure response (*left panel*) does not necessarily signify organic cardiovascular disease; it may be produced in normal subjects by volume expansion (62). *Middle panel* depicts response pattern that is qualitatively normal but quantitatively abnormal. Patients with autonomic neuropathy may develop substantial hypotension during phases 2 and 3; however, baroreflex-mediated sympathetic vasoconstriction does not occur, and therefore arterial pressure does not rise above control levels after period of straining (*right panel*).

of efferent autonomic (sympathetic and parasympathetic) activity. Heart rate changes are nearly abolished by pretreatment with atropine, but are not altered importantly by propranolol. However, the identity of the receptor populations contributing to altered autonomic outflow and the relative importance of each receptor group in mediating response patterns are conjectural.

Elevation of arterial pressure during phase 1 appears to be due to mechanical rather than neural factors. On the other hand, cardiac slowing is neurally mediated and probably is triggered primarily by stimulation of carotid arterial baroreceptors. Minor stretch of carotid arterial baroreceptors provokes cardiac slowing (36), but minor reduction of cardiac chamber size (such as may occur during the earliest phase of the Valsalva maneuver) does not (1). Aortic baroreceptors probably do not contribute to this reflex response, because increased intrathoracic pressure leads to net reduction of aortic transmural pressure. It is likely that the exertion of straining (52) and the preceding deep breath (44) oppose phase 1 cardiac slowing.

There are theoretical reasons for preferring phase 1 of the Valsalva maneuver to subsequent phases for analysis of human parasympathetic responses. The arterial pressure stimulus arises from mechanical rather than neural factors and occurs in patients with and without cardiovascular disease. This phase is very brief, however, and its quantitative analysis requires beat-by-beat measurement of arterial pressure.

Cardiac acceleration during continued straining may result from several changes of receptor input profiles. [Heart rate changes do not appear to be modified by mechanical factors, such as altered sinus node (20) or sinus node artery (64) stretch, because they can be prevented by atropine (75).] Cardiac acceleration may result from central inhibition of efferent parasympathetic activity by physical exertion (52) or by deep inspiration before the maneuver (44), as well as by lessened stimulation of arterial baroreceptors (39). The net heart rate change during phase 2 probably reflects complex central interactions between arterial baroreceptors, chemoreceptors (57), and cardiopulmonary receptors (99). Vasoconstriction, on which subsequent pressure elevation in phase 4 depends (70), is probably mediated primarily by altered activity of cardiopulmonary receptors (99, 116, 131). Although patients with transplanted hearts may experience normal reductions of arterial pressure during phases 2 and 3 of the Valsalva maneuver, they do not experience an arterial pressure overshoot during phase 4 (H. A. Kontos, unpublished observations). This may explain why some patients experience normal elevation of arterial pressure (mediated primarily by cardiopulmonary receptors) and subnormal cardiac slowing [mediated primarily by arterial baroreceptors (42)]. Although bradycardia is vagally mediated (75), it does not occur if it is not preceded by adrenergically mediated vasoconstriction.

Four phases of responses can be identified regardless of the particulars of the Valsalva maneuver used. However, if analyses of human autonomic reflex function are to be quantitative, rather than merely qualitative, responses must be related to the intensity of stimuli. Without secure knowledge of the precise contribution of each receptor area to net Valsalva responses, it may be impossible to quantitate the intensity of stimulation. Moreover, even if intra-arterial and transmural cardiac pressures are measured during the Valsalva maneuver, it still may be impossible to quantitate the stimulus. Reflex adjustments occur during the Valsalva maneuver, and these minimize measured pressure changes. This mechanism is illustrated conceptually in Figure 3. The fall of arterial pressure during phase 2 is much greater after administration of autonomic blocking drugs than before (70), and the pressure rise during phase 4 is much greater if parasympathetic reflex adjustments are prevented by atropine (46).

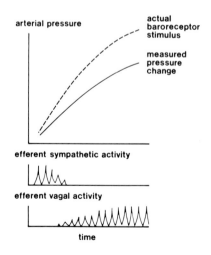

Fig. 3. Ongoing reflex adjustments reduce arterial pressure elevation that occurs during phase 4 of Valsalva maneuver (70). Therefore, measured pressure change may not accurately reflect intensity of actual baroreceptor stimulus; it is function of stimulus rather than stimulus itself. This problem is not unique to Valsalva maneuver; it arises whenever prolonged (more than a few seconds) autonomic stimuli are applied in human volunteers whose reflex responses are intact.

Therefore, ongoing reflex adjustments alter hemodynamic responses to the Valsalva maneuver, and the measured arterial pressure change is a function of the stimulus rather than the stimulus itself. (The problem that concurrent reflex adjustments modify ongoing autonomic stimuli is not unique to the Valsalva maneuver; it exists whenever autonomic stimuli last long enough for reflex adjustments to occur.)

Difficulties encountered in interpreting Valsalva responses of normal subjects are compounded when the test is used to assess autonomic function of patients with cardiovascular diseases. For an appropriate comparison to be made, stimuli must be comparable in both groups. This minimum condition is not met in almost all published studies.

First, many reports do not describe the conditions of the Valsalva maneuver used. It may be hazardous to guess what type of test was used, because there is no agreement in the medical literature on what comprises a standard Valsalva maneuver. The magnitude of hemodynamic responses to the Valsalva maneuver depends importantly on the type of test used, as reflected by the depth of inspiration before straining, the magnitude of intraoral pressure rise, and the duration of straining (70, 71, 98, 117). Therefore, valid, quantitative comparisons between studies cannot be made if the Valsalva maneuvers used were different. Also, there may be problems interpreting results within individual studies; in many (53, 65, 75, 83, 113, 114), conditions of the Valsalva maneuver were not controlled rigidly, and the intensities and durations of straining were determined by the subjects themselves. In at least one study (83) patients with heart disease were not able to maintain straining as long as normal subjects. If the Valsalva technique was not the same in patients and normal subjects, then quantitative comparisons of results are of questionable value.

Even if specific parameters of the Valsalva maneuver are controlled rigidly, the intensity of stimulation of receptor areas in normal subjects and patients with cardiovascular diseases may not be comparable. For example, abnormal arterial pressure responses of patients with atrial septal defects to the Valsalva maneuver may be due in part to a disturbed pattern of intracardiac shunting (74). In patients with congestive heart failure, arterial pressure (104) and intra-cardiac dimensions (90) may not decline during phase 2; in such patients, the Valsalva maneuver probably does not provoke the same alteration of afferent autonomic input as it does in normal subjects. If arterial pressures and cardiac chamber pressures and volumes do not change to the same extent in both groups, then the failure of patients with cardiovascular diseases to respond appropriately to the Valsalva maneuver cannot be attributed necessarily to reduced autonomic responsiveness.

The simplicity of the Valsalva maneuver is illusory. It is an exceedingly complex global test of the mechanical status of the circulation, the integrity of several popuations of autonomic receptors (including arterial baroreceptors, cardiopulmonary receptors, and chemoreceptors), complex central autonomic interactions, and efferent sympathetic and parasympathetic mechanisms. An unusual response may point toward abnormal cardiovascular hydraulics as well as toward autonomic dysfunction. In part, because of the complexity of the Valsalva maneuver, several other experimental approaches have been used that perturb the human autonomic nervous system in a more selective fashion.

Interventions That Reduce Venous Return

Upright posture, lower body negative pressure, hemorrhage, and administration of nitrates set in motion complex hemodynamic adjustments that may be triggered primarily by reduced venous return to the heart. The usual heart rate response to these interventions is cardioacceleration; heart rate slowing occurs less frequently. Both types of response can be prevented in large measure by atropine.

Upright Tilt. In healthy humans, upright posture is associated with pooling of blood in dependent portions of the body (109, 110). Central venous pressure (16, 125), right atrial pressure (106), right ventricular volume (96), left ventricular end-diastolic dimension (112) and stroke volume (2, 96), and cardiac output (85) decline. Calculated mean carotid arterial pressure declines by 15–20 mmHg, but mean aortic pressure probably declines by less than 5 mmHg (85). The pattern of neural traffic from vestibular receptors to the cerebellum and brain stem is altered (34).

There appears to be a reciprocal change of autonomic outflow when a recumbent person stands. Efferent vagal activity is reduced (102) and efferent sympathetic activity is increased (23). Recent evidence (51) suggests that the primary cause of cardioacceleration during standing is withdrawal of parasympathetic inhibition. Also, norepinephrine (60, 119), calculated peripheral vascular resistance (16, 99, 119), venous tone (16, 125), and plasma and renal vein renin levels (88) rise; arterial flow to the forearm (85), kidneys (33), and liver (30) falls.

Reflex mechanisms underlying responses to upright tilt have been examined primarily in experimental (quadruped) animals. It is clear that arterial baroreceptors contribute to the responses to upright posture; it is also clear, however, that other receptor areas are involved. During upright posture after arterial baroreceptor denervation, the magnitude of vasoconstriction and cardioacceleration is less than in intact animals (29) and arterial pressure falls to a greater extent (31). In the human, renin secretion is much greater when subjects are tilted upright than when their carotid sinuses are compressed (with neck pressure) in recumbency (81); this suggests that carotid hypotension is not solely responsible for elevation of renin levels.

Autonomic mediation of cardioacceleration during upright posture is not well understood. Heart rate speeds more (62, 80) and efferent sympathetic activity increases more (23) with the transition from sitting to standing than from lying to sitting positions, even though the change of carotid pressure is likely to be greater with the transition from lying to sitting than from sitting to standing. The importance of labyrinthine and cerebellar mechanisms in modulating autonomic responses to upright posture has been demonstrated amply in experimental animals (3, 34), but the precise role of these mechanisms in the human remains conjectural.

As with the Valsalva maneuver, the methods used to examine the effects of upright posture vary from study to study. The subject may be rotated passively on a tilt table (2, 60, 88, 112), or he may be asked to assume and maintain an upright position through his own exertions (23, 40). Hemodynamic responses to upright posture may be different when subjects are supported in the upright position by a saddle than when they stand without external support (6). The degree of rotation, which influences hemodynamic responses (85), may be 25° (112), 60° (60, 125), 70° (2), 80° (88), or 90° (23, 40). In most studies the rate of passive rotation is not specified; this factor may be an important determinant of heart rate responses (15).

Other Measures That Reduce Venous Return. Hemorrhage, lower body negative pressure, and nitrates are simpler interventions than upright posture; these reduce venous return but do not alter vestibular function or create a disparity between carotid and aortic pressures. Nevertheless, experimental evidence also points toward participation of multiple receptor areas in mediation of cardiovascular responses to these interventions. Cardioacceleration appears to be mediated primarily by reduced arterial baroreceptor afferent activity (1). On the other hand, augmentation of efferent sympathetic nerve traffic to forearm muscles (116) and forearm vasoconstriction (1) appear to be mediated primarily by cardiopulmonary receptors. Thus reduced venous return is not a simple experimental analogue of arterial hypotension.

Responses of Humans With Cardiovascular Disease to Reduced Venous Return. Patients with congestive heart failure do not have normal responses to measures that reduce venous return. In these patients, upright posture provokes minimal reductions of central venous pressure (106), left ventricular dimensions (112), cardiac output, and arterial pressure (96) and augmentations of heart rate (96). These patients also tolerate lower body negative pressure better than normal subjects; during this intervention they may experience no reduction of cardiac output and stroke volume, or, alternatively, their cardiac outputs and stroke volumes may actually rise (86).

There are probably several mechanisms underlying abnormal autonomic responses of patients with heart failure to reduced venous return. The simplest explanation is that venous pooling provoked by standard maneuvers is less in patients with heart failure than in normal subjects. Lessened venous pooling in patients may be due to a variety of factors, including reduced venous compliance (129), increased blood volume (110), and increased levels of circulating catecholamines (27). Regardless of the cause, parasympathetic responses triggered by venous pooling may be subnormal in patients because the autonomic stimulus is less intense than it is in normal subjects.

Vasovagal Reactions. The majority of healthy people (103) and some patients with cardiovascular diseases (89) respond to venous pooling by developing cardioacceleration. In a minority of healthy subjects and patients, tachycardia and mild arterial hypotension are succeeded by abrupt development of bradycardia (or by inappropriate cardiac slowing) and severe hypotension (14, 28), the vasovagal reaction (78).

Vasovagal reactions reflect apparently inappropriate central autonomic responses: bradycardia results from sudden augmentation of efferent vagal activity (78) and hypotension results from sudden reduction or cessation of sympathetic activity (23) and relaxation of arterial resistance vessels (9). Bradycardia (78), but apparently not hypotension (124), can be prevented by atropine premedication. Evidence from experimental animals suggests that receptors in the cardiopulmonary region trigger vasovagal reactions (87). It is clear that patients with cardiovascular diseases experience vasovagal reactions despite probable subnormal responsiveness of cardiopulmonary receptors (54); it is not known if patients are more or less prone to develop these responses than healthy subjects.

Selective Stimulation of Autonomic Receptors

Evidence cited above suggests that the Valsalva maneuver, upright tilt, and other interventions that reduce venous return provoke complex changes of afferent autonomic input that are mediated by various receptors. Several means are now available to stimulate receptor areas more selectively and thereby trigger autonomic responses more simply.

Arterial Baroreceptor Stimulation. Two quantitative techniques are available for stimulating human arterial baroreceptors: pharmacologic hypertension or hypotension (111) and neck suction or pressure (49). The pharmacologic technique was described by Smyth, Sleight, and Pickering in 1969 (111). An α-adrenergic agonist (usually phenylephrine, less commonly angiotensin) is injected intravenously as a bolus, and prolongation of each successive R-R interval is plotted as a function of the preceding pressure pulse. The slope of the linear regression of R-R interval on systolic arterial pressure (see Fig. 4) is considered to reflect arterial baroreceptor reflex sensitivity. Pulse-interval prolongation is mediated by augmented efferent vagal activity (42, 95, 108).

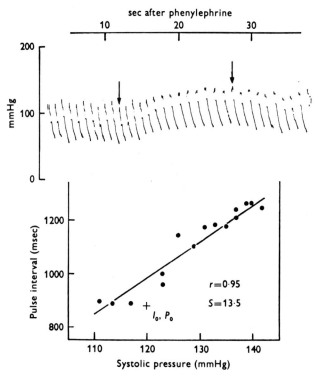

Fig. 4. Pulse-interval response of one volunteer to arterial pressure elevation caused by intravenous injection of bolus of phenylephrine. [Adapted from Bristow et al. (17).]

This technique holds several important advantages over the Valsalva maneuver. It involves no voluntary physical exertion. Participation of reflex mechanisms, in addition to the arterial baroreflex, is less. Respiration can be, but usually is not, controlled during the pressure rise. Even though left ventricular pressure rises, cardiopulmonary receptors do not appear to play an important role in mediating responses; bradycardia does not occur after sinoaortic denervation (5). Heart rate does not change if arterial pressure elevation is prevented (118); this militates against an important direct influence of the pressor drugs used on sinus node function.

There are several theoretical objections to this test. First, it treats the baroreceptor stimulus-sinus node response relation as a linear rather than as a sigmoidal (69) function. If pressure elevations occur in either threshold or saturation ranges, then calculated baroreflex slopes will be less than if pressure elevations are confined to the linear portion of the relation. The essential validity of this approach is supported by the often high correlation coefficients obtained with linear regression analyses. Moreover, a very recent study (39) suggests that in most (but not all) healthy young adults systolic pressure lies at the lowest part of the "linear" portion of this sigmoidal function. In addition, in most subjects the linear range extends well beyond the 20–30 mmHg pressure elevation that usually occurs after pressor injections. (On the basis of this study, it might be appropriate to exclude arterial pulses occurring at the beginning of the pressure rise from the regression analysis, if pulse-interval prolongation does not occur after these beats.)

A second objection is that the drugs used to raise arterial pressure may alter responsiveness of baroreceptor units through a direct action on the structures or on the receptors themselves. A preliminary study (12) indicates that this does occur; however, because naturally occurring arterial pressure elevations are preceded by bursts of efferent sympathetic activity (115), the alteration of receptor function that occurs pharmacologically may be no different from that which occurs physiologically. This seems likely, because Pickering (94) found that the slope of the regression of heart period on arterial pressure was comparable when arterial pressures were raised pharmacologically (with phenylephrine) and physiologically (during phase 4 of the Valsalva maneuver).

The technique of stimulating arterial baroreceptors with injections of pressor drugs is not uniform from study to study. First, different drugs may be used. The α-adrenergic agonist, phenylephrine, has been used in most studies (42, 108, 111), but angiotensin has been used in a minority of studies (63, 111). Phenylephrine appears to be preferable to angiotensin. One study indicates that although the initial baroreflex slope is nearly the same after injections of phenylephrine and angiotensin, the terminal pulse-interval change after angiotensin may depart from linearity in a bizarre fashion (111). Another study (63) suggests that even the initial baroreflex response may be unpredictable when angiotensin is used.

Second, different mathematical treatments may be used. The regression analysis of arterial pressure and pulse interval may begin as soon as the injection is completed (111), or it may begin with the onset of the arterial pressure elevation (42). I am not aware of any study in which analyses were begun when the regression of pulse interval on arterial pressure became linear (in subjects whose systolic arterial pressures lay below their thresholds for baroreceptor-cardiac reflex activation). Third, pressor drugs may be given as boluses or as continuous infusions. In most studies only the transients (after bolus injections) of arterial pressure are analyzed; in studies from one group, however, pressor drugs are given as a continuous infusion, and pulse-interval responses are measured after arterial pressure has reached a stable new plateau (72).

The technique of stimulating carotid baroreceptors with neck suction (one neck chamber is shown in Fig. 5) holds at least one important advantage over the pharmacological technique: baroreflex-mediated changes of arterial pressure as well as heart rate can be measured. (There is one additional advantage: neck suction does not require arterial cannulation.) This technique has been used extensively to evaluate patients with arterial hypertension (38, 82, 120); I am not aware of its use to evaluate patients with other types of cardiovascular diseases.

As with the pressor injection technique, there are pitfalls also with the neck suction technique. Certain theoretical objections have been dealt with. Altered chemoreceptor activity does not appear to contribute to sinus node responses to neck suction; responses are similar before and after inhalation of 100% oxygen (41, 79). The sinus node response does not appear to represent a nonspecific reaction to a noxious stimulus. Sinus node responses to neck suction and pressure (which are probably equally noxious) are directionally opposite (39, 79). Responses to neck suction do not appear to result from altered cerebral blood flow (79).

One unresolved question is, What contribution, if any, does stimulation of tracheal receptors by neck suction make to sinus node responses? A second question, which has been addressed by Ludbrook et al. (79), is, How accurately

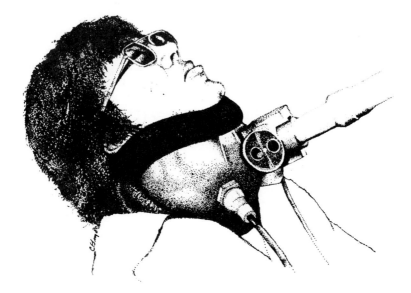

Fig. 5. Neck chamber used to alter carotid distending pressure experimentally (41). Strain gauge pressure transducer is mounted directly on chamber. White hose (right) is connected to continuous vacuum source. Pressure changes within chamber are initiated by rotation of solenoid-actuated pneumatic valve.

do pressure changes measured in the neck chamber reflect changes occurring within the carotid sinus? These authors measured pressures with a fluid-filled catheter whose tip was advanced to the vicinity of the carotid sinus in human volunteers and found that neck pressure is reduced by 14% and neck suction by 36% in transmission through neck tissues. In this study, body habitus did not influence transmission of neck chamber pressure changes into neck tissues; therefore, neck suction can be used to study different populations, such as normotensive volunteers and hypertensive patients whose average body weights are likely to be dissimilar.

As with other methods described in this report, the technique of stimulating carotid baroreceptors with neck suction is heterogeneous. Although the ultimate intensity of stimuli may be comparable, the mode of onset of stimuli may vary considerably from study to study. For example, the rate of onset of neck suction may be extremely slow [from 0 to −30 mmHg may require 10–15 s (120)] or extremely rapid [the transition from 0 to −30 may be accomplished within about 25 ms (36)]. Because the magnitude of arterial baroreflex responses is determined in part by the rate of onset of stimuli (36), it may be inappropriate to compare results from studies in which the rates of pressure change are dissimilar. Similarly the duration of baroreflex stimuli applied by different groups is highly variable; neck suction is applied commonly for about 2 min (82, 120), but it may be applied for periods as brief as 0.1 s (36). Because a striking decay of sinus node responses begins within the first 2 s after the onset of neck suction (37), the duration of the period of stimulation must exert an important influence on the magnitude of sinus node responses.

Although pressor drug infusion and neck suction seem to stimulate primarily arterial baroreceptors rather than other types of autonomic receptors, these methods share an intrinsic flaw with other techniques used to study autonomic reflexes in man. Abrupt baroreflex stimuli lead to reduction of arterial pressure within 5 s (10), and to reduction of efferent sympathetic nervous activity within 2.0 s (122). Moreover, cardioacceleration may begin as early as 0.5 s after the onset of an abrupt reduction of afferent arterial baroreceptor activity (39). These nearly instantaneous reflex changes may alter afferent arterial baroreceptor traffic during the period of baroreceptor stimulation with pressor drugs or neck suction. The findings of Evans, Knapp, and Lowery (50) support this possibility. They found that comparable doses of pressor drugs cause greater arterial pressure elevations after sinoaortic denervation than before. Therefore, as with the Valsalva maneuver (Fig. 3), concurrent reflex adjustments may modify the intensity of the baroreflex stimuli delivered; measured arterial (or neck chamber) pressure may be a function of the stimulus rather than the stimulus itself.

Responses of Patients to Arterial Baroreceptor Stimulation. Eckberg, Drabinsky, and Braunwald (42) used the pressor drug technique to evaluate arterial baroreceptor-cardiac reflexes in 22 patients with a variety of cardiovascular diseases. We found that sinus node responses to pressor injections were much smaller in patients with cardiovascular diseases than in normal subjects, even when differences in ages were considered. Responses were smallest in patients whose disease processes were symptomatically most advanced. The locus of the abnormality within the baroreflex arc was not and has not been identified; abnormalities in arterial baroreceptor areas, the central nervous system, and the sinus node all could contribute singly or in combination. Both the pressor drug technique and neck suction have been used to study baroreflex responses in hypertension. Most (18, 38, 72, 108) but not all (82, 120) studies indicate that baroreflex-mediated sinus node inhibition is reduced in patients with hypertension.

Carotid Massage. The technique of stimulating baroreceptors with carotid massage antedates both neck suction and pressor drug methods (91). Carotid massage is simple to perform but difficult to quantify; indeed, there is no standard reproducible method that can be applied during repeated tests in the same patient or during tests in different patients. Notwithstanding this substantive methodological difficulty, it seems likely that patients with heart disease (at least those with coronary heart disease) may have greater (not less) cardiac slowing during carotid massage than normal subjects (21, 107). This provides inferential evidence that the subnormal responses of patients with cardiovascular diseases to stimulation of arterial baroreceptors with pharmacologic arterial pressure elevation is due to abnormalities in baroreceptor areas rather than to abnormalities elsewhere in the reflex arc.

Cardiopulmonary Receptor Stimulation. Although it seems likely that some autonomic responses to venous pooling are mediated by receptors in the cardiopulmonary area, it is also highly probable that other receptor groups participate (see above). There are at least two circumstances in which receptors in cardiopulmonary areas appear to be stimulated selectively: during coronary arteriography and during acute inferior myocardial infarction.

Selective injection of hyperosmolar contrast medium into a coronary ostium provokes bradycardia (11), hypotension (11), reduction of cardiac output (127), and forearm vasodilatation (130) [or failure of appropriate vasoconstriction (127)]. A typical pattern of hemodynamic responses to coronary arteriography is shown in Figure 6. These changes are reduced by atropine pretreatment (11,

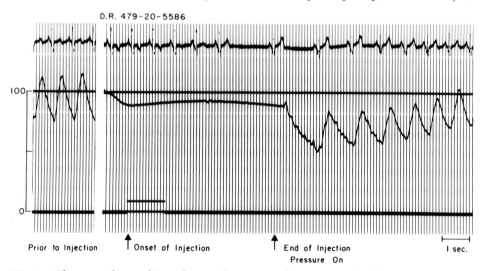

Fig. 6. Electrocardiographic and arterial pressure changes provoked by coronary arteriography. Striking bradycardia occurred despite arterial hypotension. [Data obtained in collaboration with C. W. White, J. M. Kioschos, and F. M. Abboud.]

45, 127, 130) and do no occur when contrast medium is injected into the ascending aorta above the coronary ostia (130); therefore, this hemodynamic pattern results from activation of a reflex triggered by stimulation of cardiac receptors. This reaction in the human seems to be an equivalent of the Bezold-Jarisch reflex (73) produced experimentally in laboratory animals. Bradycardia during coronary arteriography in patients with cardiovascular diseases may be comparable with or less than that in subjects with no identifiable cardiovascular disease (93). Acute inferior myocardial infarction also triggers bradycardia and arterial hypotension that are reversed by intravenous atropine (4). The occurrence of bradycardia in conjunction with hypotension (often profound) during coronary arteriography and acute inferior myocardial infarction suggests that arterial baroreceptor reflexes do not necessarily take precedence over reflexes arising in other receptor areas. This does not exclude the possibility that arterial baroreceptors modify Bezold-Jarisch responses; indeed, a recent study (26) suggests that this is the case.

Discussion

Tests available to provoke changes of parasympathetic outflow in man are multiple and varied. They may be simple or complex; they may be noninvasive or highly invasive. Most analyses rely on changes of heart rate; few are based on changes of arterial pressure or its hemodynamic determinants. Most tests of autonomic function alter heart rate by modifying parasympathetic outflow. Tests of autonomic function may provoke different responses in patients with cardiovascular diseases than in healthy volunteers. I draw several conclusions from this review.

1) Regardless of the apparent simplicity of the methods used, alterations of human autonomic function provoked by these tests may be extremely complex. The most widely used tests, the Valsalva maneuver, upright tilt, and lower body negative pressure alter afferent autonomic input from multiple receptor areas; it may be impossible to measure the precise contribution of each receptor area to integrated responses. Tests that seem to perturb autonomic receptors more selectively, such as pharmacological hypertension, neck suction, and coronary arteriography, trigger concurrent reflex changes that modify responses to the ongoing experimental interventions. Therefore, even with selective tests, it may be difficult to measure the intensity of stimuli delivered to the receptor area being studied.

I believe that several caveats should be heeded to improve the scientific credibility of tests used to assess human parasympathetic reflex responsiveness. First, the external particulars of each test should be controlled rigorously. If responses of two groups (e.g., patients with heart disease and healthy volunteers) are compared, then the tests administered to both groups must be the same. Similarly, if results from one study are compared with results from another, then the tests used in both studies must be comparable; alternatively, allowance must be made that differences between results may arise, in part, from differences between experimental techniques used. Some differences of technique, such as variable durations of straining during the Valsalva maneuver, may result in differences of responses that are merely quantitative. However, other methodological variations may account for differences of responses that

are qualitative; for example, prolonged, but not brief (10, 36), carotid baroreceptor stimulation with neck suction may be associated with cardioacceleration
rather than with cardiac slowing (120). In this context, cardioacceleration is an
experimental artifact and should not be regarded as a physiological consequence
of arterial baroreceptor stimulation.

Second, the autonomic stimuli delivered should alter function of only one
afferent receptor group; this is necessary if responses are to be related quantitatively to the intensity of stimuli [ideally, this basic scientific axiom (13) should
be as applicable to research conducted with human subjects as it is with that
conducted with experimental animals]. If this is not possible, then evidence
should be adduced that indicates that the degree of participation of other
receptor groups is negligible, or, if it is not negligible, that it can be measured
and assigned its proper role in mediation of net effector organ responses. This
may be very difficult. For example, the Valsalva maneuver is a time-honored
method that surely alters human efferent autonomic outflow, sympathetic and
parasympathetic. As this review indicates, however, the Valsalva maneuver
provokes extremely complex perturbations of afferent and efferent autonomic
activity that are difficult to isolate and quantitate.

Third, efforts should be made to control other contributing or competing
reflex mechanisms (such as those involved with respiration) that might modify
integrated parasympathetic responses. Finally, measurements of responses
should be restricted to as brief a period of time as possible after the onset of the
experimental intervention to minimize the influence of ongoing reflex hemodynamic changes. In the case of neck suction, reflex adjustments are measurable
after the first few seconds and are probably substantial after 10 s. If steady-
state responses (e.g., those occurring after 2 min of neck suction) are analyzed,
the multiplicity of contributing neurohumoral factors should be accounted for.

2) Although sinus node responses to autonomic tests can be altered or
abolished by intravenous atropine (and are therefore, by definition, parasympathetic[1]), it is extremely difficult to exclude and otherwise divorce from
consideration contributions from altered efferent sympathetic activity. Sympathetic participation may occur at several levels. First, a parasympathetic sinus
node response may depend upon prior sympathetic vasoconstriction; this is
exemplified by phase 4 of the Valsalva maneuver. Second, reflex changes of
efferent sympathetic traffic, or sympathomimetic drugs used to trigger reflex
responses, may directly alter responsiveness of autonomic receptors; an example of this possibility is the alteration of afferent baroreceptor activity that
occurs after intravenous pressor drug administration that may be independent
of changes of arterial pressure (12). Third, sympathetic reflex changes, e.g.,
baroreflex-mediated arterial hypotension, may occur concurrently with parasympathetic responses (cardiac slowing); these may modify the ongoing reflex

[1] In most studies of human autonomic cardiovascular reflex mechanisms, the efferent limb of the
reflex is identified according to changes of responses that occur after intravenous administration of
autonomic blocking drugs. This experimental approach is expedient; however, it also may be
simplistic. For example, β-adrenergic blockade appears to exert a complex influence on sinus node
baroreflex responses: it opposes changes mediated by β-adrenergic agonists, but it also enhances
responsiveness of arterial baroreceptor units to pressure changes (7).

stimulus and its vagal response. Fourth, a sympathetic-parasympathetic reflex interaction may occur (77) and may be impossible to detect in man.

3) Almost all studies of human parasympathetic cardiovascular reflexes are restricted to measurements of sinus node responses. Evidence cited above suggests that in at least two circumstances, during coronary arteriography and acute inferior myocardial infarction, arterial hypotension is in large measure cholinergically mediated. Mechanisms underlying cholinergic changes of human arterial pressure are understood poorly; they merit further investigation.

4) Although parasympathetic responses of patients with cardiovascular diseases may be different from those of healthy people, very little is known of the mechanisms responsible for these abnormalities.

First, the simplest explanation for a disparity of responses in published studies is that the autonomic stimuli delivered to patients and healthy volunteers were dissimilar. An obvious example is a Valsalva maneuver whose duration or intensity was not the same in patient and control groups. If the test used is not the same in both groups, then it is difficult to draw valid inferences regarding subnormal parasympathetic responses.

Second, a more subtle explanation is that although the autonomic tests used may have been similar in all external particulars, the actual stimuli delivered to receptor areas may have been dissimilar. Examples of this possibility include Valsalva maneuvers that do not provoke a reduction of arterial pressure and upright posture that does not cause appreciable diminution of venous return. If the actual stimuli delivered to receptor areas are not the same, then it is difficult to implicate deranged autonomic function as the sole mechanism underlying subnormal vagal responses.

Third, a reason why patients may have subnormal parasympathetic responses is that although the autonomic stimulus, e.g., diminution of venous return, may have been similar in patients and in normal subjects, the sympathetic reaction to the stimulus may have been faulty. For example, if a sympathetically mediated rise of arterial pressure does not occur in response to arterial hypotension during phases 2 and 3 of the Valsalva maneuver, then no inferences can be drawn regarding the presence or absence of subsequent parasympathetically mediated cardiac slowing.

Fourth, patients in whom abnormal parasympathetic responses have been identified are a heterogeneous lot; in some instances, cardiac anatomy may be a determinant of parasympathetic responses. For example, abnormal responses of patients with atrial septal defects to the Valsalva maneuver may be attributable in part to altered intracardiac shunting. The contributions of specific anatomic or pathophysiological defects to abnormal parasympathetic cardiovascular reflex responses have not been studied.

5) Although the mechanisms responsible for deranged parasympathetic responses in patients with cardiovascular diseases are understood incompletely, available evidence is probably sufficient to implicate one derangement at the level of the arterial baroreceptors. This hypothesis is supported by two lines of evidence. First, in patients with coronary heart disease, sinus node responses are subnormal when baroreceptors are stimulated by increased intra-arterial pressure (42); but normal (or supranormal) when receptors are deformed by direct carotid massage (21, 107). Second, patients with heart disease may develop

normal vagal bradycardia in response to other autonomic stimuli, such as coronary arteriography (93); this suggests that in such patients central nervous system and sinus node mechanisms may be intact.

Notwithstanding the formidable methodological difficulties involved with study of parasympathetic mechanisms in man, this field of enquiry is eminently worthwhile. Human pathophysiology is difficult to reproduce faithfully in experimental animals; therefore, the most unimpeachable preparation for study of the effects of human disease is the human. Moreover, although scientists who essay to study human parasympathetic cardiovascular reflex mechanisms must wrestle with certain conceptual and methodological problems (as this chapter indicates), they need not be concerned with other problems that beset scientists who conduct research with experimental animals, such as the influence of general anesthesia (which may distort the reflex mechanisms being studied) or the effects of acute or chronic surgical disruption of receptor (8) or effector (92) tissues (whose potential adverse influences have not been studied systematically).

Moreover, parasympathetic responses may figure importantly in human disease processes. Substantial fluxes of efferent cholinergic activity occur during the earliest, highly lethal stages of acute myocardial infarction (4). Other evidence from a variety of sources suggests that normal parasympathetic responses are conducive to good health. Vagal activity during experimental myocardial infarction seems to protect against catastrophic dysrhythmias (68). Patients with heart disease who have sinus arrhythmia, which is due primarily to ebb and flow of efferent vagal activity (66), live longer than patients whose heart rates during normal breathing are nearly constant (61). Drugs, such as propranolol, that among other actions enhance vagal responses (39, 95) may prolong life in patients with coronary heart disease (25).

Acknowledgments

I thank Hermes A. Kontos and David W. Richardson for their critical reviews of this chapter, Alice E. Rowe and Michelle L. Prettyman for their editorial help, and Constance Ann Lazzaro for her secretarial assistance.

This paper was presented in a symposium on the Neural Control of the Heart at the American Heart Association Meetings in Dallas in 1978.

The author is a Medical Investigator, Veterans Administration.

References

1. ABBOUD, F. M., D. L. ECKBERG, U. J. JOHANNSEN, AND A. L. MARK. Carotid and cardiopulmonary baroreceptor control of splanchnic and forearm vascular resistance during venous pooling in man. J. Physiol. London 286: 173–184, 1979.
2. ABELMANN, W. H., AND K. FAREEDUDDIN. Circulatory response to upright tilt in patients with heart disease. Clin. Aviat. Aerospace Med. 38: 60–65, 1967.
3. ACHARI, N. K., AND C. B. B. DOWNMAN. Autonomic effector responses to stimulation of nucleus fastigius. J. Physiol. London 210: 637–650, 1970.
4. ADGEY, A. A. J., J. S. GEDDES, H. C. MULHOLLAND, D. A. J. KEEGAN, AND J. F. PANTRIDGE. Incidence, significance, and management of early bradyarrhythmia complicating acute myocardial infarction. Lancet 2: 1097–1101, 1968.
5. ALEXANDER, N., AND M. DeCUIR. Loss of baroreflex bradycardia in renal hypertensive rabbits. Circ. Res. 19: 18–25, 1966.
6. ALLEN, S. C., C. L. TAYLOR, AND V. E. HALL. A study of orthostatic insufficiency by the tiltboard method. Am. J. Physiol. 143: 11–20, 1945.
7. ANGELL-JAMES, J. E., M. J. GEORGE, AND C. J. PETERS. Arterial baroreceptor function in health and disease. Jpn. Heart J. 20, Suppl. 1: 84–86, 1979.

8. ARNDT, J. O., J. KLAUSKE, AND F. MERSCH. The diameter of the intact carotid artery in man and its change with pulse pressure. *Pfluegers Arch.* 301: 230–240, 1968.

9. BARCROFT, H., O. G. EDHOLM, J. MCMICHAEL, AND E. P. SHARPEY-SCHAFER. Posthaemorrhagic fainting study by cardiac output and forearm flow. *Lancet* 1: 489–491, 1944.

10. BASKERVILLE, A. L., D. L. ECKBERG, AND M. A. THOMPSON. Arterial pressure and pulse interval responses to repetitive carotid baroreceptor stimuli in man. *J. Physiol. London* 297: 61–71, 1979.

11. BENCHIMOL, A., AND E. M. MCNALLY. Hemodynamic and electrocardiographic effects of selective coronary angiography in man. *N. Engl. J. Med.* 274: 1217–1224, 1966.

12. BERGEL, D. H., R. C. PEVELER, J. L. ROBINSON, AND P. SLEIGHT. The measurement of arterial pressure, carotid sinus radius and baroreflex sensitivity in the conscious greyhound. *J. Physiol. London* 292: 65P–66P, 1979.

13. BERNARD, C. *An Introduction to the Study of Experimental Medicine.* New York: Dover, 1957, p. 1.

14. BOSSI, M., G. CATALDO, A. COLOMBO, F. FIORISTA, D. GENTILI, AND S. PIRELLI. Ipotensione, bradicardia e prelipotimia da isosorbide dinitrato sub-linguale nell'infarto miocardico in fase iniziale. *G. Ital. Cardiol.* 7: 922–926, 1977.

15. BRÄUER, G., AND F. ROSSBERG. Zum Verhalten der Herzfrequenz des Menschen bei unterschiedlicher Geschwindigkeit des Übergangs vom Liegen zur Kopfaufwärtsposition. *Acta Biol. Med. Ger.* 34: 1153–1157, 1975.

16. BRIGDEN, W., AND E. P. SHARPEY-SCHAFER. Postural changes in peripheral blood flow in cases with left heart failure. *Clin. Sci.* 9: 93–100, 1950.

17. BRISTOW, J. D., E. B. BROWN, JR., D. J. C. CUNNINGHAM, M. G. HOWSON, M. J. R. LEE, T. G. PICKERING, AND P. SLEIGHT. The effects of raising alveolar P_{CO_2} and ventilation separately and together on the sensitivity and setting of the baroreceptor cardiodepressor reflex in man. *J. Physiol. London* 243: 401–425, 1974.

18. BRISTOW, J. D., A. J. HONOUR, G. W. PICKERING, P. SLEIGHT, AND H. S. SMYTH. Diminished baroreflex sensitivity in high blood pressure. *Circulation* 39: 48–54, 1969.

19. BROOKER, J. Z., E. L. ALDERMAN, AND D. C. HARRISON. Alterations in left ventricular volumes induced by Valsalva manoeuvre. *Br. Heart J.* 36: 713–718, 1974.

20. BROOKS, C. McC., H.-H. LU, G. LANGE, R. MANGI, R. B. SHAW, AND K. GEOLY. Effects of localized stretch of the sinoatrial node region of the dog heart. *Am. J. Physiol.* 211: 1197–1202, 1966.

21. BROWN, K. A., J. D. MALONEY, H. C. SMITH, G. O. HARTZLER, AND D. M. ILSTRUP. Carotid sinus reflex in patients undergoing coronary angiography: relationship of degree and location of coronary artery disease to response to carotid sinus massage. *Circulation* 62: 697–703, 1980.

22. BUDA, A. J., M. R. PINSKY, N. B. INGELS, JR., G. T. DAUGHTERS II, E. B. STINSON, AND E. L. ALDERMAN. Effect of intrathoracic pressure on left ventricular performance. *N. Engl. J. Med.* 301: 453–459, 1979.

23. BURKE, D., G. SUNDLÖF, AND B. G. WALLIN. Postural effects on muscle nerve sympathetic activity in man. *J. Physiol. London* 272: 399–414, 1977.

24. CANDEL, S., AND D. E. EHRLICH. Venous blood flow during the Valsalva experiment including some clinical applications. *Am. J. Med.* 15: 307–315, 1953.

25. CHAMBERLAIN, D. A. Beta-adrenergic blocking agents in prevention of sudden death. *Adv. Cardiol.* 25: 196–205, 1978.

26. CHEN, H. I. Interaction between the baroreceptor and Bezold-Jarisch reflexes. *Am. J. Physiol.* 237 (*Heart Circ. Physiol.* 6): H655–H661, 1979.

27. CHIDSEY, C. A., D. C. HARRISON, AND E. BRAUNWALD. Augmentation of the plasma norepinephrine response to exercise in patients with congestive heart failure. *N. Engl. J. Med.* 267: 650–654, 1962.

28. COME, P. C., AND B. PITT. Nitroglycerin-induced severe hypotension and bradycardia in patients with acute myocardial infarction. *Circulation* 54: 624–628, 1976.

29. COWLEY, A. W., JR., J. F. LIARD, AND A. C. GUYTON. Role of the baroreceptor reflex in daily control of arterial blood pressure and other variables in dogs. *Circ. Res.* 32: 564–576, 1973.

30. CULBERTSON, J. W., R. W. WILKINS, F. J. INGELFINGER, AND S. E. BRADLEY. The effect of the upright posture upon hepatic blood flow in normotensive and hypertensive subjects. *J. Clin. Invest.* 30: 305–311, 1951.

31. DAMPNEY, R. A. L., A. STELLA, R. GOLIN, AND A. ZANCHETTI. Vagal and sinoaortic reflexes in postural control of circulation and renin release. *Am. J. Physiol.* 237 (*Heart Circ. Physiol.* 6): H146–H152, 1979.

32. Delius, W., K.-E. Hagbarth, A. Hongell, and B. G. Wallin. Manoeuvres affecting sympathetic outflow in human muscle nerves. *Acta Physiol. Scand.* 84: 82–94, 1972.

33. De Wardener, H. E., and R. R. McSwiney. Renal haemodynamics in vaso-vagal fainting due to haemorrhage. *Clin. Sci.* 10: 209–217, 1951.

34. Doba, N., and D. J. Reis. Role of the cerebellum and the vestibular apparatus in regulation of orthostatic reflexes in the cat. *Circ. Res.* 34: 9–18, 1974.

35. Duke, P. C., J. G. Wade, R. F. Hickey, and C. P. Larson. The effects of age on baroreceptor reflex function in man. *Can. Anaesth. Soc. J.* 23: 111–124, 1976.

36. Eckberg, D. L. Baroreflex inhibition of the human sinus node: importance of stimulus intensity, duration, and rate of pressure change. *J. Physiol. London* 269: 561–577, 1977.

37. Eckberg, D. L. Adaptation of the human carotid baroreceptor-cardiac reflex. *J. Physiol. London* 269: 579–589, 1977.

38. Eckberg, D. L. Carotid baroreflex function in young men with borderline blood pressure elevation. *Circulation* 59: 632–636, 1979.

39. Eckberg, D. L. Nonlinearities of the human carotid baroreceptor-cardiac reflex. *Circ. Res.* 47: 208–216, 1980.

40. Eckberg, D. L., F. M. Abboud, and A. L. Mark. Modulation of carotid baroreflex responsiveness in man: effects of posture and propranolol. *J. Appl. Physiol.* 41: 383–387, 1976.

41. Eckberg, D. L., M. S. Cavanaugh, A. L. Mark, and F. M. Abboud. A simplified neck suction device for activation of carotid baroreceptors. *J. Lab. Clin. Med.* 85: 167–173, 1975.

42. Eckberg, D. L., M. Drabinsky, and E. Braunwald. Defective cardiac parasympathetic control in patients with heart disease. *N. Engl. J. Med.* 285: 877–883, 1971.

43. Eckberg, D. L., Y. T. Kifle, and V. L. Roberts. Phase relationship between normal human respiration and baroreflex responsiveness. *J. Physiol. London.* 304: 489–502, 1980.

44. Eckberg, D. L., and C. R. Orshan. Respiratory and baroreceptor reflex interactions in man. *J. Clin. Invest.* 59: 780–785, 1977.

45. Eckberg, D. L., C. W. White, J. M. Kioschos, and F. M. Abboud. Mechanisms mediating bradycardia during coronary arteriography. *J. Clin. Invest.* 54: 1455–1461, 1974.

46. Elisberg, E. I. Heart rate response to the Valsalva maneuver as a test of circulatory integrity. *J. Am. Med. Assoc.* 186: 200–205, 1963.

47. Elisberg, E. I., H. Goldberg, and G. L. Snider. Value of intraoral pressure as a measure of intrapleural pressure. *J. Appl. Physiol.* 4: 171–176, 1951.

48. Elisberg, E. I., G. Miller, S. L. Weinberg, and L. N. Katz. The effect of the Valsalva maneuver on the circulation. *Am. Heart J.* 45: 227–236, 1953.

49. Ernsting, J., and D. J. Parry. Some observations on the effects of stimulating the stretch receptors in the carotid artery of man. *J. Physiol. London* 137: 45P–46P, 1957.

50. Evans, J. M., C. F. Knapp, and T. R. Lowery. Pressor response buffering by β-adrenergic and cholinergic vasodilation in tranquilized dogs. *Am. J. Physiol.* 236 (*Heart Circ. Physiol.* 5): H165–H173, 1979.

51. Ewing, D. J., L. Hume, I. W. Campbell, A. Murray, J. M. Nielson, and B. F. Clarke. Autonomic mechanisms in the initial heart rate to standing. *J. Appl. Physiol.: Respirat. Environ. Exercise Physiol.* 49: 809–814, 1980.

52. Freyschuss, U. Cardiovascular adjustment to somatomotor activation. *Acta Physiol. Scand. Suppl.* 342: 1–63 1970.

53. Goldberg, H., E. I. Elisberg, and L. N. Katz. The effects of the Valsalva-like maneuver upon the circulation in normal individuals and patients with mitral stenosis. *Circulation* 5: 38–47, 1952.

54. Greenberg, T. T., W. H. Richmond, R. A. Stocking, P. D. Gupta, J. P. Meehan, and J. P. Henry. Impaired atrial receptor responses in dogs with heart failure due to tricuspid insufficiency and pulmonary artery stenosis. *Circ. Res.* 32: 424–433, 1973.

55. Greene, D. G., and I. L. Bunnell. The circulatory response to the Valsalva maneuver of patients with mitral stenosis with and without autonomic blockade. *Circulation* 8: 264–268, 1953.

56. Greenfield, J. C., Jr., R. L. Cox, R. R. Hernandez, C. Thomas, and F. W. Schoonmaker. Pressure-flow studies in man during the Valsalva maneuver with observations on the mechnical properties of the ascending aorta. *Circulation* 35: 653–661, 1967.

57. Gross, P. M., B. J. Whipp, J. T. Davidson, S. N. Koyal, and K. Wasserman. Role of the carotid bodies in the heart rate response to breath holding in man. *J. Appl. Physiol.* 41: 336–340, 1976.

58. Hagbarth, K.-E., and Å. B. Vallbo. Pulse and respiratory grouping of sympathetic impulses in human muscle nerves. *Acta Physiol. Scand.* 74: 96–108, 1968.

59. HAMILTON, W. F., R. A. WOODBURY, AND H. T. HARPER, JR. Physiologic relationships between intrathoracic, intraspinal and arterial pressures. *J. Am. Med. Assoc.* 107: 853–856, 1936.

60. HICKLER, R. B., R. E. WELLS, JR., H. R. TYLER, AND J. T. HAMLIN III. Plasma catechol amine and electroencephalographic responses to acute postural change. *Am. J. Med.* 26: 410–423, 1959.

61. HINKLE, L. E., JR., S. T. CARVER, AND A. PLAKUN. Slow heart rates and increased risk of cardiac death in middle-aged men. *Arch. Intern. Med.* 129: 732–750, 1972.

62. HOSHI, T., T. KOJIMA, S. KAMEYAMA, AND K. MATSUDA. On the influence of postural change upon the cardiac rate in man. *Tohoku J. Exp. Med.* 62: 221–234, 1955.

63. ISMAY, M. J. A., E. R. LUMBERS, AND A. D. STEVENS. The action of angiotensin II on the baroreflex response of the conscious ewe and the conscious fetus. *J. Physiol. London* 288: 467–479, 1979.

64. JAMES, T. N., AND R. A. NADEAU. Sinus bradycardia during injections directly into the sinus node artery. *Am. J. Physiol.* 204: 9–15, 1963.

65. JUDSON, W. E., J. D. HATCHER, AND R. W. WILKINS. Blood pressure responses to the Valsalva maneuver in cardiac patients with and without congestive failure. *Circulation* 11: 889–899, 1955.

66. KATONA, P. G., AND F. JIH. Respiratory sinus arrhythmia: noninvasive measure of parasympathetic cardiac control. *J. Appl. Physiol.* 39: 801–805, 1975.

67. KATONA, P. G., J. W. POITRAS, G. O. BARNETT, AND B. S. TERRY. Cardiac vagal efferent activity and heart period in the carotid sinus reflex. *Am. J. Physiol.* 218: 1030–1037, 1970.

68. KENT, K. M., E. R. SMITH, D. R. REDWOOD, AND S. E. EPSTEIN. Electrical stability of acutely ischemic myocardium. *Circulation* 47: 291–298, 1973.

69. KOCH, E. Die reflektorische Selbststeuerung des Kreislaufes. In: *Ergebnisse der Kreislaufforschung*. Dresden: Steinkopff, 1931, p. 211–215.

70. KORNER, P. I., A. M. TONKIN, AND J. B. UTHER. Reflex and mechanical circulatory effects of graded Valsalva maneuvers in normal man. *J. Appl. Physiol.* 40: 434–440, 1976.

71. KORNER, P. I., A. M. TONKIN, AND J. B. UTHER. Valsalva constrictor and heart rate reflexes in subjects with essential hypertension and with normal blood pressure. *Clin. Exp. Pharmacol. Physiol.* 6: 97–110, 1979.

72. KORNER, P. I., M. J. WEST, J. SHAW, AND J. B. UTHER. 'Steady-state' properties of the baroreceptor-heart rate reflex in essential hypertension in man. *Clin. Exp. Pharmacol. Physiol.* 1: 65–76, 1974.

73. KRAYER, O. The history of the Bezold-Jarisch effect. *Arch. Exp. Pathol. Pharmacol.* 240: 361–368, 1961.

74. KRONIK, G., J. SLANY, AND H. MOESSLACHER. Contrast M-mode echocardiography in diagnosis of atrial septal defect in acyanotic patients. *Circulation* 59: 372–378, 1979.

75. LEON, D. F., J. A. SHAVER, AND J. J. LEONARD. Reflex heart rate control in man. *Am. Heart J.* 80: 729–739, 1970.

76. LEVIN, A. B. A simple test of cardiac function based upon the heart rate changes induced by the Valsalva maneuver. *Am. J. Cardiol.* 18: 90–99, 1966.

77. LEVY, M. N. Sympathetic-parasympathetic interactions in the heart. *Circ. Res.* 29: 437–445, 1971.

78. LEWIS, T. Vasovagal syncope and the carotid sinus mechanism. *Br. Med. J.* 1: 873–876, 1932.

79. LUDBROOK, J., G. MANCIA, A. FERRARI, AND A. ZANCHETTI. Factors influencing the carotid baroreceptor response to pressure changes in a neck chamber. *Clin. Sci. Mol. Med.* 51: 347s–349s, 1976.

80. MACWILLIAM, J. A. Postural effects on heart-rate and blood-pressure. *Q. J. Exp. Physiol.* 23: 1–33, 1933.

81. MANCIA, G., G. LEONETTI, L. TERZOLI, AND A. ZANCHETTI. Reflex control of renin release in essential hypertension. *Clin. Sci. Mol. Med.* 54: 217–222, 1978.

82. MANCIA, G., J. LUDBROOK, A. FERRARI, L. GREGORINI, AND A. ZANCHETTI. Baroreceptor reflexes in human hypertension. *Circ. Res.* 43: 170–177, 1978.

83. MCINTOSH, H. D., J. F. BURNUM, J. B. HICKAM, AND J. V. WARREN. Circulatory changes produced by the Valsalva maneuver in normal subjects, patients with mitral stenosis, and autonomic nervous system alterations. *Circulation* 9: 511–520, 1954.

84. MELLANDER, S., AND B. JOHANSSON. Control of resistance, exchange and capacitance functions in the peripheral circulation. *Pharmacol. Rev.* 20: 117–196, 1968.

85. MENGESHA, Y. A., AND G. H. BELL. Forearm and finger blood flow responses to passive body tilts. *J. Appl. Physiol.: Respirat. Environ. Exercise Physiol.* 46: 288–292, 1979.

86. MURRAY, R. H., L. J. THOMPSON, J. A. BOWERS, E. F. STEINMETZ, AND C. D. ALBRIGHT. Hemodynamic effects of hypovolemia in normal subjects and patients with congestive heart failure. *Circulation* 39: 55–63, 1969.

87. ÖBERG, B., AND S. WHITE. The role of vagal cardiac nerves and arterial baroreceptors in the circulatory adjustments to hemorrhage in the cat. *Acta Physiol. Scand.* 80: 395–403, 1970.

88. OPARIL, S., C. VASSAUX, C. A. SANDERS, AND E. HABER. Role of renin in acute postural homeostasis. *Circulation* 41: 89–95, 1970.

89. PACKER, M., J. MELLER, N. MEDINA, R. GORLIN, AND M. V. HERMAN. Rebound hemodynamic events after the abrupt withdrawal of nitroprusside in patients with severe chronic heart failure. *N. Engl. J. Med.* 301: 1193–1197, 1979.

90. PARISI, A. F., J. J. HARRINGTON, J. ASKENAZI, R. C. PRATT, AND K. M. MCINTYRE. Echocardiographic evaluation of the Valsalva maneuver in healthy subjects and patients with and without heart failure. *Circulation* 54: 921–927, 1976.

91. PARRY, C. H. *An Inquiry into the Symptoms and Causes of the Syncope Angiosa, Commonly Called Angina Pectoris.* Bath, England: Cruttwell, 1799, p. 123–125.

92. PÉREZ, J. E., M. YOKOYAMA, J. NELSON, AND P. D. HENRY. Decreased coronary arterial contractility and regional myocardial catecholamine depletion after chronic instrumentation (Abstract). *Circulation* 60 Suppl. II: 261, 1979.

93. PEREZ-GOMEZ, F., AND A. GARCIA-AGUADO. Origin of ventricular reflexes caused by coronary arteriography. *Br. Heart J.* 39: 967–973, 1977.

94. PICKERING, T. G. *Baroreceptor Reflex in Man in Health and Disease* (thesis). Oxford, UK: University of Oxford, 1970, p. 17–23.

95. PICKERING, T. G., B. GRIBBIN, E. S. PETERSEN, D. J. C. CUNNINGHAM, AND P. SLEIGHT. Effects of autonomic blockade on the baroreflex in man at rest and during exercise. *Circ. Res.* 30: 177–185, 1972.

96. RAPAPORT, E., M. WONG, E. E. ESCOBAR, AND G. MARTINEZ. The effect of upright posture on right ventricular volumes in patients with and without heart failure. *Am. Heart J.* 71: 146–152, 1966.

97. ROBERTSON, D., G. A. JOHNSON, R. M. ROBERTSON, A. S. NIES, D. G. SHAND, AND J. A. OATES. Comparative assessment of stimuli that release neuronal and adrenomedullary catecholamines in man. *Circulation* 59: 637–643, 1979.

98. ROBERTSON, D., R. M. STEVENS, G. C. FRIESINGER, AND J. A. OATES. The effect of the Valsalva maneuver on echocardiographic dimensions in man. *Circulation* 55: 596–602, 1977.

99. RODDIE, I. C., AND J. T. SHEPHERD. Receptors in the high-pressure and low-pressure vascular systems. *Lancet* 1: 493–496, 1958.

100. RUSKIN, J., A. HARLEY, AND J. C. GREENFIELD, JR. Pressure flow studies in patients having a pressor response to the Valsalva maneuver. *Circulation* 38: 277–281, 1968.

101. SARNOFF, S. J., E. HARDENBERGH, AND J. L. WHITTENBERGER. Mechanism of the arterial pressure response to the Valsalva test: the basis for its use as an indicator of the intactness of the sympathetic outflow. *Am. J. Physiol.* 145: 316–327, 1948.

102. SCHER, A. M., W. W. OHM, K. BUMGARNER, R. BOYNTON, AND A. C. YOUNG. Sympathetic and parasympathetic control of heart rate in the dog, baboon and man. *Federation Proc.* 31: 1219–1225, 1972.

103. SCHLANT, R. C., T. S. TSAGARIS, AND R. J. ROBERTSON, JR. Studies on the acute cardiovascular effects of intravenous sodium nitroprusside. *Am. J. Cardiol.* 9: 51–59, 1962.

104. SHARPEY-SCHAFER, E. P. Effects of Valsalva's manoeuvre on the normal and failing circulation. *Br. Med. J.* 1: 693–695, 1955.

105. SHARPEY-SCHAFER, E. P. Effect of respiratory acts on the circulation. In: *Handbook of Physiology. Circulation.* Washington, DC: Am. Physiol. Soc., 1965, sect. 2, vol. III, chapt. 52, p. 1875–1886.

106. SIEKER, H. O., AND O. H. GAUER. A study of postural effects on pressure relationships in the venous circulation (Abstract). *Clin. Res. Proc.* 5: 102–103, 1957.

107. SIGLER, L. H. The hyperactive cardioinhibitory carotid-sinus reflex as an aid in the diagnosis of coronary disease. *N. Engl. J. Med.* 226: 46–51, 1942.

108. SIMON, A. C., M. E. SAFAR, Y. A. WEISS, G. M. LONDON, AND P. L. MILLIEZ. Baroreflex sensitivity and cardiopulmonary blood volume in normotensive and hypertensive patients. *Br. Heart J.* 39: 799–805, 1977.

109. SJÖSTRAND, T. The regulation of the blood distribution in man. *Acta Physiol. Scand.* 26: 312–327, 1952.

110. SJÖSTRAND, T. Volume and distribution of blood and their significance in regulating the circulation. *Physiol. Rev.* 33: 202–228, 1953.

111. SMYTH, H. S., P. SLEIGHT, AND G. W. PICKERING. Reflex regulation of arterial pressure during sleep in man. *Circ. Res.* 24: 109–121, 1969.

112. STEFADOUROS, M. A., M. EL SHAHAWY, F. STEFADOUROS, AND A. C. WITHAM. The effect of upright tilt on the volume of the failing human left ventricle. *Am. Heart J.* 90: 735–743, 1975.

113. STONE, D. J., A. F. LYON, AND A. S. TEIRSTEIN. A reappraisal of the circulatory effects of the Valsalva maneuver. *Am. J. Med.* 39: 923–933, 1965.

114. STUCKI, P., J. D. HATCHER, W. E. JUDSON, AND R. W. WILKINS. Studies of circulation time during the Valsalva test in normal subjects and in patients with congestive heart failure. *Circulation* 11: 900–908, 1955.

115. SUNDLÖF, G., AND B. G. WALLIN. Human muscle nerve sympathetic activity at rest. Relationship to blood pressure and age. *J. Physiol. London* 274: 621–637, 1978.

116. SUNDLÖF, G., AND B. G. WALLIN. Effect of lower body negative pressure on human muscle nerve sympathetic activity. *J. Physiol. London* 278: 525–532, 1978.

117. VALENTINUZZI, M. E., L. E. BAKER, AND T. POWELL. The heart rate response to the Valsalva manoeuvre. *Med. Biol. Eng.* 12: 817–822, 1974.

118. VARMA, S., S. D. JOHNSEN, D. E. SHERMAN, AND W. B. YOUMANS. Mechanisms of inhibition of heart rate by phenylephrine. *Circ. Res.* 8: 1182–1186, 1960.

119. VENDSALU, A. Studies on adrenaline and noradrenaline in human plasma. *Acta Physiol. Scand. Suppl.* 173 49: 70–77, 1960.

120. WAGNER, J. V., J. WACKERBAUER, AND H. H. HIGLER. Arterielles Blutdruck- und Herzfrequenzverhalten bei Hypertonikern unter Änderung des transmuralen Druckes im Karotissinusbereich. *Z. Kreislauf.* 57: 701–712, 1968.

121. WALLIN, G., W. DELIUS, AND K. E. HAGBARTH. Regional control of sympathetic outflow in human skin and muscle nerves. In: *Central Rhythmic and Regulation,* edited by W. Umbach and H. P. Koepchen. Stuttgart: Hippokrates, 1974, p. 190–195.

122. WALLIN, B. G., AND D. L. ECKBERG. Sympathetic transients caused by abrupt alterations of carotid baroreceptor activity in man. *Am. J. Physiol.* 242 (*Heart Circ. Physiol.* 11): February, 1982.

123. WATSON, W. E. Some circulatory responses to Valsalva's manoeuvre in patients with polyneuritis and spinal cord disease. *J. Neurol. Neurosurg. Psychiatr.* 25: 19–23, 1962.

124. WEISS, S., R. W. WILKINS, AND F. W. HAYNES. The nature of circulatory collapse induced by sodium nitrite. *J. Clin. Invest.* 16: 73–84, 1937.

125. WEISSLER, A. M., J. J. LEONARD, AND J. V. WARREN. Effects of posture and atropine on the cardiac output. *J. Clin. Invest.* 36: 1656–1662, 1957.

126. WEXLER, L., D. H. BERGEL, I. T. GABE, G. S. MAKIN, AND C. J. MILLS. Velocity of blood flow in normal human venae cavae. *Circ. Res.* 23: 349–359, 1968.

127. WHITE, C. W., D. L. ECKBERG, J. M. KIOSCHOS, AND F. M. ABBOUD. A study of coronary artery reflexes in man (Abstract). *Circulation* 48 *Suppl.* IV: 65, 1973.

128. WILKINS, R. W., AND J. W. CULBERTSON. The effects of surgical sympathectomy upon certain vasopressor responses in hypertensive patients. *Trans. Assoc. Am. Physicians* 60: 195–207, 1947.

129. WOOD, J. E., J. LITTER, AND R. W. WILKINS. Peripheral venoconstriction in human congestive heart failure. *Circulation* 13: 524–527, 1956.

130. ZELIS, R., C. C. CAUDILL, K. BAGGETTE, AND D. T. MASON. Reflex vasodilation induced by coronary angiography in human subjects. *Circulation* 53: 490–493, 1976.

131. ZOLLER, R. P., A. L. MARK, F. M. ABBOUD, P. G. SCHMID, AND D. D. HEISTAD. The role of low pressure baroreceptors in reflex vasoconstrictor responses in man. *J. Clin. Invest.* 51: 2967–2972, 1972.

Index